Structural Equation Modeling

A Second Course

a volume in
Quantitative Methods in Education and the Behavioral Sciences:
Issues, Research, and Teaching

Series Editor:
Ronald C. Serlin
University of Wisconsin–Madison

Structural Equation Modeling

A Second Course

edited by

Gregory R. Hancock
University of Maryland

and

Ralph O. Mueller
The George Washington University

Greenwich, Connecticut • www.infoagepub.com

Library of Congress Cataloging-in-Publication Data

Structural equation modeling : a second course / edited by Gregory R. Hancock and Ralph O. Mueller.
p. cm. -- (Quantitative methods in education and the behavioral sciences)
Includes bibliographical references and index.
ISBN 1-59311-015-4 (hardcover) -- ISBN 1-59311-014-6 (pbk.)
1. Structural equation modeling. 2. Analysis of covariance.
3. Linear models (Statistics) 4. Multilevel models (Statistics)
I. Hancock, Gregory R. II. Mueller, Ralph O. III. Series.
QA278.3.S77 2006
519.5'3--dc22

2005036850

Printed in the United States of America

CONTENTS

Series Introduction
Ronald C. Serlin *ix*

Preface
Richard G. Lomax *xi*

Acknowledgments *xv*

1. Introduction
Gregory R. Hancock and Ralph O. Mueller *1*

PART I. FOUNDATIONS

2. The Problem of Equivalent Structural Models
Scott L. Hershberger *13*

3. Reverse Arrow Dynamics: Formative Measurement and Feedback Loops
Rex B. Kline *43*

4. Power Analysis in Covariance Structure Modeling
Gregory R. Hancock *69*

PART II. EXTENSIONS

5. Evaluating Between-Group Differences in Latent Variable Means
Marilyn S. Thompson and Samuel B. Green *119*

6. Using Latent Growth Models to Evaluate Longitudinal Change
Gregory R. Hancock and Frank R. Lawrence *171*

7. Mean and Covariance Structure Mixture Models
Phill Gagné *197*

8. Structural Equation Models of Latent Interaction and Quadratic Effects
Herbert W. Marsh, Zhonglin Wen, and Kit-Tai Hau 225

PART III. ASSUMPTIONS

9. Nonnormal and Categorical Data in Structural Equation Modeling
Sara J. Finney and Christine DiStefano 269

10. Analyzing Structural Equation Models with Missing Data
Craig K. Enders 315

11. Using Multilevel Structural Equation Modeling Techniques with Complex Sample Data
Laura M. Stapleton 345

12. The Use of Monte Carlo Studies in Structural Equation Modeling Research
Deborah L. Bandalos 385

About the Authors 427

To our students, from whom we learn
far more than we could ever teach

SERIES INTRODUCTION

Quantitative Methods in Education and the Behavioral Sciences: Issues, Research, and Teaching is a unique book series sponsored by the American Educational Research Association's Special Interest Group Educational Statisticians. Motivated by the group's central purpose—to increase interaction among educational researchers interested in the theory, applications, and teaching of statistics in the social sciences—the new series is devoted to didactically oriented presentations that introduce, extend, and clarify state-of-the-art quantitative methods for students and researchers in the social and behavioral sciences. As such, the current series aims to present selected topics in a technically sophisticated yet didactically oriented format. This allows for the individual volumes to be used to enhance the teaching of quantitative methods in course sequences typically taught in graduate behavioral science programs. Although other series and journals exist that routinely publish new quantitative methodologies, this book series is dedicated to both the teaching and applied research perspectives of specific modern quantitative research methods through each volume's relevant and accessible topical treatments. In line with this educational commitment, royalties from the sale of series' volumes will be used in large part to support student participation in annual special interest group conference activities.

Ronald C. Serlin
University of Wisconsin–Madison
Series Editor

PREFACE

This is the first volume in the series, *Quantitative Methods in Education and the Behavioral Sciences: Issues, Research, and Teaching,* sponsored by the American Educational Research Association's Special Interest Group/ Educational Statisticians (SIG/ES). I served as an officer of SIG/ES during the development of this volume, as program chair, as president, and now as past president. Thus, I am pleased to see this volume completed, both from the perspective of our SIG, and as a structural equation modeling (SEM) researcher.

The purpose of this volume is to provide a user-friendly, applied, intermediate coverage of SEM to complement the introductory textbooks of the same perspective. Following an overview of the field of SEM by editors Hancock and Mueller, the volume is divided into three parts, Foundations, Extensions, and Assumptions. The *Foundations* part deals with three topics that are typically not thoroughly covered in a first course in SEM, but are crucial topics for the field. These three topics are equivalent models, reverse arrow dynamics, and power analysis.

The equivalent models chapter (Hershberger) is important because any theoretical model has many equivalent models to consider, which differ somewhat in structure but provide the exact same fit. While knowledge of such models has been around since the birth of SEM, it is only recently that they have been systematically studied. The chapter on reverse arrow dynamics (Kline) considers feedback loops and reciprocal relations in measurement and structural models. This represents another topic not usually given the attention it deserves in an introductory course, despite such models becoming more and more prevalent. Power analysis (Hancock) has featured prominently in the social sciences in recent years, but mainly in general linear models of manifest variables. With journals and funding agencies asking researchers to conduct their own power analyses (either a priori or post hoc), this chapter is important information for SEM researchers to follow suit with latent variable models.

The second part of the volume is entitled *Extensions*, and considers the more advanced topics of latent variable means models, latent growth models, mixture models, and nonlinear models. SEM software and models have been capable of dealing with multiple samples for many years, although their usage in applied work is only now becoming popular. The first chapter (Thompson & Green) represents an important guide to the many possible forms of multiple sample models in comparing latent variable means. Latent growth models have been popular for over a decade in terms of modeling growth longitudinally over time. The latent growth chapter (Hancock & Lawrence) presents a nice overview on testing such models. Returning again to multiple group modeling, the chapter on mixture models (Gagné) deals with situations where data arise from multiple or mixtures of populations. When models are tested with mixtures of populations, whether membership can be determined or not, there are critical methodological challenges that the field of SEM has recently begun to address. The final chapter in this section is on nonlinear models (Marsh, Wen, & Hau), that is, models with latent interaction or quadratic effects. While models of this type are common in the manifest variable context, it is only recently that such models have begun to appear in the latent variable context (effects due to mediators, moderators, interactions, etc.). This is one area in SEM that is ready to explode with interest.

The third and final part of the volume treats a critical area, *Assumptions*, with chapters on non-normal and categorical data, missing data, multilevel data, and simulation studies. More often than not, researchers use manifest variables that are non-normal and/or categorical within their models (Finney & DiStefano), and thus we need to know how to treat such data when relying up normal theory. As well, missing data are very prevalent in the applied context (Enders). While there are a number of new missing data methods available in SEM software, we need to know which strategies are best to use in particular situations. Multilevel models (Stapleton) have been around for some time with manifest variable models (e.g., HLM), but only recently for multilevel latent variable models. This chapter considers such models with complex sample designs, popularized by the multitude of large-scale datasets gathered at the national level. The final chapter provides an overview of how to systematically study assumption violations, known as simulation or Monte Carlo studies (Bandalos). Knowledge of the important issues in designing such studies is crucial to the further development of SEM.

I hope that you find this volume to be as useful as I have already. I believe that this volume represents a vital contribution to the field of SEM beyond the introductory level. Finally, I want to thank the editors and the authors for doing such a fine job of presenting highly technical SEM information at an applied level. Happy modeling!

Richard G. Lomax
Department of Educational Studies in Psychology, Research Methodology, and Counseling, University of Alabama

ACKNOWLEDGMENTS

We wholeheartedly thank Richard Lomax, past president of the American Educational Research Association's Special Interest Group Educational Statisticians, and members of its editorial board, for their trust both in the project and in us as coeditors.

We acknowledge the support throughout from George Johnson at Information Age Publishing and thank Frank Aguirre from Hypertext Book and Journal Services for typesetting the manuscript and realizing our design for the cover.

A very special thanks and recognition goes to the members of the "Breakfast Club" who, for over a year, spent Wednesday mornings sharing breakfast, discussing and critiquing every word of every chapter draft, and offering their comprehensive feedback, both of a technical and pedagogical nature: Jaehwa Choi, Weihua Fan, Phill Gagné, Jennifer Hamilton, Marc Kroopnick, Roy Levy, and Chin-fang Weng. Indeed, you all earned your Mickey ears!

Finally, and most importantly, we give our deepest appreciation to Goldie, Sydney, and Quinn, for all their love, patience, and support throughout this project, and for perspective on what truly matters most.

CHAPTER 1

INTRODUCTION

Gregory R. Hancock and Ralph O. Mueller

The origins of modern structural equation modeling (SEM) are usually traced to biologist Sewall Wright's development of path analysis (e.g., Wright, 1921, 1934; see Wolfle, 1999, for an annotated bibliography of Wright's work). With respect to the social and behavioral sciences, path analysis lay largely dormant until the 1960s when Otis Duncan (1966) and others introduced the technique in sociology. Simultaneously, statistical developments by Karl Jöreskog (e.g., 1966, 1967) articulated a method for confirmatory factor analysis (CFA), an application of normal theory maximum likelihood estimation to factor models with specific a priori hypothesized theoretical latent structures. A milestone in the development of modern SEM was Jöreskog's provision for a formal χ^2-test comparing the observed pattern of relations among measured variables to that implied by an a priori specified factor model, thereby allowing for the disconfirmation (or tentative confirmation) of such an hypothesized model. Soon after, quite unceremoniously it seems, the fusion of Wright's measured variable path analysis and Jöreskog's CFA occurred and SEM was quietly born (see Wolfle, 2003, for an annotated bibliography of the introduction of SEM to the social sciences).

Despite its tremendous potential, SEM remained generally inaccessible to researchers in the social and behavioral sciences until well into the 1970s. Not only did it require access to a special statistical software pack-

Structural Equation Modeling: A Second Course, 1–9

age, LISREL,[1] but utilizing this package required knowledge of matrix algebra. However, by the 1980s and 1990s, examples of measured variable path models, CFAs, and latent variable models started to become increasingly common, largely due to the vastly improved user-friendliness of SEM software. EQS (Bentler, 1985) was the first program to offer non-matrix-based syntax, followed by the SIMPLIS command language of LISREL 8 (Jöreskog & Sörbom, 1993), and later Mplus (Muthén & Muthén, 1998). Social scientists could now use rather intuitive commands that mirrored the structural equations themselves. Versatile graphic interfaces were also developed within some SEM programs to further simplify the modeling process; in fact, a primarily graphics-based SEM program, AMOS, was created (see the AMOS 4 User's Guide; Arbuckle & Wothke, 1999). By the end of the 1990s, SEM had ascended to the ranks of the most commonly used multivariate techniques within the social sciences.[2]

In addition, also responsible for SEM's tremendous increase in popularity was the increase in accessibility of the materials used to train growing numbers of social science graduate students in SEM. In the 1970s and 1980s a first wave of influential SEM texts relied heavily on matrix formulations of SEM and/or emphasized measured variable path analysis (e.g., Bollen, 1989; Byrne, 1989; Duncan, 1975; Hayduk, 1987; James, Mulaik, & Brett, 1982; Kenny, 1979). Authors of a second wave of books in the 1990s, some of which are now in subsequent editions, moved toward a less mathematical and more accessible treatment. Their focus was much more applied in nature, trading matrix-based explanations for coverage of a broader range of topics, tending to include at a minimum such model types as measured variable path analysis, CFA, latent variable path analysis, and multigroup analyses, as well as applications with SEM software packages (e.g., Byrne, 1994, 1998, 2001; Dunn, Everitt, & Pickles, 1993; Kline, 2004; Loehlin, 2004; Maruyama, 1998; Mueller, 1996; Raykov & Marcoulides, 2000; Schumacker & Lomax, 2004).

With regard to more advanced SEM topics, a host of books are also now available. Specific topics covered by such works include latent growth models (Duncan, Duncan, Strycker, Li, & Alpert, 1999), multilevel models (Heck & Thomas, 2000; Reise & Duan, 2003), and latent variable interactions (Schumacker & Marcoulides, 1998), in addition to other advanced topic compendia (e.g., Marcoulides & Moustaki, 2002; Marcoulides & Schumacker, 1996, 2001) and texts covering both introductory and some more advanced topics (Kaplan, 2000). Most of these resources, however, serve researchers with highly specialized interests, and tend to be more appropriate for a technically sophisticated audience of methodologists. Less common are texts addressing advanced topics while achieving the accessibility that is characteristic of the second wave of introductory texts. The present volume is intended to parallel the movement toward more accessible introductory texts, serving as a didactically oriented resource covering a broad range of more advanced topics often not discussed in an introductory course. Such topics are important in furthering the understanding of foundations and assumptions of SEM as well as in exploring SEM as a potential tool to address new types of research questions that might not have arisen during a first encounter with SEM.

THE PRESENT VOLUME: GOALS, CONTENT, AND NOTATION

In a second course in SEM the treatment of new topics can, arguably, be approached in two ways: (1) emphasizing topics' mathematical and statistical underpinnings and/or (2) stressing their relevance, conceptual foundations, and appropriate applications. As much as we value the former as methodologists, in keeping with the goal of maximizing accessibility to the widest audience possible, we have tried to emphasize the latter to a much greater degree. Accordingly, with regard to the selection of chapter authors, all certainly have established technical expertise in their respective areas; however, the most salient professional attribute for our purposes was that their previous work was characterized by a notable didactic clarity. Drawing on these skills in particular, selected authors were asked to present a thorough, balanced, and illustrated treatment of the state-of-knowledge in the assigned topic area; they were asked *not* to set forth a specific agenda or to propose new methods. Where appropriate, authors focused on the clear explanation and application of topics rather than on analytical derivations, and provided syntax files and excerpts from EQS, SIMPLIS/LISREL, and/or Mplus output.[3]

Regarding the topics themselves, we assembled chapters targeted toward graduate students and professionals who have had a solid but introductory exposure to SEM. In doing so we intended to provide the reader with three important things: (1) the necessary foundational knowledge to fill in possible gaps left during a first course on SEM, (2) an extension of the range of research questions that might be addressed with modern SEM techniques, and (3) ways to analyze data that violate traditional assumptions. As such, the remaining chapters in the book have been organized into three sections. The first, *Foundations*, complements introductory texts, filling in some foundational gaps by addressing the areas of equivalent models (Hershberger), formative measurement and feedback loops (Kline), and power analysis (Hancock). The second section, *Extensions*, presents methods that allow researchers to address applied research questions that introductory SEM methods cannot satisfactorily address. These extensions include latent means models (Thompson & Green), latent growth models (Hancock & Lawrence), mixture models (Gagné), and latent interaction and quadratic effects (Marsh, Wen, & Hau). The third and last section, *Assumptions*, acknowledges that real-world data rarely resemble the textbook ideals. Rather, they are replete with complexities often serving as obstacles to the real substantive questions of interest. The topics in this section are non-normal and categorical data (Finney & DiStefano), missing data (Enders), multilevel models (Stapleton), and Monte Carlo methods (Bandalos), with the last chapter providing researchers with a means to investigate for themselves the consequences of assumption violations in their models.

Finally, in order to maximize the utility of this text as a teaching tool, we strived to ensure as much homogeneity in chapter structure and notation as possible. As illustrated in Figure 1.1, the Bentler-Weeks "VFED" system is used for labeling variables and residuals wherever possible in the chapters. These designations, and their more traditional Jöreskog–Keesling–Wiley (JKW) notational analogs in parentheses, are shown below:

V = observed Variable (X, Y)
F = latent Factor (ξ, η)
E = Error/variable residual (δ, ε)
D = Disturbance/factor residual (ζ)

Chapter figures also generally follow the convention of showing measured variables (V) in squares/rectangles and latent variables (F) in circles/ellipses. Measured or latent residuals (E, D) are not typically enclosed in any shape (although the latent nature of residuals could warrant an ellipse). For mean structure models, the pseudovariable (1) is enclosed in a triangle.

As for the parameters relating the VFED elements, we herein propose a parameter notation system that we call the "*a-b-c*" system. While the JKW system is very useful for mathematical representations of SEM, for didactic expositions we find the *a-b-c* system generally more intuitive. First, drawing from regression notation, in mean structure models the symbol a is used to designate an intercept (or mean) parameter, which is represented diagrammatically as a path from the pseudovariable (1) to a measured or latent variable. A subscript on a indicates the variable whose intercept (or mean) is being modeled, such as a_{V1} or a_{F1}, shown at the bottom of Figure 1.1. Analogs to a in the JKW system include subscripted τ (measured variable intercept), κ (exogenous factor mean), and α (endogenous factor intercept). Sample-based estimates of an a parameter in the *a-b-c* system have an additional hat designation, such as $\hat{a}_{F1}$.

Second, also drawing from regression notation, the symbol b is used to designate a structural (causal) relation hypothesized between any VFED

Covariance structure model (not all parameters shown)

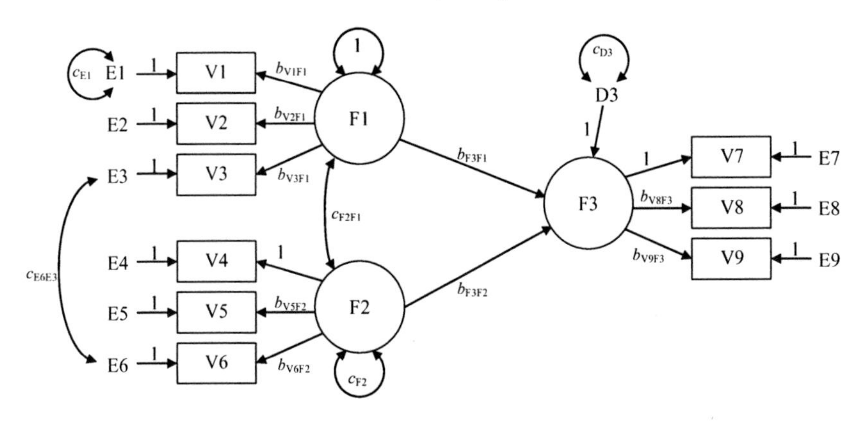

Mean structure model (not all parameters shown)

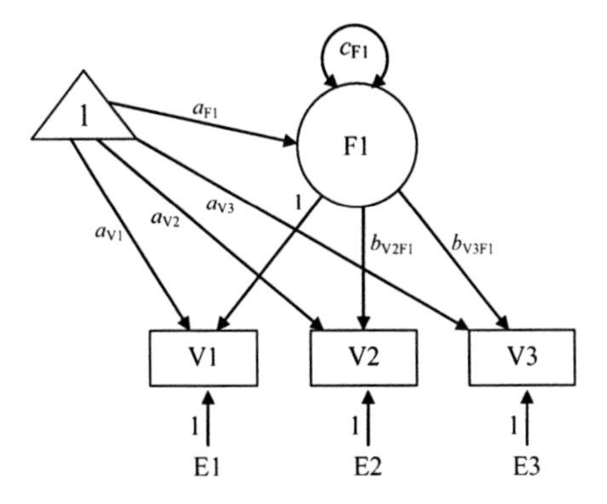

Figure 1.1. Diagrammatic and notational conventions for this volume

elements; this relation is represented diagrammatically as a single-headed (usually straight) arrow. Subscripts on b follow the "*to–from*" convention, indicating first where the arrow is going *to* and second where the arrow is coming *from*; examples include the factor-loading b_{V2F1} or the latent structural relation b_{F3F2} shown at the top of Figure 1.1. In measured variable path models, where no latent variables exist, the subscripts may be shortened to include variable numbers only (e.g., b_{V4V2} becomes b_{42}). Analogs to b in the JKW system include subscripted λ (measurement model loading), γ (structural path from exogenous variable to endogenous variable), and β (structural path from endogenous variable to endogenous variable). Sample-based estimates of a b parameter in the *a-b-c* system have an additional hat designation, such as $\hat{b}_{F3F2}$.

Third, the symbol c is used to designate a covariance (or variance) parameter for any exogenous VFED elements. This relation is represented diagrammatically as a two-headed (usually curved) arrow; a covariance parameter is depicted by an arrow connecting two elements, while a variance parameter is depicted by a circular arrow connecting an element to itself (as a variance is a covariance of a variable with itself). Subscripts on c indicate the elements being connected, where two subscripts (in arbitrary order) indicate the covarying exogenous elements and a single subscript indicates the exogenous element whose variance is a parameter. Covariance examples in Figure 1.1 include the c_{E6E3} and c_{F2F1}, while variance examples include c_{E1}, c_{F2}, and c_{D3}. In measured variable path models, where no latent variables exist, the subscripts may be shortened to include variable numbers only (e.g., c_{V1V2} becomes c_{12}, and c_{V3} becomes c_3). Analogs to c in the JKW system include subscripted ϕ (exogenous variable co/variance), θ (measured residual co/variance), and ψ (latent residual co/variance). Sample-based estimates of a c parameter in the *a-b-c* system have an additional hat designation, such as $\hat{c}_{F2F1}$ and $\hat{c}_{D3}$.

To summarize the *a-b-c* notational system:

a = intercept or mean path from a pseudovariable (1) to V or F;
b = structural path connecting any VFED;
c = co/variance associated with exogenous VFED elements.

LOOKING FORWARD

Over the last 30 years or so, SEM has matured into a methodological framework that allows researchers to address theory-driven causal research questions at both the observed variable and latent variable levels. In our view, the integration of structure and measurement facilitated by this analytical paradigm stands as one of the most significant methodological developments of the 20th century. More and more, we observe institutions of higher education offering SEM instruction, as well as a need in the professional social/behavioral science community for SEM tutelage. We conceived this volume out of a mutual passion for teaching SEM, both to beginning and more experienced modelers, and indeed learned much about SEM ourselves from selecting and editing the chapters in this textbook. Our hope is for you to share in our excitement about SEM as you fill in some of the holes that may be in your foundational knowledge (Part I), extend the kind of research questions that you can answer with SEM (Part II), and learn what can be done when assumptions commonly made about data do not necessarily hold true in yours (Part III).

NOTES

1. According to Sörbom (2001), LISREL III became available in 1974.
2. Between 1994 and 2001, the number of both substantive and technical SEM journal publications more than doubled to over 340 articles; over 60% of the total number of journals published by the American Psychological Association during the same time period contained articles using SEM (Hershberger, 2003; for the 1985–1994 period, see Tremblay & Gardner, 1996).
3. AMOS, which is an excellent SEM software package as well, was not emphasized in the current volume only because its graphic nature made it more difficult for authors to provide sample syntax for readers.

REFERENCES

Arbuckle, J. L., & Wothke, W. (1999). *Amos 4.0 user's guide*. Chicago: SPSS.

Bentler, P. M. (1985). *Theory and implementation of EQS, A structural equations program*. Los Angeles: BMDP Statistical Software.

Bollen, K. A. (1989). *Structural equations with latent variables*. New York: Wiley.

Byrne, B. M. (1989). *A primer of LISREL: Basic applications and programming for confirmatory factor analytic models*. New York: Springer-Verlag.

Byrne, B. M. (1994). *Structural equation modeling with EQS and EQS/Windows*. Thousand Oaks, CA: Sage.

Byrne, B. M. (1998). *Structural equation modeling with LISREL, PRELIS, and SIMPLIS*. Mahwah, NJ: Erlbaum.

Byrne, B. M. (2001). *Structural equation modeling with AMOS*. Mahwah, NJ: Erlbaum.

Duncan, O. D. (1966). Path analysis: Sociological examples. *American Journal of Sociology, 72*, 1–16.

Duncan, O. D. (1975). *Introduction to structural equation models*. New York: Academic Press.

Duncan, T. E., Duncan, S. C., Strycker, L. A., Li, F., & Alpert, A. (1999). *An introduction to latent variable growth curve modeling: Concepts, issues, and applications*. Mahwah, NJ: Erlbaum.

Dunn, G., Everitt, B., & Pickles, A. (1993). *Modelling covariances and latent variables using EQS*. London: Chapman & Hall.

Hayduk, L. A. (1987). *Structural equation modeling with LISREL*. Baltimore: Johns Hopkins University Press.

Heck, R. H., & Thomas, S. L. (2000). *An introduction to multilevel modeling techniques*. Mahwah, NJ: Erlbaum.

Hershberger, S. L. (2003). The growth of structural equation modeling: 1994–2001. *Structural Equation Modeling: A Multidisciplinary Journal, 10,* 35–46.

James, L. R, Mulaik, S. A., & Brett, J. (1982). *Causal analysis: Models, assumptions, and data.* Beverly Hills, CA: Sage.

Jöreskog, K. G. (1966). Testing a simple structure hypothesis in factor analysis. *Psychometrika, 31,* 165–178.

Jöreskog, K. G. (1967). Some contributions to maximum likelihood factor analysis. *Psychometrika, 32,* 443–482.

Jöreskog, K. G., & Sörbom, D. (1993). *LISREL 8: Structural equation modeling with the SIMPLIS command language.* Chicago: Scientific Software International.

Kaplan, D. (2000). *Structural equation modeling: Foundations and extensions.* Thousand Oaks, CA: Sage.

Kenny, D. A. (1979). *Correlation and causation.* New York: Wiley.

Kline, R. B. (2004). *Principles and practice of structural equation modeling* (2nd ed.). New York: Guilford Press.

Loehlin, J. C. (2004). *Latent variable models* (4th ed.). Hillsdale, NJ: Erlbaum.

Marcoulides, G. A., & Moustaki, I. (Eds.) (2002). *Latent variable and latent structure models.* Mahwah, NJ: Erlbaum.

Marcoulides, G. A., & Schumacker, R. E. (Eds.). (1996). *Advanced structural equation modeling.* Mahwah, NJ: Erlbaum.

Marcoulides, G. A., & Schumacker, R. E. (Eds.). (2001). *New developments and techniques in structural equation modeling.* Mahwah, NJ: Erlbaum.

Maruyama, G. M. (1998). *Basics of structural equation modeling.* Thousand Oaks, CA: Sage.

Mueller, R. O. (1996). *Basic principles of structural equation modeling: An introduction to LISREL and EQS.* New York: Springer.

Muthén, L. K., & Muthén, B. (1998). *Mplus user's guide.* Los Angeles: Authors.

Raykov, T., & Marcoulides, G. A. (2000). *A first course in structural equation modeling.* Mahwah, NJ: Erlbaum.

Reise, S. P., & Duan, N. (Eds.). (2003). *Multilevel modeling: Methodological advances, issues, and applications.* Mahwah, NJ: Erlbaum.

Schumacker, R. E., & Lomax, R. G. (2004). *A beginner's guide to structural equation modeling* (2nd ed.). Hillsdale, NJ: Erlbaum.

Schumacker, R. E., & Marcoulides, G. A. (Eds.). (1998). *Interaction and nonlinear effects in structural equation modeling.* Mahwah, NJ: Erlbaum.

Sörbom, D. (2001). Karl Jöreskog and LISREL: A personal story. In R. Cudeck, S. Du Toit, & D. Sörbom (Eds.), *Structural equation modeling: Present and future* (pp. 3–10). Lincolnwood, IL: Scientific Software International.

Tremblay, P. F., & Gardner, R. C. (1996). On the growth of structural equation modeling in psychological journals. *Structural Equation Modeling: A Multidisciplinary Journal, 3,* 93–104.

Wolfle, L. M. (1999). Sewall Wright on the method of path coefficients: An annotated bibliography. *Structural Equation Modeling: A Multidisciplinary Journal, 6,* 280–291.

Wolfle, L. M. (2003). The introduction of path analysis to the social sciences, and some emergent themes: An annotated bibliography. *Structural Equation Modeling: A Multidisciplinary Journal, 10,* 1–34.

Wright, S. (1921). Correlation and causation. *Journal of Agricultural Research, 20,* 557–585

Wright, S. (1934). The method of path coefficients. *Annals of Mathematical Statistics, 5,* 161–215.

PART I

FOUNDATIONS

CHAPTER 2

THE PROBLEM OF EQUIVALENT STRUCTURAL MODELS

Scott L. Hershberger

This chapter provides a general review of *equivalent models* from a structural equation modeling perspective. Broadly defined, equivalent models differ in causal structure, but are identical in fit to the data. Narrowly defined, they are a set of models that yield identical (a) implied covariance, correlation, and other moment matrices when fit to the same data; (b) residuals and fitted moment matrices; (c) fit function and chi-square values; and (d) goodness-of-fit indices based on fit functions and chi-square. The social sciences' strong interest in equivalent models dates from Stelzl (1986); since then, psychometricians have studied model equivalence as a source of difficulty for inferring causal relations from structural equation models (McDonald, 2002). However, concurrent with work on statistical identification, econometricians were the first to investigate the properties of equivalent models formally (Koopmans, Rubin, & Leipnik, 1950); we can still see the influence of their work from the many contemporary econometric investigations that note the presence of equivalent models (e.g., Chernov, Gallant, Ghysels, & Tauchen, 2003). Arguably the most significant recent developments in model equivalence have

Structural Equation Modeling: A Second Course, 13–41

arisen from graph theorists' attempts to formulate a mathematical basis for causal inferences (Pearl, 2000). This is not surprising given the many shared characteristics of graph models and structural equation models. Yet one suspects that interest in model equivalence has been limited to only a few quantitatively oriented researchers; for example, two reviews of published studies found few researchers show any concern or even any awareness of the problem (Breckler, 1990; Roesch, 1999). In the following sections, I identify different types of equivalent models, and how their presence can be detected from the structure of a model.

GENERAL DEFINITION OF EQUIVALENT MODELS

Understanding the idea of equivalent models first requires knowing how model fit is evaluated in structural equation models. The basis for most fit indices used in structural equation modeling is the discrepancy between a sample covariance matrix of the observed variables (**S**) and the covariance matrix of these variables implied by the model $\hat{\boldsymbol{\Sigma}}$. Using numerical algorithms, parameter estimates describing the structure of a model are generated that produce an implied covariance matrix with least discrepancy from the observed sample covariance matrix. In the most uncommon situation, the implied covariance matrix is identical to the sample covariance matrix; thus, the discrepancy between **S** and $\hat{\boldsymbol{\Sigma}}$ (i.e., $\mathbf{S} - \hat{\boldsymbol{\Sigma}}$) is a matrix of zeros, **0**, implying a perfect fit of the data to the model (in this case, the model has zero residuals). Because the implied covariance matrix is bound by the restrictions imposed by the model, and because of the inevitable presence of sampling error, the sample and implied covariance matrices are almost never identical; that is, $\mathbf{S} - \hat{\boldsymbol{\Sigma}} \neq \mathbf{0}$ in most modeling situations. This (nonzero) value of the discrepancy yielded during parameter estimation is used in the computation of most fit indices. For example, the contribution of $\mathbf{S} - \hat{\boldsymbol{\Sigma}}$ to the goodness-of-fit index (GFI; Jöreskog & Sörbom, 1993) is clear:

$$\mathrm{GFI} = 1 - \frac{tr\left[(\mathbf{s} - \hat{\boldsymbol{\sigma}})'\mathbf{W}(\mathbf{s} - \hat{\boldsymbol{\sigma}})\right]}{tr\left[\mathbf{s}'\mathbf{W}\mathbf{s}\right]},$$

where *tr* is the trace operator, **s** and $\hat{\boldsymbol{\sigma}}$ refer to vectors of the nonduplicated elements of **S** and $\hat{\boldsymbol{\Sigma}}$, respectively, and **W** is a weight matrix that the method of parameter estimation defines (e.g., **W** is an identity matrix for unweighted least squares; **W** derives from $\mathbf{S}^{-1}$ for generalized least squares; and **W** derives from $\hat{\boldsymbol{\Sigma}}^{-1}$ for reweighted least squares, the asymptotic equivalent of maximum likelihood). The GFI is scaled to range between 0 and 1, with higher values signifying better fit. In those cases where the discrepancy value does not appear in the formula for a fit index, other terms are used that reflect the difference between the observed and implied matrices.

Equivalent models may be defined as a set of models that, regardless of the data, yield identical (a) implied covariance, correlation, and other moment matrices when fit to the same data, which in turn imply identical (b) residuals and fitted moment matrices, (c) fit functions and chi-square values, and (d) goodness-of-fit indices based on fit functions and chi-square. One most frequently thinks of equivalent models as described in (a) above. To be precise, consider two alternative models, denoted $M1$ and $M2$, each of which is associated with a set of estimated parameters (contained in vectors $\hat{\boldsymbol{\theta}}_{M1}$ and $\hat{\boldsymbol{\theta}}_{M2}$) and a covariance matrix implied by those parameter estimates (denoted as $\hat{\boldsymbol{\Sigma}}_{M1}$ and $\hat{\boldsymbol{\Sigma}}_{M2}$). Models $M1$ and $M2$ are considered equivalent if, for any sample covariance matrix **S**, the implied matrices $\hat{\boldsymbol{\Sigma}}_{M1} = \hat{\boldsymbol{\Sigma}}_{M2}$, or alternatively, $(\mathbf{S} - \hat{\boldsymbol{\Sigma}}_{M1}) = (\mathbf{S} - \hat{\boldsymbol{\Sigma}}_{M2})$. That is to say, model equivalence is not defined by the data, but rather by an algebraic equivalence between the models' parameters. In turn, because of this equivalence, the values of statistical tests of global fit (e.g., test of exact fit, tests of close fit) will be identical, as will goodness-of-fit indices that are based on the discrepancy between the sample covariance matrix and the model-implied covariance matrix. Thus, even when a hypothesized model fits well according to multiple fit indices, there may be equivalent models with identical fit—even if the theoretical implications of those models are very different.

As an example, I define a model involving three observed variables, V1, V2, and V3. This model is shown in Figure 2.1 as *Model 1*. The covariance matrix for the three variables (**S**), the set of parameter estimates obtained from fitting the model to the covariance matrix ($\hat{\boldsymbol{\theta}}_{M1}$) using maximum likelihood estimation, and the covariance matrix implied by the parameter estimates ($\hat{\boldsymbol{\Sigma}}_{M1}$) are

$$\mathbf{S} = \begin{bmatrix} 2.000 & & \\ 0.940 & 1.800 & \\ 1.030 & 0.650 & 1.500 \end{bmatrix},$$

$$\hat{\boldsymbol{\theta}}_{M1} = \begin{bmatrix} \hat{b}_{V2V1} = 0.470 \\ \hat{b}_{V3V2} = 0.361 \\ \hat{c}_{V1} = 2.000 \\ \hat{c}_{E2} = 1.358 \\ \hat{c}_{E3} = 1.265 \end{bmatrix},$$

$$\hat{\boldsymbol{\Sigma}}_{M1} = \begin{bmatrix} 2.000 & & \\ 0.940 & 1.800 & \\ 0.339 & 0.650 & 1.500 \end{bmatrix}.$$

A chi-square goodness-of-fit test suggests that *Model 1* is consistent with the data, but the process of model confirmation has not been successfully completed. There exist two models equivalent to *Model 1*: *Model 2* and *Model 3*, also shown in Figure 2.1. Fitting *Model 2* and *Model 3* to the covariance matrix **S** results in two new sets of maximum likelihood parameter estimates:

$$\hat{\boldsymbol{\theta}}_{M2} = \begin{bmatrix} \hat{b}_{V2V1} = 0.522 \\ \hat{b}_{V3V2} = 0.433 \\ \hat{c}_{V1} = 1.500 \\ \hat{c}_{E2} = 1.509 \\ \hat{c}_{E3} = 1.518 \end{bmatrix}, \qquad \hat{\boldsymbol{\theta}}_{M3} = \begin{bmatrix} \hat{b}_{V2V1} = 0.522 \\ \hat{b}_{V3V2} = 0.361 \\ \hat{c}_{V1} = 1.800 \\ \hat{c}_{E2} = 1.509 \\ \hat{c}_{E3} = 1.265 \end{bmatrix},$$

Yet, despite the different parameterizations of *Models 1*, *2*, and *3*, all have the same model-implied covariance matrix:

$$\hat{\boldsymbol{\Sigma}}_{M1} = \hat{\boldsymbol{\Sigma}}_{M2} = \hat{\boldsymbol{\Sigma}}_{M3} = \begin{bmatrix} 2.000 & & \\ 0.940 & 1.800 & \\ 0.339 & 0.650 & 1.500 \end{bmatrix}.$$

Therefore, because *Models 1*, *2*, and *3* are equivalent for all **S**, we observe for this specific example $(\mathbf{S} - \hat{\boldsymbol{\Sigma}}_{M1}) = (\mathbf{S} - \hat{\boldsymbol{\Sigma}}_{M2}) = (\mathbf{S} - \hat{\boldsymbol{\Sigma}}_{M3})$.

Model 1:

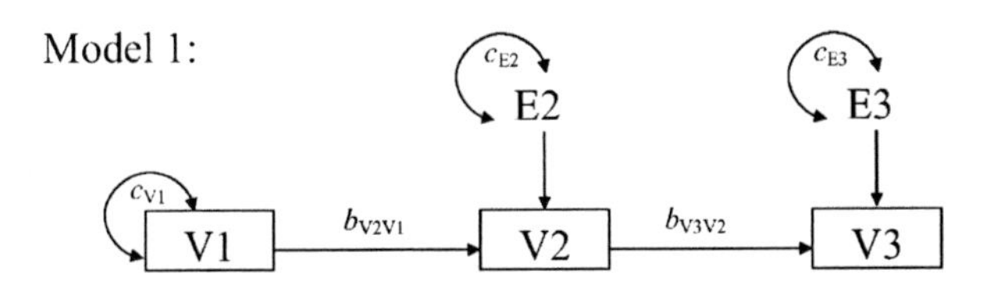

Model 2:

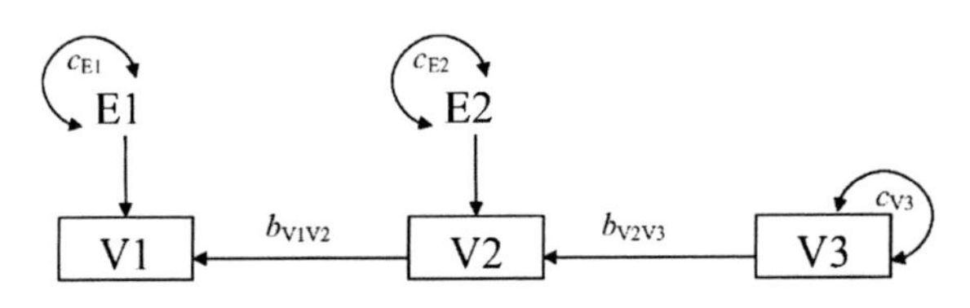

Model 3:

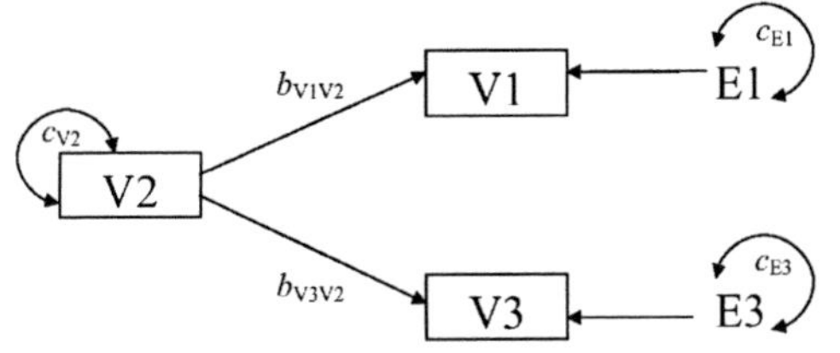

Figure 2.1. Three equivalent models.

WHY EQUIVALENT MODELS ARE IMPORTANT TO IDENTIFY

Admittedly, model equivalence can be a very dry subject, exciting to only a few. And as described later, the entire family of equivalent models can also be very difficult to identify, using statistical techniques that are far from easy to apply. Considering these unattractive attributes, we require a compelling reason for model equivalence to be a significant factor in drawing causal inferences from models. A compelling reason does exist: Model equivalence is methodologically significant because it dramatically exposes the limitations of structural equation modeling to test theories. Equivalent models remind us that we never test *a* model but rather a whole class of models from which the hypothesized model cannot be distinguished by any statistical means.

Although it is true that (overidentified) structural equation models have testable implications, those implications are only one part of what a model represents: a set of claims, assumptions, and implications. Equivalent models can cause damage by not only offering false causal relations; they may also lead one to accept their equally false implications. For example, Williams, Bozdogan, and Aiman-Smith (1996) discussed many hypothesized models of job satisfaction and its correlates; for each model, at least one model equivalent to it altered both the size and direction of the job satisfaction relationships. The impressions received about what factors contribute to job satisfaction very much depended on which equivalent model was selected.

Most would agree that alternative models should always be considered. Yet why are equivalent models considered a special threat to causal inference, much more so than alternative but nonequivalent models? After all, we could propose alternative but nonequivalent models that fit the data almost as well as the hypothesized model, alternative models so close in fit as to render trivial the choice among the models based solely on fit. The reason equivalent models are so special is that one of them can *never* be supported without *all* of them being supported. On the other hand, alternative models that fit almost as well as a hypothesized model in one sample could be discarded if their fit decreases substantially in a new sample.

In the remainder of this chapter, different types of model equivalence are first discussed, and then strategies for identifying them are presented. I describe important statistical and conceptual similarities between model equivalence and model identification next. The final section of the chapter offers some suggestions for selecting among equivalent models.

TYPES OF MODEL EQUIVALENCE

Observationally Equivalent/Covariance Equivalent

Model equivalence is a generic term that may refer to one of several situations of equivalence. Hsiao (1983), Luijben (1991), McDonald (2002), Pearl (2000), and Raykov and Penev (1999) distinguish among several different types of model equivalence. Two models are *observationally equivalent* only if one model can generate every probability distribution that the other model can generate. Observational equivalence is model equivalence in the broadest sense, and can be shown using data of any type. In contrast, models are *covariance equivalent* if every covariance matrix generated by one can be generated by the other. Thus, observational equivalence encompasses covariance (model) equivalence; that is, observational equivalence requires the identity of individual data values, whereas covariance equivalence requires the identity of summary statistics such as covariances and variances. Thus, observationally equivalent models are always covariance equivalent; on the other hand, covariance equivalent models might not be observationally equivalent. For the purposes of the current chapter, the term *equivalent* will be used specifically to refer to models that are covariance equivalent, as well as *globally equivalent* as described next.

Globally Equivalent/Locally Equivalent

A distinction can also be drawn between *globally equivalent* models and *locally equivalent* models. For two models to be globally equivalent, *a function* must exist that translates *every* parameter of one model into the parameters of another model. The result of this mapping of one parameter set into another is an identity between the implied covariance matrices of the two models. On the other hand, if only a *subset* of one model's parameter set is translatable into the parameter set of another model, the models are *locally equivalent*. Local equivalence does not guarantee that the models' implied covariance matrices will be the same.

Global Equivalence Example. This example involves a factor analysis model (*M0*) with one common factor F1 for three observed variables, V1, V2, and V3, specified as

$$\begin{bmatrix} \text{V1} \\ \text{V2} \\ \text{V3} \end{bmatrix} = \begin{bmatrix} b_{\text{V1F1}} \\ b_{\text{V2F1}} \\ b_{\text{V3F1}} \end{bmatrix} \text{F1} + \begin{bmatrix} \text{E1} \\ \text{E2} \\ \text{E3} \end{bmatrix}.$$

The parameters b_{V1F1}, b_{V2F1}, and b_{V3F1} are factor loadings; E1, E2, and E3 are error terms with variances c_{E1}, c_{E2}, and c_{E3}, and covariances c_{E1E2}, c_{E1E3}, and c_{E2E3}; and c_{F1} is the variance of F1. Three constraints have been placed on the parameters of *M0*: (1) $c_{\text{F1}} = 1$, (2) $c_{\text{E1}} = c_{\text{E2}} = c_{\text{E3}} = c_{\text{E}}$, and (3) $c_{\text{E2E1}} = c_{\text{E3E1}} = c_{\text{E3E2}} = c_{\text{EE}}$. Because of these constraints, *M0* is overidentified with one *df* (i.e., six variances and covariances minus the five parameter estimates: b_{V1F1}, b_{V2F1}, b_{V3F1}, the common error variance c_{E}, and the common error covariance c_{EE}).

Now consider two other models, *M1* and *M2*, both derived from *M0*. In *M1* the covariance c_{E3E2} is constrained to be zero; in *M2* the covariance c_{E2E1} is constrained to be zero. The implied covariance matrix $\hat{\Sigma}_1$ for *M1* is

$$\begin{bmatrix} b^2_{\text{V1F1}} + c_{\text{E}} & & \\ b_{\text{V2F1}} b_{\text{V1F1}} + c_{\text{EE}} & b^2_{\text{V2F1}} + c_{\text{E}} & \\ b_{\text{V3F1}} b_{\text{V1F1}} + c_{\text{EE}} & b_{\text{V3F1}} b_{\text{V2F1}} & b^2_{\text{V3F1}} + c_{\text{E}} \end{bmatrix},$$

and the implied covariance matrix $\hat{\Sigma}_2$ for *M2* is

$$\begin{bmatrix} b_{\mathrm{V1F1}}^2 + c_{\mathrm{E}} & & \\ b_{\mathrm{V2F1}} b_{\mathrm{V1F1}} & b_{\mathrm{V2F1}}^2 + c_{\mathrm{E}} & \\ b_{\mathrm{V3F1}} b_{\mathrm{V1F1}} + c_{\mathrm{EE}} & b_{\mathrm{V3F1}} b_{\mathrm{V2F1}} + c_{\mathrm{EE}} & b_{\mathrm{V3F1}}^2 + c_{\mathrm{E}} \end{bmatrix}.$$

Both models are fit to

$$\mathbf{S} = \begin{bmatrix} s_{\mathrm{V1}}^2 & & \\ s_{\mathrm{V2V1}} & s_{\mathrm{V2}}^2 & \\ s_{\mathrm{V3V1}} & s_{\mathrm{V3V2}} & s_{\mathrm{V3}}^2 \end{bmatrix}.$$

If a function exists that maps the parameters of *M1* onto the parameters of *M2*, thereby yielding the same model-implied covariance matrix (i.e., $\hat{\Sigma}_1 = \hat{\Sigma}_2$) for all possible observed covariance matrices **S**, then they are globally equivalent. In contrast, if a function cannot be found that transforms every parameter, then, at best, the two models are locally equivalent.

For example, in *M1* the parameter b_{V1F1} can be shown to have an implied value of

$$\frac{(s_{\mathrm{V1}}^2 - s_{\mathrm{V3V1}})\sqrt{s_{\mathrm{V1}}^2 - s_{\mathrm{V2V1}} - s_{\mathrm{V3V2}}}}{s_{\mathrm{V3V2}}},$$

while in *M2* it has an implied value of

$$-\frac{s_{\mathrm{V2V1}}}{\sqrt{s_{\mathrm{V2}}^2 - s_{\mathrm{V3V2}}}}.$$

The global equivalence of the two models is confirmed by noting that implied values of b_{V1F1} in *M1* and *M2* can be shown algebraically to be equal for all **S**; that is, $b_{\mathrm{V1F1}}(M1) = b_{\mathrm{V1F1}}(M2)$. Similarly, the implied values of the other four parameters in *M1* (i.e., b_{V2F1}, b_{V3F1}, c_{E}, and c_{EE}) are algebraically equivalent to their counterparts in *M2*. Furthermore, at one *df*, both *M1* and *M2* are overidentified, with each of their parameters having more than one implied value. These implied alternatives are also algebraically identical in *M1* and *M2*.

The strategies described later in this chapter identify globally equivalent models; fewer strategies have been proposed for identifying locally equivalent models. The methods of identifying local equivalence depend on a model's *Jacobian matrix*. What a Jacobian matrix is and how it is used to identify local equivalence are complex topics, well outside the purview of this chapter. We refer the reader to Bekker, Merckens, and Wansbeek (1994), an excellent source of information on local equivalence.

STRATEGIES FOR IDENTIFYING EQUIVALENT MODELS

The best time to detect models equivalent to one's theoretical model is before a significant amount of resources have been spent on a research project. An ideal time would be before data collection begins, when the study is still in the planning stage. Forewarned is forearmed: Becoming aware of the presence of equivalent models should motivate the researcher to (a) acknowledge the potentially severe inferential limitations regarding a model achieving satisfactory data–model fit; (b) revise the model so that the number of equivalent models is substantially decreased; (c) replace structural equation modeling as the method of statistical analysis with a technique that will not be compromised by the presence of equivalent models; or (d) cancel the study. Alternative (a), although important, is rather minimal alone, whereas (b) could be considered highly desirable; but first, a method must be used to find or detect the presence of equivalent models without recourse to data. This section describes strategies proposed for determining the number and nature of equivalent models.

The Replacing Rule and Structural Models

The four rules developed by Stelzl (1986) for completely recursive structural models may be simplified (and extended to models with nonrecursive relationships) by use of the *replacing rule* for locating covariance equivalent models. This more general rule was shown by Lee and Hershberger (1990) to subsume Stelzl's four rules. Before the replacing rule is explained, however, it is first necessary to define several terms. A structural model can be divided into three blocks: a *preceding block*, a *focal block*, and a *succeeding block*. Figure 2.2 shows the location of the three blocks in the "initial model." The focal block includes the relations we are interested in altering to produce an equivalent model; the preceding block consists of all the variables in the model that causally precede the variables in the focal block; and the succeeding block consists of all the variables in the model that causally succeed the variables in the focal block. Within the econometrics literature, the division of a structural model into blocks, where recursiveness exists between and within blocks, is termed a *block recursive system*. For the application of the replacing rule, *limited block recursiveness* is required. In the limited block recursive model, the relations between the blocks and within the focal block are recursive; however, relations within the preceding and succeeding blocks may be nonrecursive. (Under certain conditions discussed below, relations within the focal block may be nonrecursive. For now, we will assume a fully recursive focal block.)

The replacing rule is defined as follows. Let V*i* and V*j* represent two variables in a focal block, where V*i* and V*j* within the focal block stand in the relationship V*i*→V*j*; V*i* is a *source variable* and V*j* is an *effect variable* in the focal block. Both V*i* and V*j* receive paths from other variables in the preceding and focal blocks, and they may also send paths to other variables in the succeeding block. Let E*i* and E*j* represent the residuals of V*i* and V*j*. According to the replacing rule, the directed path between V*i* and V*j* (i.e., V*i*→V*j*) may be replaced by their residual covariance (i.e., E*i*↔E*j*)

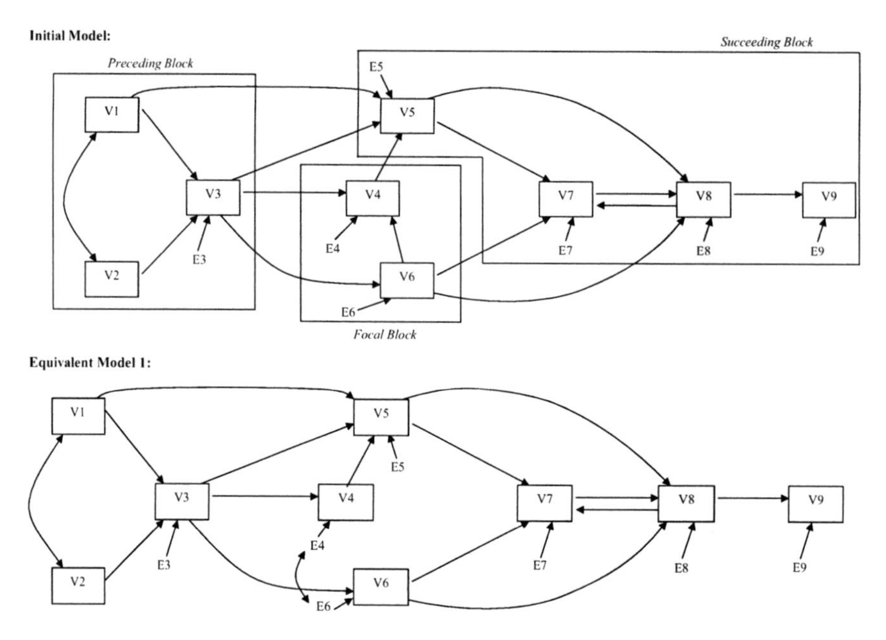

Figure 2.2. An example of the generation of equivalent structural models using the replacing rule.

if the predictors of the effect variable V*j* are the same as or include those of the source variable V*i*. The reverse application of the replacing rule will also produce an equivalent model; that is, a residual covariance, E*i*↔E*j*, may be replaced by a directed path, V*i*→V*j* or V*i*←V*j*, when the predictors of the effect variable are the same as or include those of the source variable following the replacement.

An example of a model to which the replacing rule can be applied is shown in Figure 2.2. In the initial model of Figure 2.2, V1, V2, and V3 are defined as a preceding block, V4 and V6 as a focal block, and V5, V7, V8, and V9 as a succeeding block. Within the focal block, the directed path V6→V4 may be replaced by the residual covariance E6↔E4; the predictor V3 influences both the source (V6) and effect (V4) variables in the focal block. This application of the replacing rule is shown in *Model 1*, a model equivalent to the initial model. This example also illustrates an important special case of the replacing rule, wherein both the source and effect variables have the same predictors. In this situation, the variables in the focal block are said to be *symmetrically determined*, and it is a matter of statistical indifference if the directed path between the source and effect variables is reversed or replaced by a residual covariance. The source and effect variables are said to be in a *symmetric focal block*.

Frequently, applications of the replacing rule will create equivalent models, models that were not equivalent prior to application of the replacing rule. For example, let a new focal block be defined as consisting of V4 and V5, with V4 as the source variable and V5 as the effect variable. Within the initial model, application of the replacing rule to this focal block is not possible because both V4 and V5 have one predictor not shared with each other. After the alteration to the relation between V4 and V6 shown in *Model 1*, however, the effect variable V5 has as one of its predictors the only predictor (V3) of the source variable V4. In equivalent *Model 2* of Figure 2.3, the directed path V4→V5 has been replaced by the residual covariance E4↔E5. The replacing rule will not only generate a model equivalent to a hypothesized model, but on occasion, a model that has other models equivalent to it—but that are *not* equivalent to the hypothesized model. Such was the case with *Model 1*, generated from the initial model by replacing V4→V6 with E4↔E6. *Model 2* can be generated by replacing V4→V5 of *Model 1* with E4↔E5. However, *Model 2* is not equivalent to the initial model, suggesting that model equivalence is not transitive. That *Model 2* is not equivalent to the initial model can be shown by fitting the two models to the same covariance matrix: the fit of the two models will not be identical. Because the initial model is not equivalent to *Model 2*, they may differ in the number of models equivalent to them.

Equivalent Model 2:

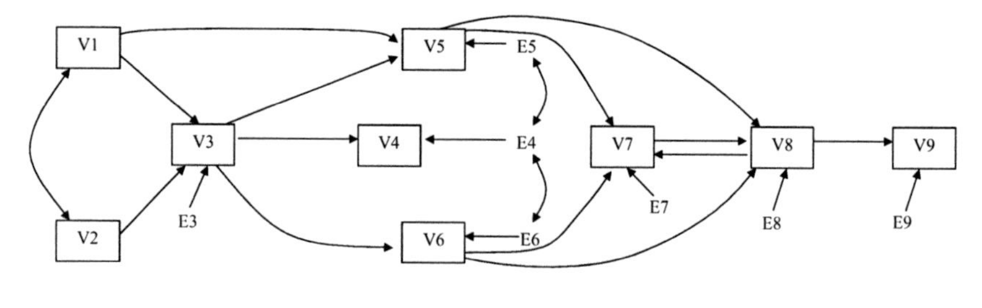

Equivalent Model 3:

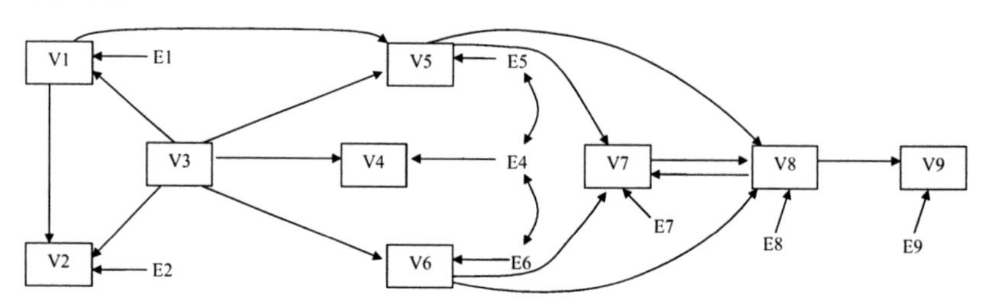

Figure 2.3. An example of equivalent models generated by the replacing rule, not equivalent to the initial model.

This example also illustrates that focal blocks are not invariantly defined within a structural model: Any two variables within the same model may be designated as part of a focal block if the focal block meets the requirements of the replacing rule. For instance, neither V3→V4, V5→V7, nor V6→V7 may be defined as a focal block for the application of the replacing rule: The source variables in all three cases have predictors not shared with the effect variable, and in addition, defining V5→V7 or V6→V7 as the focal block would require defining a succeeding block (V8 and V9) nonrecursively connected to a focal block.

Another special case of the replacing rule occurs when a preceding block is completely saturated, and all variables are connected by either directed paths or residual covariances. In this case, the preceding block is defined as a focal block, and any residual covariance/directed path replacement yields an equivalent model. The replacing rule may be applied whenever the necessary conditions are met; the existence of a preceding or a succeeding block is not absolutely necessary. In the initial model (Figure 2.2), V1, V2, and V3 can be designated as a focal block. The results of applying the replacing rule to this saturated focal block are shown in *Model 3* in Figure 2.3, where V1↔V2 has been replaced by V1→V2, V1→V3 has been replaced by V1←V3, and V2→V3 has been replaced by V2←V3. Furthermore, application of the replacing rule to

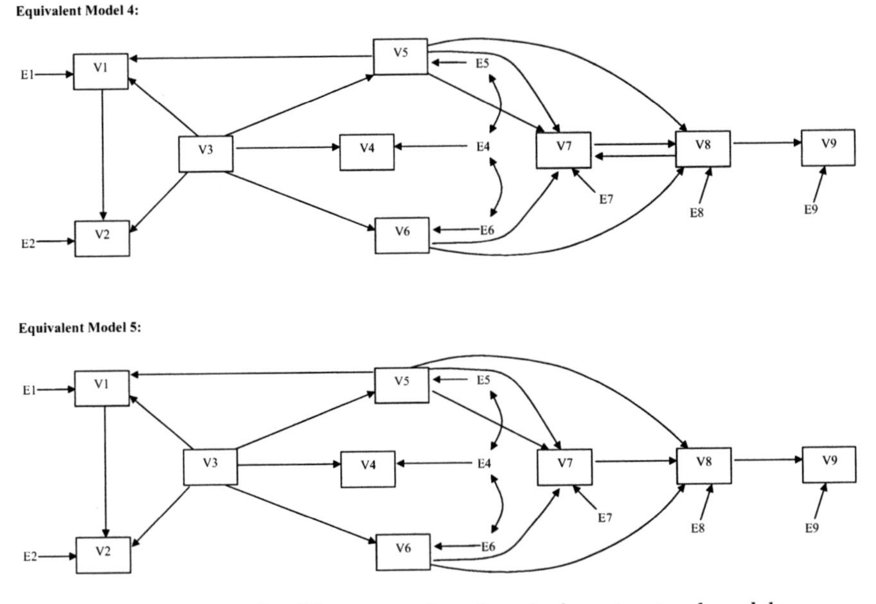

Figure 2.4. An example of the generation of equivalent structural models appling the replacing rule to a symmetric focal block.

this focal block has provided yet another heretofore unavailable opportunity to generate an equivalent model. Variables V1 and V5 now define a symmetric focal block, thus permitting the substitution of V1→V5 by E1↔E5 or V1←V5, of which the latter substitution is shown in *Model 4* in Figure 2.4.

Symmetric focal blocks also provide a basis for generating equivalent models when two variables are nonrecursively connected. Whenever two nonrecursively connected variables define a symmetric focal block (i.e., have all their predictors in common), the nonrecursive path between them may be replaced by a single directed path or a residual covariance. Importantly, the nonrecursive paths must be specified as equal due to the impossibility of uniquely identifying each path when both variables have the same predictors. The converse of this rule holds as well: Directed paths or residual covariances within a symmetric focal block may be replaced by equated nonrecursive paths (symbolized herein by ⇔). In *Model 4*, let V7⇔V8 be a focal block where V7→V8 = V7←V8; then V7⇔V8 may be replaced by either V7→V8, V7←V8, or E7↔E8. *Model 5* presents a model equivalent to all of the prior models where now V7→V8. Thus, under the special circumstance of a symmetric focal block, the replacing rule can be applied to nonrecursive paths.

It is interesting to note that if, in *Model 4*, V8←V9 was specified instead of V8→V9, then V7 and V8 would not have defined a symmetric focal block, thus preventing application of the replacing rule. In the alternative circumstance where V9 is an additional predictor of V8, V9 would have held the position of an *instrumental variable* allowing the unconstrained estimation of the nonrecursive paths between V7 and V8. Instrumental variables are an example of a more general strategy within structural equation modeling for the reduction of equivalent models. Instrumental variables are predictor variables added to a model in order to help identify the model. To the degree that variables in a model do not share predictors, the number of equivalent models will be reduced. Symmetric focal blocks exemplify the model equivalence difficulties that arise when focal block variables share all their predictors, and just-identified preceding blocks illustrate the problem even more so. In general, constraining the number of free parameters within a model will reduce the number of equivalent models.

The Reversed Indicator Rule and Measurement Models

Measurement models are factor analysis models, but with an important difference from traditional (exploratory) factor analysis models: The traditional factor analysis model is underidentified, with the resulting factor pattern subject to an infinite number of rotational solutions. Therefore, an infinite number of equivalent solutions exist. On the other hand, the specific (confirmatory) pattern of factor loadings obtained from structural equation modeling is at least just-identified (or should be) and is unique up to and including rotation. For a given set of variables, and a hypothesized factor structure, equivalent measurement models would appear not to be problematic. However, model equivalence is still very much an issue for two reasons.

First, the pattern of correlations among a set of indicators may indicate the existence of two or more distinct factors nested within the general common factor. For example, the basic measurement model in Figure 2.5 depicts a single common factor defined by five indicators. Assume that this model was tested on a set of data and rejected, the rejection apparently stemming from a relatively high correlation between V1 and V2 (i.e., higher than their correlation with the other three variables). In order to modify the model so as to accommodate the relatively high correlation between V1 and V2, any one of four mutually equivalent measurement models could be specified. These four equivalent models are shown in Figure 2.5. If tested against a set of data, identical fit statistics would be obtained for each model. Whenever a model is modified to accommodate

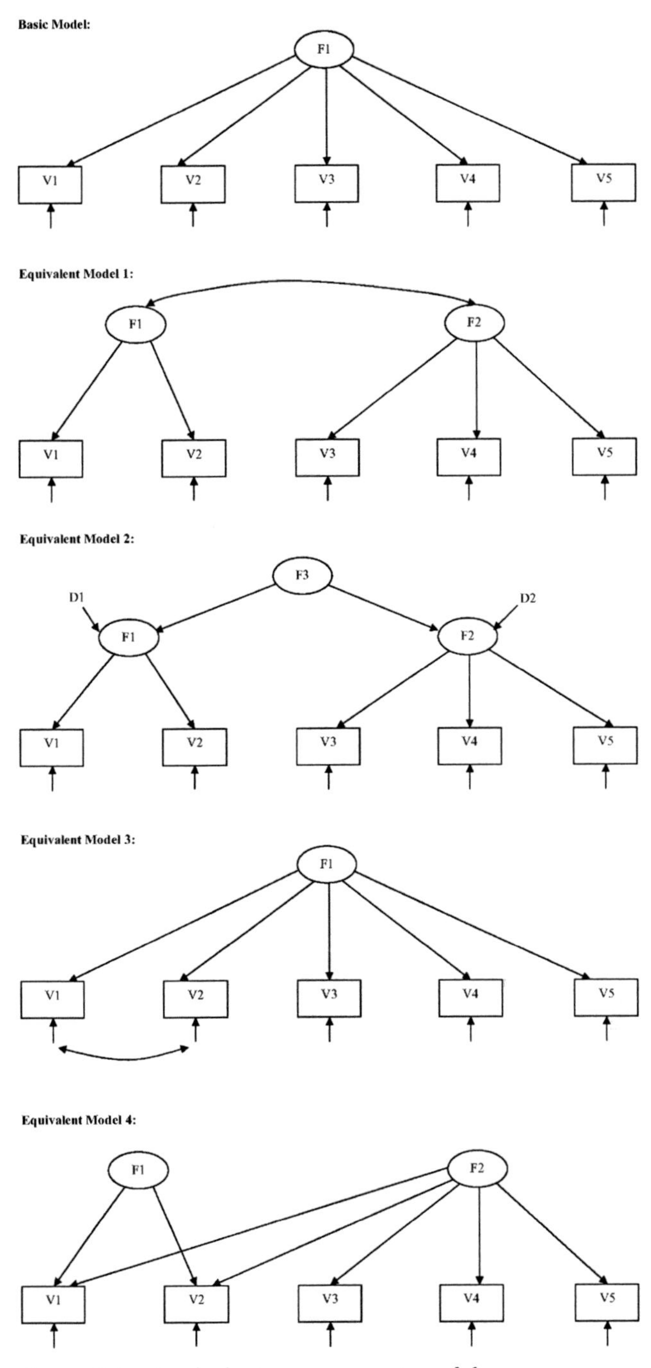

Figure 2.5. Equivalent measurement models.

a sizable but unmodeled correlation between two variables, the potential for model equivalence increases due to the many ways in which the source of the correlation may be represented. Another factor may be specified (*Models 1, 2,* and *4*), or the correlation between the two variables may be left as a residual covariance (*Model 3*). If the model with a second factor is selected, the choice then is whether the association between the two factors will be represented by an unanalyzed correlation (*Model 1*), a higher order factor (*Model 2*, with, e.g., $b_{F1F3} = b_{F2F3}$), or shared indicators (*Model 4*).

The second reason why model equivalence will be a problem for measurement models is described by the *reversed indicator rule* (Hershberger, 1994). According to the reversed indicator rule, the direction of a path between a latent variable and an indicator is arbitrary for one and only one latent-indicator path in the measurement model of an exogenous factor, and therefore, this path may be reversed. A single recursive path may also be replaced by equated nonrecursive paths. Reversing the direction of the path, or replacing the recursive path with equated nonrecursive paths, results in an equivalent model. Two requirements exist, in addition to the reversal of only one indicator, for application of the reversed indicator rule: (1) the measurement model should be completely exogenous or affected by only one indicator before and after the rule's application, and (2) the exogenous latent variable must be uncorrelated with other exogenous latent variables before and after the rule's application.

In Figure 2.6, three equivalent measurement models are shown. These three models, equivalent by the reversed indicator rule, imply very different conceptualizations. In Figure 2.6a, a latent variable exerts a common effect on a set of indicators. In contrast, in Figure 2.6b, a latent variable is influenced by an indicator. Figure 2.6c shows another equivalent model, in which the latent variable influences, and is influenced by, an indicator.

Summary of Rules for Specifying Equivalent Structural Models

The Replacing Rule—General Case. Given a directed path between two variables, V1→V2 in a focal block, the directed path may be replaced by the residual covariance, E1↔E2, if

1. Only recursive paths exist between blocks;
2. Only recursive paths exist within a focal block (condition may be relaxed when the block is a symmetric focal block); and
3. The predictors of the effect variable are the same as or include those for the source variable.

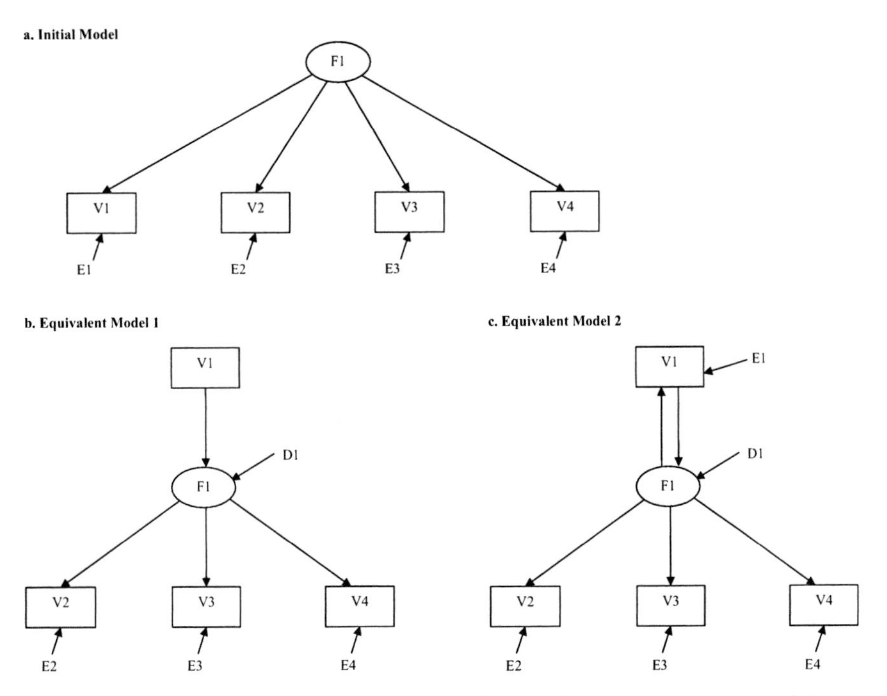

Figure 2.6. An example of the generation of equivalent measurement models using the reversed indicator rule.

Replacing Rule—Symmetric Focal Blocks. Given a directed path between two variables, V1→V2 in a focal block, the directed path may be replaced by the residual covariance, E1↔E2, or V1←V2, or V1⇔V2, if

1. Only recursive paths exist between blocks;
2. When the original specification is replaced by a nonrecursive relation, the two paths are equated; and
3. The source and effect variables have the same predictors.

Replacing Rule—Just-Identified Preceding Block. Given a directed path between two variables, V1→V2, in a focal block that is a just-identified preceding block,[1] the directed path may be replaced by the residual covariance E1↔E2, by V1←V2, or by V1⇔V2, if

1. Only recursive paths exist between the preceding (focal) block and succeeding block; and
2. When the original specification is replaced by a nonrecursive relation, the two paths are equated.

Summary of Rules for Specifying Equivalent Measurement Models

The Reversed Indicator Rule. Given a directed path between a latent variable and an indicator, F1→V1, the directed path may be replaced by F1←V1, or by F1⇔V1, if

1. Only the directed path for one indicator for each measurement model is altered;
2. The latent variable is affected by a single indicator, or is completely exogenous, before and after application of the rule;
3. The latent variable is uncorrelated with other exogenous latent variables; and
4. When the original specification is replaced by a nonrecursive relation, the two paths are equated.

For any model consisting of both measurement and structural portions, the model equivalence rules may be applied to both portions, as long as the requirements of the rules for both are met.

Other Strategies

The replacing rule and the reversed indicator rule are hardly the only strategies available for identifying equivalent models based on their structure alone, without resort to data. I mention two others below.

Of late, advances in graph theory have helped to detect equivalent models. These advances have resulted from efforts to provide causal inferences with a more objective, mathematical foundation than had been available in the past. The strategy providing this foundation involves translating the causal relationships implied by graph models (which are very similar to structural equation models) into statistical relationships. The logic here is that we expect certain statistical relationships only if a proposed causal relationship is true. Zero-order partial correlations are a statistical relationship important here because a causal relationship implies a particular pattern of partial correlations in the model. Furthermore, for recursive models at least, partial correlations are sufficient: a set of zero-order partial correlations can fully characterize the statistical content of a structural model that the causal relationships embodied in the model entail. This should come as no surprise, because we found that the replacing rule worked because the only changes in a model implied by the rule did not alter the model's partial correlation structure. Algorithms, most of which can only be carried out by computer, have been developed for finding models that imply the same partial correlations but differ in causal structure. Some of these models turn out to be equivalent.[2] The generality of the algorithms vary; some are restricted to completely recursive latent variable models (Spirtes, Glymour, & Scheines, 1993), others are suitable for nonrecursive latent variable models, complete with feedback loops and correlated errors (Richardson, 1997).

The graph theory approach to identifying equivalent models is far too complex for further discussion in this chapter. To learn more, the two excellent introductory texts are by Glymour, Scheines, Spirtes, and Kelly (1987) and Shipley (2000); a more difficult but clear and comprehensive account is by Pearl (2000).

A second approach to identifying equivalent models uses the rank of the matrix of correlations among the parameters of the model (Bekker et al., 1994).[3] To use this method, one must combine the matrices from two models *suspected* of equivalence. If the rank of the combined matrix is less than the sum of the ranks of the separate matrices, the two models are minimally *locally equivalent*. Although with the correct software the rank method is straightforward in application, it does have three disadvantages: (1) it is conceptually very difficult to understand without advanced knowledge of linear algebra; (2) only the local equivalence of two models is tested; and (3) models suspected of equivalence must be known beforehand.

MODEL EQUIVALENCE AND MODEL IDENTIFICATION

Rigorous mathematical treatment of the identification problem began in the 1950s with the work of Koopmans and Reiersøl (1950). Econometricians contributed much to the development of identification; probably the best-known contribution was the introduction of criteria for the identification of simultaneous equation systems (e.g., the full rank condition). Fisher (1966) provided a thorough treatment of identification for linear equation systems and Bowden (1973) extended identification theory to embrace nonlinear models. Hannan and Deistler (1988) generalized the theory of identification for linear structural equations with serially correlated residual errors, establishing conditions for identifiability for the important case of auto-regressive moving average error terms. Some more recent, important contributions to the identification of structural equation models have been made by Rigdon (1995) and Walter and Pronzato (1997).

Identification refers to the uniqueness of the parameterization that generated the data. Hendry (1995) specified three aspects of identification:

(1) uniqueness, (2) correspondence to a "desired entity," and (3) a satisfying interpretation of a theoretical model. As an analogy, the beta weight determined by regressing the quantity of a product sold on the price charged is uniquely determined by the data, but need not correspond to any underlying economic behavior, and may be incorrectly interpreted as a supply schedule due to rising prices.

We categorize a model's identifiability in one of two ways: (1) *Global identification*, in which *all* of a model's parameters are identified, or in contrast, (2) *local identification*, in which at least one—but not all—of a model's parameters is identified. Globally identified models are locally identified, but locally identified models may or may not be globally identified. Global identification is a prerequisite for drawing inferences about an entire model. When a model is not globally identified, local identification of some of its parameters permits inferential testing in only that section of the model. At least in theory—but debatable in practice—nonidentified parameters do not influence the values of ones that are identified. Thus, the greater the local identifiability of a model, the stronger the inferences that can be drawn from the subset of identified equations.

Besides categorizing models as either globally or locally identified, models can be (1) *underidentified*, (2) *just-identified*, or (3) *overidentified*. Underidentification occurs when not enough relevant data are available to obtain unique parameter estimates. Note that when the degrees of freedom of a model is negative, at least one of its parameters is underidentified. However, positive degrees of freedom does not prove that each parameter is identified. Thus, negative degrees of freedom is a sufficient but not a necessary criterion for showing that a model is globally underidentified.

Just-identified models are always identified in a trivial way: Just-identification occurs when the number of data elements equals the number of parameters to be estimated. If the model is just-identified, a solution can always be found for the parameter estimates that will result in perfect fit—a discrepancy function equal to zero. This implies that both the chi-square goodness-of-fit test and its associated degrees of freedom will be zero. As such, there is no way one can really test/confirm the veracity of a just-identified model. In fact, all just-identified models are equivalent, but not all equivalent models are just-identified.

If theory testing is one's objective, the most desirable identification status of a model is *overidentification*, where the number of available data elements is more than what we need to obtain a unique solution. In an overidentified model, the degrees of freedom are positive so that we can explicitly test model fit. An overidentified model implies that, for at least one parameter, there is more than one equation the estimate of a parameter must satisfy: Only under these circumstances—the presence of multiple solutions—are models provided with the opportunity to be rejected by the data.

Though a family of equivalent models may be statistically identified, equivalent models are underidentified in their causal assumptions, suggesting that an intimate association is present among them. To begin, both underidentification and model equivalence are sources of indeterminacy: Underidentification requires the selection of a parameter's best estimate from an infinite number of estimates, while model equivalence requires the selection of the best model from among many (if not infinite; Raykov & Marcoulides, 2001) models. In one case, we cannot determine the correct population value of a parameter, and in the other we cannot determine the correct population model. A comparison of the mapping functions of identified parameter and equivalent model sets illustrates the similarity, and subtle difference, between identification and model equivalence. Recall that for models to be globally equivalent, a function must result in a one-to-one mapping associating (reparameterizing) each parameter of one model with every parameter of a second model, *without remainder* (see Figure 2.7). One-to-one mappings that exhaust every element of two sets are *bijective*. On the other hand, for a model to be identified, a one-to-one mapping of each parameter onto a unique estimate of that parameter must be *injective*. We illustrate injective one-to-one mapping in Figure 2.8a. An injective mapping associates two sets so that all members of set A are paired with a unique member of set B, *but with remainder*. The set A elements do not exhaust the elements of set B; there are some elements within set B that are not paired with a member of set A. For a model to be identified, all of a model's parameters within set A must be paired with a unique parameter estimate in set B. Figure 2.8b shows

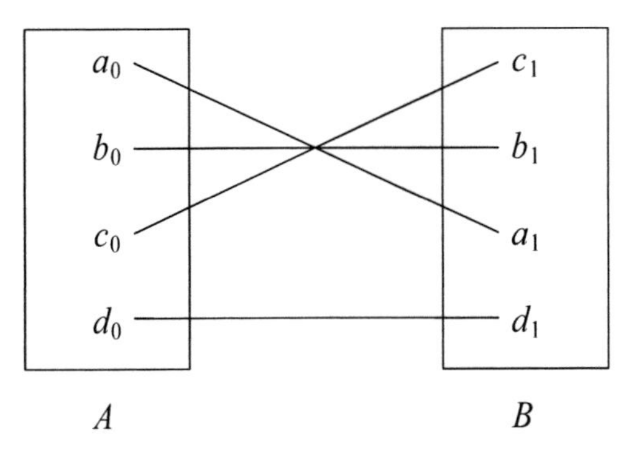

Figure 2.7. One-to-one mapping function of two models' parameters required for global equivalence.

a. Identified

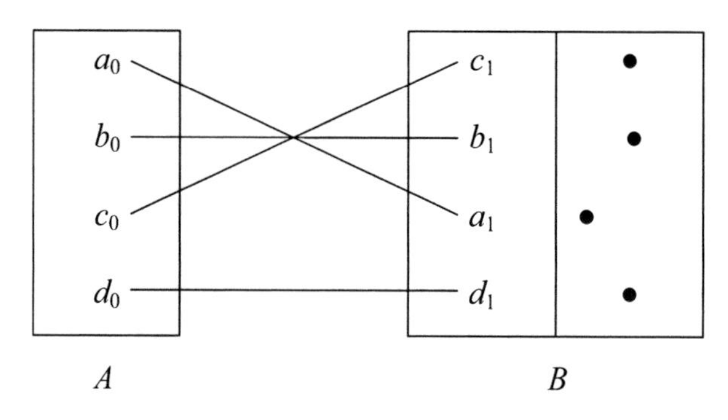

b. Underidentified

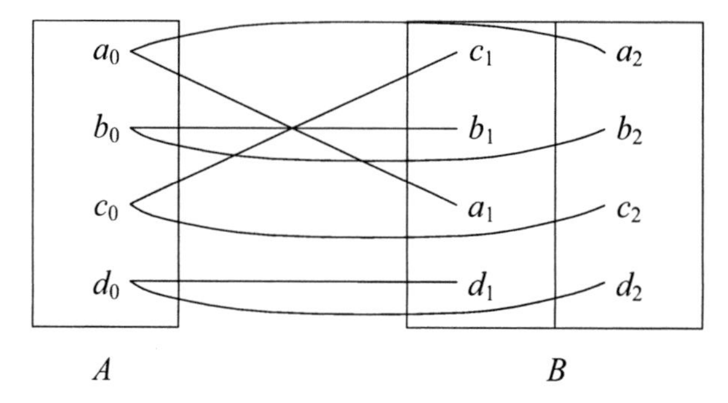

Figure 2.8. Mapping functions relating parameters (*A*) to estimates (*B*) for identified and underidentified models.

the mapping that would occur if a model were underidentified: Each parameter within set *A* is potentially paired with each member of set *B*, which, though not shown, now has an infinite number of members that are potential parameter estimates.

Another association between underidentification and observational equivalence concerns the *uniqueness* of the one-to-one mapping function linking the parameters of two observationally equivalent models. If only one function can be found that reparameterizes model *A* into model *B*, then the parameters from both models are identified; that is, there is only one set of parameter estimates for each model. On the other hand, if more than one function exists, both models are underidentified, each having an infinite number of parameter estimates for each parameter.[4]

This discussion concerning similarities between model equivalence and model identification does not imply the ideas are simply "two sides of the same coin." Model equivalence and model identification are, in many respects, different. For example, a globally underidentified model is forever uninterpretable in its current formulation, worthless for theory confirmation. In contrast, a model with many equivalent models can potentially confirm a theory, at least to the extent that the alternative hypotheses posed by the equivalent models can be ruled out on other (e.g., theoretical) grounds.

SELECTING FROM AMONG EQUIVALENT MODELS

Before Data Collection

Before data collection begins, getting a sense of how many models are equivalent to a hypothesized model, and understanding one's preventative options, are important. Even when only one equivalent model is possible, causal inferences from the hypothesized model are compromised. Unfortunately, unless the hypothesized model is significantly overidentified due to imposing restrictions on the model, we should expect at least some equivalent models to exist. However, the causal models typically specified in the behavioral and social sciences can hardly be characterized as significantly overidentified; many are only a few degrees of freedom away from just-identified status. In the case of just-identification, the number of models equivalent to a hypothesized model is at a maximum. If the number of equivalent models is "large," we should consider revising the hypothesized model to reduce the number of equivalent models. Revising the hypothesized model may in turn require revising the original data collection design by, for example, requiring the collection of data on an additional set of variables.

Collecting information on additional variables need not change one's original theory: the role of the additional variables can be comparable to the role of covariates introduced when random assignment has not been fully implemented. For example, in traditional regression modeling, to control for confounding by a variable one must include it in the structural part of the statistical model. The confounding variable is not itself of theoretical interest. We could also introduce the new variables into the model as instruments serving a purpose similar to that of instrumental variables when a model is just- or underidentified. If introducing the new variables results in a positive gain in the model's degrees of freedom, and if each of the new variables is selectively associated with only a few of the original variables, the number of potentially equivalent models will decrease substantially. Of course, the revised model should retain the most important causal relations specified in the original model, or minimally, should not have specified causal relations theoretically inconsistent with the causal relations of the original model. In practice, reducing the number of equivalent models while at the same time retaining the integrity of the original model will be difficult. Consider also that although there may be many equivalent models initially, many will have nonsensical, illogical, or factually inconsistent causal relations. The presence of time going backward is a common illogical specification found in equivalent models.

After Data Collection

If the hypothesized model (and its equivalent models) fit, one can (a) retain the hypothesized model, (b) decide to select a particular equivalent model as the best model, or (c) assume a position of agnosticism with respect to which model should be selected. Most researchers would prefer (a), which, after all, was the purpose of the study. At this point, they would try "somehow" to justify, using quantitative criteria, retaining the hypothesized model. The four strategies described below were formulated to help researchers rationalize retaining their hypothesized model when model equivalence is an issue.

Strategy 1. An information complexity criterion (ICOMP). An information complexity criterion (*ICOMP*) has been proposed for selecting among equivalent models (Williams et al., 1996). The idea behind *ICOMP*, and complexity criteria in general, is that the value of a model should be based on the fit of the model to the data and the *complexity* of the parameter estimates: *ICOMP* = *model misfit* + *complexity*. Model misfit is defined in the usual way as a function of the difference between the observed and model-implied covariance matrices. *Complexity* is a measure of the correlations among the parameter estimates—higher correlations mean more complexity. These correlations are obtained from a model's information matrix, which is computed during the process of parameter estimation. While small discrepancies between the observed and model-implied covariance matrices suggest a good fit of the model to the data, large correlations among the parameter estimates do not. Correlations among the parameter estimates are not expected because, ideally, the estimation of one parameter should be independent of the estimation of the others. Problems begin when these correlations become high, a sign indicative of insufficient information in the data to estimate each parameter uniquely. Under these conditions, the same information (data) is used to estimate more than one parameter, thus inducing correlations among them. At a minimum, the model is locally underidentified. *ICOMP* is scaled so the model with the lowest value is selected, since a low *ICOMP* arises with a small discrepancy between **S** and $\hat{\boldsymbol{\Sigma}}$ (i.e., small misfit) and low correlations among the parameter estimates (i.e., small complexity). Among a set of equivalent models, which have identical misfit, the one selected with the lowest *ICOMP* will be the least complex of those models.

Strategy 2. Tortured iteration histories, negative variances, and constrained boundary values. Although equivalent models are all destined to arrive at the same model-implied covariance matrix, they travel over different roads. And some of these roads are bumpier than others. Indications of a difficult trip include an unusual number of iterations, characterized by haphazard improvement, negative parameter estimates for variances, and parameter estimates constrained by a modeling program to zero or one to avoid improper parameter estimates (e.g., constraining a variance to zero when, if freely estimated, would assign a negative parameter estimate to a variance). Both empirical research and anecdotal observation agree that these estimation difficulties frequently result from highly misspecified models (e.g., Green, Thompson, & Poirier, 1999).

The computational problems associated with misspecification can be used to distinguish among equivalent models. Equivalent models are equivalent in fit, but almost necessarily not equivalent in structure. In fact, a good chance exists that most equivalent models have nonsensical causal relations. Thus, the task of selecting the best model from a family of equivalent models becomes less ambiguous and arbitrary when models of dubious structure also show poor estimation histories. A model with a theoretically unlikely causal structure and with unlikely parameter estimates is, justifiably, unlikely to be selected as the "true" model.

Strategy 3. Comparing R^2 values among models. Jöreskog and Sörbom (1993) have suggested that R^2 values for specific structural equations might be used to select the optimal model from a set of equivalent alternatives. Because the R^2 values of structural equations are independent, they can be aggregated to reflect an overall R^2 for a model. While sensible, differences in aggregate R^2 as a criterion for selecting a model among a set of equivalent models has not been systematically investigated. One well-known disadvantage of using an R^2 selection criterion is the tendency for its value to fluctuate dramatically across samples. Such volatility may lead to the selection of different models for different samples, not at all a desirable situation. In contrast to the sample dependency of R^2 values, the values of information criteria such as *ICOMP* are very stable, suggesting that a numerical index to be used for the identification of an optimal model among equivalent alternatives should be based on characteristics like model parsimony and complexity instead of the amount of explained variance.

Strategy 4. Extended individual case residuals. Because the model-implied covariance matrix is identical among equivalent models, the residual covariance matrix denoting the discrepancy between the observed and model-implied covariance matrices is identical as well. Although these residuals are not helpful in discriminating among equivalent models, residuals computed at the level of the individual can be helpful in doing so. Raykov and Penev (2001) proposed the use of *extended individual case residuals (EICR)* to help in this discrimination. An *EICR* for an individual, h_i, is the difference between an observed data point and its model-predicted value:

$$h_i = \mathbf{\Theta}^{-1/2}[\mathbf{I} - \mathbf{\Lambda}(\mathbf{\Lambda}'\mathbf{\Theta}^{-1}\mathbf{\Lambda})^{-1}\mathbf{\Lambda}'\mathbf{\Theta}^{-1}]y_i ,$$

where $\mathbf{\Theta}$ is an error covariance matrix, $\mathbf{\Lambda}$ is a matrix of factor loadings, $\mathbf{I}$ is an identity matrix, and y_i is the *i*th subject's score on an observed variable.

The *EICR* for an individual will differ across equivalent models if (a) is different for each equivalent model, (b) $\mathbf{\Lambda}$ is full rank in each model, and (c) the equivalent models have the same number of factors specified. An *EICR* is computed for each individual. While none of the individual *EICR* values will be identical across the equivalent models, the sum of squares of the *EICR* values will be. If the *EICR* values are available, we must decide how they can be used to select among equivalent models. Raykov and Penev (2001) suggested selecting the model that has the smallest average standardized *EICR* relative to its equivalent model alternatives.

CONCLUSION

Model equivalence should convince us of the limitations of confirming structural equation models. Arguably, model equivalence is one of the most compelling demonstrations available for showing why confirming a structural equation model, and ultimately the causal relations the model implies, is so difficult: No matter how much better a hypothesized model fits the data in comparison with competing models, no possibility exists of unequivocally confirming the model if there are models equivalent to it. This is not to suggest that it is impossible to rule out equivalent models based on plausibility alone. But there will always be doubt. The best method of reducing the number of potential equivalent models is to be as parsimonious as possible during model specification. Parsimonious specification acknowledges the inverse relationship between the number of model constraints and the number of equivalent models.

But it is unlikely that consideration of equivalent models will become a standard part of the process of defining, testing, and modifying models. Despite nearly two decades of reports arguing convincingly for the importance of model equivalence to the model-fitting enterprise, efforts to identify equivalent models are still rare. Not helping the situation is the absence of readily available software capable of identifying equivalent models. To my knowledge, TETRAD (Scheines, Spirtes, Glymour, & Meek, 1994) is the only commercially available program that does so. None of the popular structural equation modeling programs do. There really is no excuse for the omission of model equivalence. Rules and algorithms available for finding equivalent models are available and programmable; these rules and algorithms typically identify no less than 99% of the possible equivalent models. Researchers should take the initiative, and routinely report the presence of equivalent models. If reporting equivalent models becomes common practice, software companies will soon respond by including automated methods for detecting equivalent models in their SEM programs.

NOTES

1. By definition, the variables of a just-identified block are symmetrically determined; a just-identified preceding (focal) block is a symmetric focal block.
2. All equivalent models have identical partial correlation structures but not all models with identical partial correlation structure are equivalent. The partial correlation structure of a model refers to the partial correlations that are equal to zero as implied by the model. For example, if one model implies the partial correlation $r_{xy.z} = 0$, then all of the models equivalent to it must also imply $r_{xy.z} = 0$.
3. It is not strictly true that this matrix, the Jacobian matrix referred to earlier, is a matrix of correlations among the model's parameters, but it is highly related to the matrix that is (the inverted information matrix).
4. Econometricians view the model equivalence–model identification relationship as a deterministic: *A model is identified if there is no observationally equivalent model*. In other words, whenever a model has models equivalent to it, the model must be underidentified. This view, which is not shared by most psychometricians, is based on the following reasoning. For a parameter to be identified, complete knowledge of the joint distribution of the random variables must provide enough information to calculate parameters uniquely. Furthermore, every independent distribution of the same variables must also provide unique estimates. Yet, observationally equivalent models, by definition, always fit the same distribution equally well, *but do so with different parameter sets*. Although the distribution of the random variables is the same, the parameters of the equivalent models are different. Recall that observationally equivalent models are linked by a function that translates one model's configuration of parameters into another

model's configuration exactly; the two models' parameter sets are simply transformations of each other and are therefore the same parameters, only configured (organized) differently. A model cannot be identified and have parameters with two solutions. Thus, an observationally equivalent model is always underidentified.

REFERENCES

Bekker, P. A., Merckens, A., & Wansbeek, T. J. (1994). *Identification, equivalent models, and computer algebra*. Boston: Academic Press.

Bowden, R. (1973). The theory of parametric identification. *Econometrica, 41*, 1069–1074.

Breckler, S. J. (1990). Applications of covariance structure modeling in psychology: Cause for concern? *Psychological Bulletin, 107*, 260–273.

Chernov, M., Gallant, A. R., Ghysels, E., & Tauchen, G. (2003). Alternative models for stock price dynamics. *Journal of Econometrics, 116*, 225–257.

Fisher, F. M. (1966). *The identification problem in econometrics*. New York: McGraw-Hill.

Glymour, C., Scheines, R., Spirtes, R., & Kelly, K. (1987). *Discovering causal structure: Artificial intelligence, philosophy of science, and statistical modeling*. Orlando, FL: Academic Press.

Green, S. B., Thompson, M. S., & Poirier, J. (1999). Exploratory analyses to improve model fit: Errors due to misspecification and a strategy to reduce their occurrence. *Structural Equation Modeling: A Multidisciplinary Journal, 6*, 113–126.

Hannan, E. J., & Deistler, M. (1988). *The statistical theory of linear systems*. New York: Wiley.

Hendry, D. F. (1995). *Dynamic econometrics*. New York: Oxford University Press.

Hershberger, S. L. (1994). The specification of equivalent models before the collection of data. In A. von Eye & C. Clogg (Eds.), *The analysis of latent variables in developmental research* (pp. 68–108). Beverly Hills, CA: Sage.

Hsiao, C. (1983). Identification. In Z. Griliches & M. D. Intriligator (Eds.), *Handbook of econometrics* (Vol. 1, pp. 224–283). Amsterdam: Elsevier Science.

Jöreskog, K., & Sörbom, D. (1993). *LISREL 8: User's reference guide*. Chicago: Scientific Software.

Koopmans, T. C., & Reiersøl, O. (1950). The identification of structural characteristics. *Annals of Mathematical Statistics, 21*, 165–181.

Koopmans, T. C., Rubin, H., & Leipnik, R. B. (1950). Measuring the equation system of dynamic economics. In T. C. Koopmans (Ed.), *Statistical inference in dynamic economic models*. New York: Wiley.

Lee., S., & Hershberger, S. L. (1990). A simple rule for generating equivalent models in covariance structure modeling. *Multivariate Behavioral Research, 25*, 313–334.

Luijben, T. C. W. (1991). Equivalent models in covariance structure analysis. *Psychometrika, 56*, 653–665.

McDonald, R. P. (2002). What can we learn from the path equations?: Identifiability, constraints, equivalence. *Psychometrika, 67*, 225–249.

Pearl, J. (2000). *Causality: Models, reasoning, and inference*. New York: Cambridge University Press.

Raykov, T., & Marcoulides, G. A. (2001). Can there be infinitely many models equivalent to a given covariance structure model? *Structural Equation Modeling: A Multidisciplinary Journal, 8*, 142–149.

Raykov, T., & Penev, S. (1999). On structural equation model equivalence. *Multivariate Behavioral Research, 34*, 199–244.

Richardson, T. (1997). A characterization of Markov equivalence for directed cyclic graphs. *International Journal of Approximate Reasoning, 17*, 107–162.

Rigdon, E. E. (1995). A necessary and sufficient identification rule for structural models estimated in practice. *Multivariate Behavioral Research, 30*, 359–383.

Roesch, S. C. (1999). Modeling stress: A methodological review. *Journal of Behavioral Medicine, 22*, 249–269.

Scheines, R., Spirtes, P., Glymour, C., & Meek, C. (1994). *TETRAD II: Tools for discovery.* Hillsdale, NJ: Erlbaum.

Shipley, B. (2000). *Cause and correlation in biology.* New York: Cambridge University Press.

Spirtes, P., Glymour, C., & Scheines, R. (1993). *Causation, prediction, and search.* New York: Springer-Verlag.

Stelzl, I. (1986). Changing a causal hypothesis without changing the fit: Some rules for generating equivalent path models. *Multivariate Behavioral Research, 21*, 309–331.

Walter, E., & Pronzato, L. (1997). *Identification of parametric models from experimental data*. London: Springer-Verlag.

Williams, L. J., Bozdogan, H., & Aiman-Smith, L. (1996). Inference problems with equivalent models. In G. A. Marcoulides & R. E. Schumacker (Eds.), *Advanced structural equation modeling: Issues and techniques* (pp. 279–314). Mahwah, NJ: Erlbaum.

CHAPTER 3

REVERSE ARROW DYNAMICS

Formative Measurement and Feedback Loops

Rex B. Kline

This chapter is about structural equation models where some arrows (paths) are "backwards" compared with standard models. These include (1) formative measurement models (also called emergent variable systems) and (2) structural models with feedback loops where some variables are specified as causes and effects of each other. In standard measurement models—referred to below as reflective measurement models (also called latent variable systems)—all observed variables (indicators) are specified as *effects* of latent variables (factors) that represent hypothetical constructs. Reflective measurement is based on classical measurement theory. In contrast, the indicators in a formative measurement model are specified as *causes* of latent variables. When all indicators of a factor are so specified, the factor is actually a latent composite. Formative measurement of composites requires quite different assumptions compared with reflective measurement of constructs.

Standard structural models are recursive, which means that (1) all causal effects are represented as unidirectional—that is, no two variables

Structural Equation Modeling: A Second Course, 43–68

measured at the same time are specified as direct or indirect causes of each other—and (2) there are no disturbance correlations between endogenous variables with direct effects between them. These assumptions greatly simplify the analysis of a recursive model, but they are very restrictive. For example, many "real world" causal processes, especially dynamic ones, are based on cycles of mutual influence; that is, feedback. A feedback loop involves the specification of mutual causation among two or more variables measured at the same time, and the presence of such a loop in a structural model automatically makes it nonrecursive. The estimation of feedback effects with cross-sectional data requires several special considerations, but the alternative is a much more expensive longitudinal design. These points are elaborated later.

Discussed next are types of hypotheses that can be tested through the specification of models with "backwards" arrows and some special problems that can crop up in their analysis. It is hoped that this presentation informs the reader about additional ways to represent hypotheses about measurement or causality in his or her structural equation models.

FORMATIVE MEASUREMENT MODELS

Reflective Versus Formative Measurement

A standard measurement model assumes that all indicators are endogenous and caused by the factor(s) they are specified to measure, plus a residual term that reflects unique sources of variability such as measurement error. A standard model is thus a *reflective measurement model*, and observed variables in such models are *effect indicators* or *reflective indicators*. In structural equation modeling (SEM), factors in reflective measurement models are continuous latent variables. The relation between factor F1, its indicators V1 through V3, and residual terms E1 through E3 in a reflective measurement model is illustrated in Figure 3.1(a).[1] All direct effects in this model are from latent to observed variables. The *b* terms in the figure are path coefficients (i.e., factor loadings) for causal effects of F1 on V1 through V3. The unstandardized loading of V1 on F1 is fixed to 1.0 to assign a scale (metric) to F1. This scale is related to that of the explained variance of V1, and this specification makes V1 the *reference variable* for F1. Assuming that scores on each of V1 through V3 are equally reliable, the choice of which indicator is to be the reference variable is generally arbitrary. Otherwise it makes sense to select the indicator with the most reliable scores as the reference variable. The equations for this reflective measurement model are presented below:

(a) Reflective Measurement Model

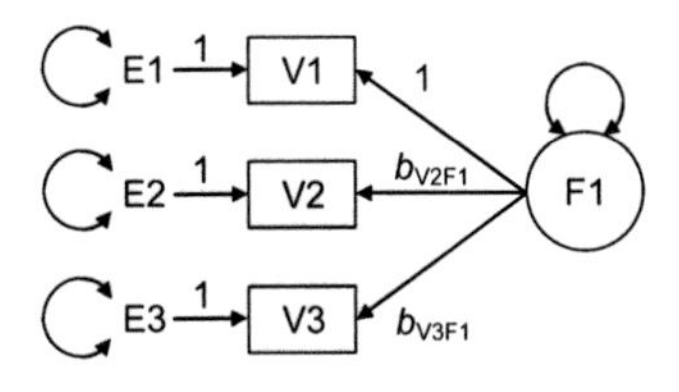

(b) Formative Measurement Model

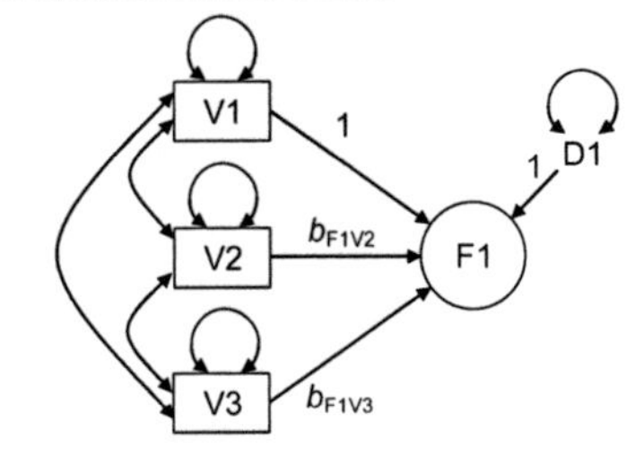

(c) Measurement Model with MIMIC Factor

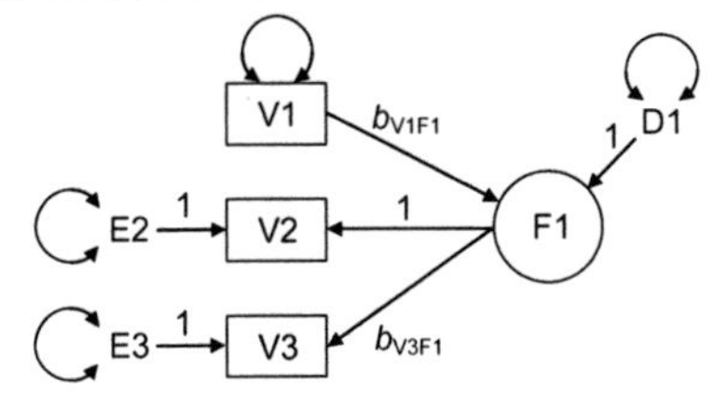

Figure 3.1. Examples of factors with (a) all effects indicators, (b) all causal indicators, and (c) both kinds of indicators. MIMIC: multiple indicators and multiple causes.

$$
\begin{aligned}
\mathrm{V1} &= \mathrm{F1} + \mathrm{E1} \\
\mathrm{V2} &= b_{\mathrm{V2F1}}\,\mathrm{F1} + \mathrm{E2} \\
\mathrm{V3} &= b_{\mathrm{V3F1}}\,\mathrm{F1} + \mathrm{E3}
\end{aligned}
\tag{1}
$$

The rationale that underlies the specification of a reflective measurement model is based on classical test theory in general and the domain sampling model in particular (e.g., Nunnally & Bernstein, 1994, chap. 6). In this view of measurement, effect indicators like V1 through V3 in Figure 3.1(a) should as a set be internally consistent. This means that their intercorrelations should be moderately high in magnitude. If there is more than one factor and assuming that each indicator depends (loads) on only one factor, within-factor correlations among indicators should be greater than between-factor correlations. The patterns of indicator intercorrelations just described correspond to, respectively, convergent validity and discriminant validity in construct measurement. It is also assumed in the domain sampling model that equally reliable effect indicators of a unidimensional construct are interchangeable (Bollen & Lennox, 1991), which implies that they can be substituted for another without appreciably affecting construct measurement.

The assumption that scores on indicators are caused by underlying factors is not always appropriate. That is, some indicators may be viewed as *cause indicators*—also called *formative indicators*—that affect a latent variable instead of the reverse. A typical example is socioecomonic status (SES), which is often measured with variables such as income, education, occupation, and place of residence. In a reflective measurement model, these variables would be represented as effect indicators each caused by an underlying SES factor (and a residual term). But we usually think of SES as being determined by all these variables, not the other way around. For example, a change in any one of these indicators (e.g., a salary increase) may affect SES even if the other indicators did not change. Conversely, a change in SES does not necessarily imply a change in all four indicators. From the perspective of *formative measurement*, SES is a composite (index) variable that is caused by its indicators. There are many examples of composite variables in the economics literature, such as various cost-of-living indexes. Formative measurement is also especially relevant in the study of organizational or social constructs when the unit of analysis is groups or companies instead of individuals (Diamantopoulos & Winklhofer, 2001).

A formative measurement model is illustrated in Figure 3.1(b). Note in the figure that F1 is specified as caused by its indicators, V1 through V3. Because F1 is endogenous in this model, it has a disturbance, D1. It is the presence of D1 that makes F1 a *latent composite*. With no disturbance, F1 would be just a linear combination of V1 through V3, and in this sense would be similar to a principal component in exploratory factor analysis or a predicted criterion score in multiple regression. Latent composites require a scale, just like latent variables in reflective measurement models. For this reason, the unstandardized direct effect of one of F1's cause indicators, V1, is specified in Figure 3.1(b) as fixed to 1.0. When all indicators are cause indicators, the choice of which direct effect of the cause indicators to fix to 1.0 in order to scale the factor is generally arbitrary if their scores are equally reliable. Note in the figure that V1 through V3 are represented as observed exogenous variables. This means that (1) their variances and covariances are model parameters and (2) analysis of the model in Figure 3.1(b) would *not* take direct account of the measurement errors of V1 through V3. Instead, measurement error in V1 through V3 would be manifested in the factor disturbance, D1.[2] Thus, a relatively large estimated disturbance variance for a latent composite may indicate poor score reliability among its cause indicators. If the latent composite F1 were "embedded" in a model where it is specified that variables other than V1 through V3 have direct effects on F1, then a relatively large estimated disturbance variance would also reflect low predictive power of those other variables. The equation for the formative measurement model of Figure 3.1(b) is:

$$\text{F1} = \text{V1} + b_{\text{F1V2}}\,\text{V2} + b_{\text{F1V3}}\,\text{V3} + \text{D1} \qquad (2)$$

Compare Equation 2 for Figure 3.1(b) with Equation 1 for Figure 3.1(a). See Edwards and Bagozzi (2000) for more examples of formative measurement models.

The traditional assumptions of classical test theory do not apply to formative measurement. Diamantopoulos and Winklhofer (2001) noted that the origins of formative measurement lie in the "operational definition" model. An older, stricter form of operationalism views constructs as synonymous with the single measure that corresponds to its operational definition. More contemporary forms of operationalism allow for both multiple measures (indicators) and disturbance terms for the composite variables. Unlike effect indicators, cause indicators are *not* generally interchangeable. This is because the removal of a cause indicator is akin to removing a part of the underlying construct (Bollen & Lennox, 1991). Internal consistency is *not* generally required for a set of cause indicators. In fact, cause indicators are allowed to have any pattern of intercorrelations, including positive versus negative or higher versus lower in magnitude. Because cause indicators are exogenous, their variances and covariances are not explained by a formative measurement model. This can make it more difficult to assess the validity of cause indicators (Bollen, 1989). In contrast, the observed variances and covariances of effect indicators can be compared against those implied by a reflective measurement model. Finally, note that the formative measurement model of Figure 3.1(b) is *not* identified. Its parameters could be uniquely estimated only if it is placed in a larger model with specific patterns of direct effects on other variables. (This point is discussed momentarily.) In contrast, the reflective measurement model of Figure 3.1(a) is identified (specifically, just-identified).

There is a compromise between specifying that the indicators of a factor are all either effect or causal. It is achieved by specifying a measurement model with a MIMIC (multiple indicators and multiple causes) factor with both effect and cause indicators. An example of a MIMIC factor is presented in Figure 3.1(c). Note in the figure that V1 is represented as a cause indicator, but V2 and V3 are each specified as effect indicators. Consequently, V2 and V3 have residual terms (the measurement errors E2 and E3), but V1 does not. In theory, a scale could be assigned to F1 by fixing the direct effect of cause indicator V1 to 1.0 or either of the factor loadings of effect indicators V2 or V3 to 1.0. In practice, it would make the most sense to select either V2 or V3 as the reference variable and fix its loading to 1.0. This specification would leave the direct effect of V1 on F1 as a free parameter, and the unstandardized estimate of this effect can be tested for statistical significance. A MIMIC factor is always endogenous, so F1 in Figure 3.1(c) has a disturbance term (D1). This disturbance reflects measurement error in its cause indicator V1. If the MIMIC factor F1 were "embedded" in a model where it is specified to have direct causes on it from variables other than V1, the disturbance D1 would also reflect the overall predictive power of those other variables. The MIMIC model of Figure 3.1(c) is identified (specifically, just-identified).[3] The equations for this measurement model of Figure 3.1(c) are presented below:

$$\begin{aligned} \text{F1} &= b_{\text{V1F1}}\,\text{V1} + \text{D1} \\ \text{V2} &= \text{F1} + \text{E2} \\ \text{V3} &= b_{\text{V3F1}}\,\text{F1} + \text{E3} \end{aligned} \tag{3}$$

Compare Equation 3 with Equation 1 and Equation 2 for all models represented in Figure 3.1.

Briefly, there are several examples in the SEM literature of the analysis of MIMIC factors. For example, Hershberger (1994) described a MIMIC depression factor with indicators that represented various symptoms. Some of these indicators, such as "crying" and "feeling depressed," were specified as effect indicators because they are classical symptoms of depression. However, another indicator, "feeling lonely," was specified as a cause indicator of depression. This is because "feeling lonely" may be something that causes depression rather than vice versa. Kaplan (2002, pp. 70–74) described the specification of MIMIC models with categorical cause indicators that represent group membership as an alternative method for estimating group differences on latent variables. Kano (2001) described the analysis of MIMIC models for experimental designs that are latent-variable analogs of techniques such as the analysis of variance (ANOVA) and the analysis of covariance (ANCOVA); see also Thompson and Green (Chapter 5, this volume). Beckie, Beckstead, and Webb (2001) analyzed a MIMIC model of women's quality of life after suffering a cardiac event correcting for measurement error in all cause indicators. The analysis of measurement models with MIMIC factors is discussed in more detail next.

Identification Issues in Formative Measurement

The main stumbling block to the analysis of measurement models where some factors have only cause indicators is identification. This is because it can be difficult to specify such a model that reflects the researcher's hypotheses about measurement and is identified. Two necessary but insufficient requirements for identification are ones that apply to any type of structural equation model: The number of free parameters cannot exceed the number of observations (i.e., the model degrees of freedom are at least zero), and every unobserved variable must be assigned a scale. Recall that the number of observations is *not* the sample size (N). It is instead the number of unique pieces of information in the variance–covariance matrix, which is calculated as $v(v+1)/2$ where v is the number of observed variables and means are not analyzed (i.e., the model has just a covariance structure) or as $v(v+3)/2$ when means are analyzed (i.e., the model has both a covariance structure and a mean structure). The assignment of a scale to a factor with cause or effect indicators was discussed earlier. MacCallum and Browne (1993) showed that an additional requirement for identification of the disturbance variance of a latent composite is that the composite must have at least two direct effects on other endogenous factors measured with effect indicators. If a factor measured with cause indicators only emits just a single path (i.e., it is specified to have a direct effect on just one endogenous variable), its disturbance variance will be underidentified. This means that it is not theoretically possible for the computer to derive a unique estimate for this term.

One way to deal with the identification requirement just mentioned is to fix the disturbance variance for the latent composite to zero, which drops the disturbance from the model. However, this is not a very good solution for a few reasons. Recall that the disturbance of a latent composite reflects measurement error in its cause indicators. Dropping the disturbance is akin to assuming that the cause indicators are measured without error, which is very unlikely. It also makes the latent composite equivalent to a simple linear combination of the cause indicators. MacCallum and Browne (1993) showed that dropping from the model a latent composite that emits a single path and converting the indirect effects of its cause indicators on other endogenous variables to direct effects results in an equivalent model. An equivalent structural equation model has the same overall fit to data as the researcher's original model but has a different configuration of paths. An equivalent model thus provides an account of the data that is statistically indistinguishable from that of the researcher's original model (MacCallum, Wegener, Uchino, & Fabrigar, 1993; see also Hershberger, Chapter 2, this volume).

Another complication concerning identification was noted by MacCallum and Browne (1993) for measurement models with two or more latent composites. Suppose that factors F1 and F4 are each measured by effect indicators only and that factors F2 and F3 are each measured by cause indicators only (i.e., they are latent composites). A researcher specifies a structural model where F1 has two different indirect effects on F4, including (1) F1 → F3 → F4 and (2) F1 → F2 → F3 → F4. In other words, there are multiple indirect effects from F1 to F4 through different subsets of latent composites (F2 and F3). In this case, one of the structural paths (direct effects) may be underidentified.

The model in Figure 3.2, based on an example by MacCallum and Browne (1993), illustrates some of the identification problems just discussed. Suppose that a researcher specifies a measurement model with two factors measured by effect indicators only, F1 and F4, and two factors measured by cause indicators only, F2 and F3. To save space, not all

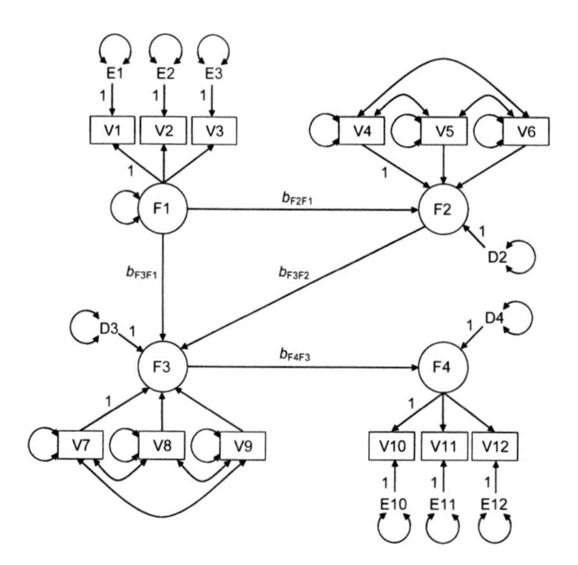

Figure 3.2. Example of a nonidentified latent variable path model where some factors have cause indicators only. Note: Unanalyzed associations among V4 through V6, V7 through V9, and F1 are not shown but assumed.

unanalyzed associations among the cause indicators V4 through V6, V7 through V9, and the exogenous factor F1 are shown in the figure, but they are assumed. The structural paths represented in Figure 3.2 reflect the researcher's hypotheses about direct and indirect effects among the factors. MacCallum and Browne showed that the variances of the three factor disturbances (D2 through D4) are underidentified as are the path coefficients for the two direct effects of F1 (b_{F2F1}, b_{F3F1}). Unfortunately, it is not possible to remedy the nonidentification of the disturbance variances by specifying that the latent composites F2 and F3 each emits two direct effects. This is because doing so would result in an underidentified structural model. It would also require a change in the researcher's hypotheses about structural relations. The indeterminacy in the estimation of b_{F2F1} and b_{F3F1} results from the fact that F1 has two different indirect effects on F4 that each run through two different latent composites. Fixing either b_{F2F1} or b_{F3F1} to zero may identify the structural model, but such a respecification would not reflect the researcher's original hypotheses about causality.

Example of a MIMIC Factor as an Alternative to a Latent Composite

MacCallum and Browne (1993) noted that one way to remedy many of the identification problems described earlier is to add effect indicators for latent composites represented in the original model as measured with cause indicators only; that is, specify a MIMIC factor. This specification requires a theoretical rationale; that is, a cause indicator should not arbitrarily be respecified as an effect indicator just to achieve identification. An example follows.

The data for this example come from Worland, Weeks, Janes, and Strock (1984), who administered measures of the cognitive and achievement status of 158 adolescents. They also collected teacher reports about classroom adjustment and measured family SES and the degree of parental psychiatric disturbance (e.g., one parent diagnosed as schizophrenic). The correlations among these variables are reported in Table 3.1. Note that Worland and colleagues did not report standard deviations, which means that it is not possible to analyze a covariance matrix for this example. This is a potential problem for the default estimation methods of many SEM computer programs such as maximum likelihood, which assume the analysis of unstandardized variables. It can happen that such methods give incorrect results when the variables are standardized (Cudeck, 1989). Although there are methods to correctly analyze a correlation matrix (e.g., Steiger, 2002), it is beyond the scope of this chapter to

Table 3.1. Input Data (Correlations, Standard Deviations) for the Analysis of a Model of Risk with Cause Indicators (Figure 3.3)

Variable	*1*	*2*	*3*	*4*	*5*	*6*	*7*	*8*	*9*
Risk indicators									
1. Parental Psychiatric	1.00								
2. Low Family SES	−.42	1.00							
3. Verbal IQ	−.43	−.50	1.00						
Achievement indicators									
4. Reading	−.39	−.43	.78	1.00					
5. Arithmetic	−.24	−.37	.69	.73	1.00				
6. Spelling	−.31	−.33	.63	.87	.72	1.00			
Classroom adjustment indicators									
7. Motivation	−.25	−.25	.49	.53	.60	.59	1.00		
8. Harmony	−.25	−.26	.42	.42	.44	.45	.77	1.00	
9. Stability	−.16	−.18	.33	.36	.38	.38	.59	.58	1.00
SD	13.00	13.50	13.10	12.50	13.50	14.20	9.50	11.10	8.70

Note: Correlations from Worland, Weeks, Janes, and Strock (1984); $N = 158$.

describe them in any detail. Thus, for didactic purposes, the author assigned plausible standard deviations to each of the variables listed in Table 3.1. Taking this pedagogical license does not affect the overall fit of the model described next. It also allows the reader to reproduce this analysis with basically any SEM computer program.

Suppose that the construct of *risk* is conceptualized for this example as a latent composite with cause indicators family SES, parental psychiatric disturbance, and adolescent verbal IQ. That is, high risk is indicated by any combination of low family SES, a high degree of parental psychopathology, or low verbal IQ. The intercorrelations among these three variables are not all high or positive (see Table 3.1), but this does not matter for cause indicators. Presented in Figure 3.3 is an example of a three-factor latent variable path model with a single latent composite—in this case risk—that is identified (MacCallum & Browne, 1993). Note in the figure that the risk composite emits two direct effects onto factors each measured with effect indicators only, achievement and classroom adjustment. This specification identifies the residual (disturbance) term for the risk composite. It also reflects the assumption that the association between achievement and classroom adjustment is spurious due to a common cause (risk). This assumption may not be very plausible. For example, it seems reasonable to expect that achievement is a direct cause of adjustment; specifically, students with better scholastic skills may be better

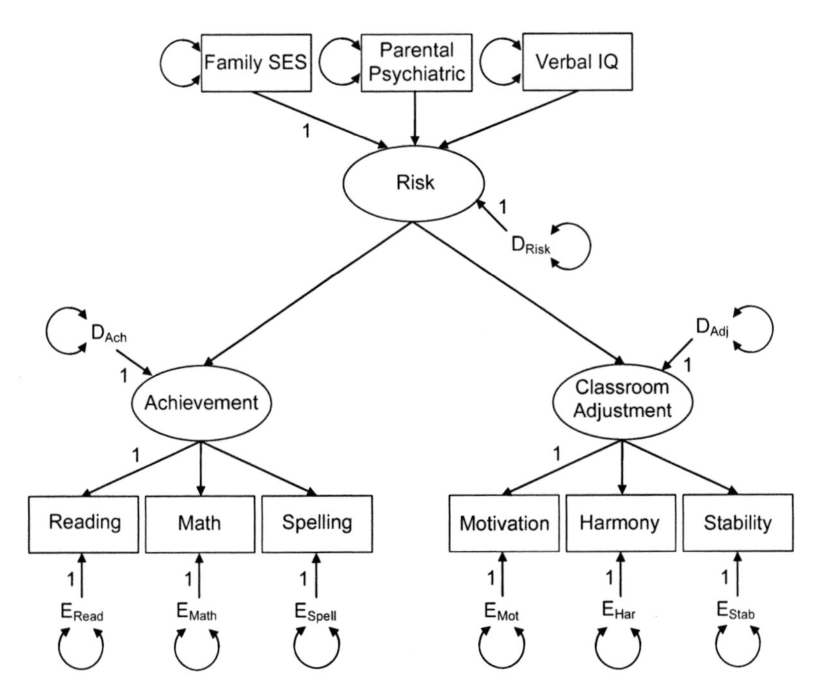

Figure 3.3. An identified latent variable path model with a latent composite of the relation between risk, achievement, and classroom adjustment.

adjusted in the classroom. The direct effect just mentioned is estimated later.

The model of Figure 3.3 with just two direct effects in the structural model (Risk → Achievement, Risk → Adjustment) was fitted to the covariance matrix "assembled" from the correlations and standard deviations in Table 3.1 with the maximum likelihood method of EQS 6 (Bentler, 2003). The analysis converged to an admissible solution. Values of selected fit indexes are χ^2 (22) = 65.467 (p < .001), Bentler Comparative Fit Index (CFI) = .951, Standardized Root Mean Square Residual (SRMR) = .039, and Steiger-Lind Root Mean Square Error of Approximation (RMSEA) = .112 with the 90% confidence interval .081–.143. The latter result (i.e., RMSEA > .08) suggests poor overall fit of the data to the model in Figure 3.3. Inspection of the correlation residuals (differences between observed and predicted correlations) verifies this conclusion: Several residuals are close to or greater than .10 in absolute value, which suggests that the model does not adequately explain the corresponding correlations. These relatively high correlation residuals generally occurred between indicators of the achievement factor and the classroom adjustment factor (see Figure 3.3).

The pattern of correlation residuals for the model of Figure 3.3 is consistent with the possibility raised earlier that the coefficient for the path Achievement → Adjustment is not zero.[4] However, MacCallum and Browne (1993) showed that the model in Figure 3.3 with the path Achievement → Adjustment is not identified. However, this identification problem is resolved if the risk composite of Figure 3.3 is respecified as a MIMIC factor with at least one effect indicator. For example, suppose that it is reasonable to view adolescent verbal IQ as affected by family SES and parental psychopathology through the latent risk composite. This respecification of risk as a MIMIC factor instead of as a latent composite also changes the identification status of the model. Specifically, we can add the path Achievement → Adjustment to this model with the MIMIC risk factor and wind up with an identified model. The respecified model just described was fitted to the same data (Table 3.1) with EQS. This second analysis converged to an admissible solution. As expected, the magnitude of the direct effect of achievement on classroom adjustment is relatively large: the standardized path coefficient equals .533, and the unstandardized coefficient (.392) is statistically significant at the .01 level. Values of selected fit indexes are $\chi^2(23) = 65.543$ ($p < .001$), CFI = .952, SRMR = .038, and RMSEA = .109 with the 90% confidence interval .078–.139.[5] The latter result suggests poor overall fit of the revised model. Inspection of the correlation residuals indicate relatively high values for some associations between indicators of the achievement factor and indicators of the classroom adjustment factor. This pattern of correlation residuals suggests that it may be possible to substantially improve overall model fit by adding a few theoretically sensible measurement error covariances between indicators of these different factors. Although these possibilities are not explored here, the point is that there may be greater flexibility in hypothesis testing when a MIMIC factor with both effect indicators and cause indicators is substituted for a latent composite with cause indicators only, again when it makes sense to do so.

STRUCTURAL MODELS WITH FEEDBACK LOOPS

Theoretical Issues

A structural model reflects the researcher's hypotheses about *why* observed variables, such as V1 and V2, correlate with each other. This explanation can reflect two kinds of causal relations or two kinds of noncausal associations. The first kind of causal relation reflects presumed unidirectional influence and includes direct effects of one variable on another (e.g., V1 → V2) or indirect effects through at least one mediating variable (e.g., V1 → V3 → V2). The second kind concerns feedback loops, either direct or indirect: *direct feedback* involves only two variables in a reciprocal relation (e.g., V1 ⇆ V2); *indirect feedback* involves three or more variables (e.g., V1 → V2 → V3 → V1). Feedback loops represent mutual influence among variables that are concurrently measured. As an example of a possible feedback relation, consider vocabulary breadth and reading: a bigger vocabulary may facilitate reading, but reading may result in an even bigger vocabulary.

The two kinds of noncausal relations are *spurious associations* and *unanalyzed associations*. Spurious associations are represented by specifying common causes. For example, if V3 is specified as a direct cause of both V1 and V2 (i.e., V3 → V1, V3 → V2), then at least part of the observed correlation between V1 and V2 is presumed to be spurious. If the model contains no other direct or indirect causal effects between these variables (e.g., V1 → V2), the entire association between them is presumed to be spurious. Spurious associations can also involve multiple common causes.

Most structural models with more than one exogenous variable assume unanalyzed associations between them. In diagrams of structural models, unanalyzed associations are typically represented with a curved line with two arrowheads. Although unanalyzed associations are estimated by the computer, they are unanalyzed in the sense that no prediction is put forward about why the two variables covary (e.g., Does one cause the other? Do they have a common cause?). It is also possible to represent in structural models unanalyzed associations between disturbances. Recall that a disturbance represents all omitted causes of the corresponding endogenous variable. An unanalyzed association between a pair of disturbances is called a *disturbance correlation* (for standardized variables) or a *disturbance covariance* (for unstandardized variables)—the former term is used from this point regardless of whether the variables are standardized or not. A disturbance correlation reflects the assumption that the corresponding endogenous variables share at least one common omitted cause. Accordingly, the *absence* of the symbol for an unanalyzed association between two disturbances reflects the presumption of independence of unmeasured causes. It also represents the hypothesis that the observed correlation between that pair of endogenous variables can be entirely explained by other variables in the structural model.

The disturbances of variables involved in feedback loops are often specified as correlated. This specification also often makes sense because if variables are presumed to mutually cause one another, then it seems plausible to expect that they may have common omitted causes. However, the specification of correlated disturbances for variables involved in feedback loops is *not* required. It is also the general practice in the behavioral sciences *not* to specify correlated disturbances unless there are substantive reasons for doing so. However, in other disciplines, such as econometrics, the specification of correlated residuals is more routine (i.e., it is a standard assumption). The presence of disturbance correlations in particular patterns in nonrecursive models also helps to determine their identification status, a point that is elaborated below.

There are two basic kinds of structural models. In *recursive models*, (1) all causal effects are unidirectional and (2) there are no disturbance correlations for endogenous variables with direct effects between them. *Nonrecursive models* have feedback loops or disturbance covariances for endogenous variables with direct effects between them (e.g., Kline, 2004, pp. 102–105; see also Brito & Pearl, 2003). As mentioned, a feedback loop represents reciprocal causal effects among endogenous variables measured at the same time; that is, the data are cross-sectional rather than longitudinal. A direct feedback loop between V1 and V2 is represented in Figure 3.4(a) without disturbances or other variables. An alternative way to estimate reciprocal causal effects requires a longitudinal design where V1 and V2 are each measured at two or more different points in time. For example, the symbols V11 and V21 in the *panel model* shown in Figure 3.4(b) without disturbances or other variables represent, respectively, V1 and V2 at the first measurement occasion. Likewise, the symbols V12 and V22 represent scores from the same two variables at the second measurement. Reciprocal causation is represented in Figure 3.4(b) by the *cross-lag* direct effects between V1 and V2 measured at different times, such as V11 → V22 and V21 → V12. A complete panel model may be recursive or nonrecursive depending on its pattern of disturbance correlations (e.g., Maruyama, 1998, Chap. 6).

Panel models for longitudinal data offer some potential advantages over models with feedback loops for cross-sectional data. These include

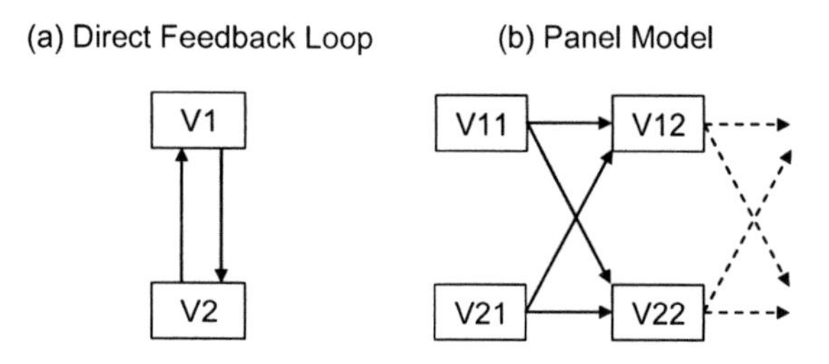

Figure 3.4. Reciprocal causal effects between V1 and V2 represented with (a) a direct feedback loop based on a cross-sectional design and (b) a cross-lag effect based on a longitudinal design (panel model).

the explicit representation of a finite causal lag that corresponds to the measurement occasions and the ability to measure stability versus change in rank order or variability over time. However, the analysis of a panel model is not a panacea for estimating reciprocal causal effects (Kenny, 1979). Not only are longitudinal designs more expensive and subject to loss of cases over time, it can also be difficult to specify measurement occasions that match actual finite causal lags. Panel designs are also *not* generally useful for resolving causal priority between reciprocally related variables—for example, does V1 cause V2 or vice versa?—unless some highly restrictive (i.e., unrealistic) assumptions are met. One of these is *stationarity*, the requirement that the causal structure does not change over time. The complexity of panel models can also increase rapidly as more variables in the form of repeated measurements are added to the model (e.g., Finkel, 1995). Maruyama (1998) noted that the general requirement that there are no omitted causes correlated with those in the model is even more critical for panel models because of the repeated sampling of variables over time. For many researchers, the estimation of reciprocal causation between variables measured at the same time is the only realistic alternative to a longitudinal design.

Kaplan, Harik, and Hotchkiss (2001) and others have reminded us that data from a cross-sectional design give only a "snapshot" of an ongoing dynamic process. Therefore, the estimation of reciprocal effects with cross-sectional data requires the assumption of *equilibrium*. This means that any changes in the system that underlies a feedback relation have already manifested their effects and that the system is in a steady state. This does *not* mean that the two direct effects that make up the feedback loop, such as V1 → V2 and V2 → V1, are equal. Instead, it means that the estimates of their values do not depend upon the particular time point of data collection. Heise (1975) described equilibrium from a somewhat different perspective: It means that a dynamic system has completed its cycles of response to a set of inputs and that the inputs do not vary over time. Kenny (1979) noted another reason for the equilibrium assumption: Causation is ordinarily assumed to occur within a finite amount of time period, also known as the *lag period*. That is, the cause precedes the effect in time, and consequences of causation require a certain amount of time to manifest themselves, although in some cases the lag period may approach zero. Thus, a feedback loop theoretically implies effects that last forever as two variables keep influencing each other, but estimation of these effects with cross-sectional data requires that the causal process has basically dampened out and is not just beginning.

It is important to realize that there is generally no statistical way to directly evaluate whether the equilibrium assumption holds when the data are cross-sectional. Therefore, it must be argued *substantively* in a cross-sectional design that the equilibrium assumption is plausible. However, Kaplan and colleagues (2001) noted that rarely is this assumption explicitly acknowledged in the literature on applications of SEM where feedback effects are estimated with cross-sectional data. This is unfortunate because the results of some computer simulation studies indicate that violation of the equilibrium assumption can lead to severely biased results. In one such study, Kaplan and colleagues specified a true dynamic system that was perturbed at some earlier time and was headed toward but had not attained yet equilibrium in either a smooth or oscillatory manner. By oscillatory it is meant that the sign of a direct effect in a feedback loop changes from positive to negative and back again over time as the system moves back toward stability after being perturbed. A smooth manner means that the sign of the direct effect is constant, but its value tends to increase or decrease over time toward an asymptotic value as the system approaches stability. In both cases, the coefficient for the direct effect eventually heads toward its initial value. The computer-generated datasets of Kaplan and colleagues simulated cross-sectional studies conducted at different numbers of cycles before equilibrium is reached. Kaplan and colleagues found that estimates of direct effects in feedback loops can vary dramatically depending on when the simulated data are collected. As expected, this phenomenon was more striking for dynamic systems in oscillation before reaching equilibrium.

Kaplan and colleagues (2001) also found that the *stability index* did not accurately measure the degree of bias due to lack of equilibrium. The stability index is printed in the output of some SEM computer programs, such as Amos 5 (Arbuckle, 2003), when a nonrecursive model is analyzed. It is based on certain mathematical properties of the matrix of coefficients for direct effects among all the endogenous variables in the structural model, not just those involved in feedback loops. These properties concern whether estimates of the direct effects would get infinitely larger over time. If so, the system is said to "explode" because it may never reach equilibrium, given the observed direct effects among the endogenous variables. The mathematics of the stability index are quite complex (e.g., Bentler & Freeman, 1983; Kaplan et al., 2001, pp. 317–322), and it is beyond the scope of this chapter to describe them in detail. A standard interpretation of the stability index is that values less than 1.0 are taken as positive evidence for equilibrium, but values greater than 1.0 suggest the lack of equilibrium. However, this interpretation is not generally supported by the computer simulation results of the Kaplan and colleagues study. This finding again emphasizes the need to evaluate the equilibrium assumption on rational rather than statistical grounds.

Problems in the Analysis of Models with Feedback Loops

There are two general kinds of problems that arise in the estimation of feedback effects. The first concerns identification. Assuming an identified measurement model, a recursive structural model is always identified. However, it is not always easy to determine whether a nonrecursive structural model is identified. Briefly summarized, if a nonrecursive structural model with feedback loops has either (1) all possible disturbance covariances or (2) all possible disturbance covariances within blocks of endogenous variables that are recursively related to each other (Kline, 2004, Chap. 9), it must satisfy two conditions in order to be identified. One of these conditions is necessary, but the other is sufficient. The necessary condition is the *order condition*. The order condition requires that the number of variables in the structural model (not counting disturbances) excluded from the equation of each endogenous variable (i.e., there is no direct effect) equals or exceeds the total number of endogenous variables minus 1. Suppose that a path model has two endogenous variables, V1 and V2, and two exogenous variables, V3 and V4. The endogenous are specified as involved in a direct feedback loop (V1 $\leftrightarrows$ V2) and their disturbances are specified as correlated. The direct effects V3 $\rightarrow$ V1 and V4 $\rightarrow$ V2 are also included in the model. The order condition for this model requires that there be 2 – 1, or 1 variable excluded from the equation of each endogenous variable, which is here true: Exogenous variable V4 is not specified as a direct cause of V1, so V4 is excluded from the equation for V1. Likewise, V3 is not specified as a direct cause of V2, so V3 is excluded from the equation for V2. Because one variable is excluded from the equation of each endogenous variable, the order condition is satisfied for this model. However, we still do not know whether this nonrecursive path model is identified because the order condition is only necessary. We can say, though, that a nonrecursive model that fails to meet the order condition is not identified and thus must be respecified.

The other requirement, the *rank condition*, provides a more stringent test of a nonrecursive structural model's identification status because it is a sufficient condition. That is, nonrecursive models with all possible disturbance covariances in the patterns outlined in the previous paragraph that meet the rank condition are in fact identified. The rank condition is usually described in matrix terms (e.g., Bollen, 1989, pp. 98–103), which is fine for those with knowledge of linear algebra. Berry (1984) devised an algorithm for checking the rank condition that does not require extensive knowledge of matrix operations, a simpler version of which is described in Kline (2004, pp. 243–247). It is beyond the scope of this chapter to demonstrate the evaluation of the rank condition, but several examples can be found in Kline (2004, Chap. 9) and Berry (1984).

There may be no sufficient conditions that are relatively straightforward to evaluate the identification status of nonrecursive structural models with patterns of disturbance covariances (including none) that do not match the two patterns described earlier. However, there are empirical checks that can be conducted to evaluate the uniqueness of a converged solution for such models (e.g., Bollen, 1989, pp. 246–251; Kline, 2004, pp. 172–174). These tests are only necessary conditions for identification, however. That is, a solution that passes them is not guaranteed to be unique. Some of these empirical tests are briefly summarized below:

1. Analyze the predicted covariance matrix from the original analysis as the input data for a second analysis of the same model. If the same solution is obtained in the second analysis, then the model may be identified. Note that this empirical check works only for overidentified models (i.e., there are more observations than free parameters).
2. Conduct a second analysis using different start values than in the first analysis. If estimation converges to the same solution working from different initial estimates, then the model may be identified.
3. Inspect the matrix of estimated correlations among the parameter estimates. (Some SEM computer programs generate this matrix as a user-specified option.) An identification problem is indicated if any of these correlations are close to 1.0 in absolute value. Correlations so high literally say that one parameter estimate cannot be derived independently of another estimate. Otherwise, the model may be identified.

The other kind of problem in the estimation of feedback effects concerns technical difficulties in the analysis. For example, iterative estimation of models with feedback loops may fail to converge unless the start values are quite accurate. This is especially true for the direct effects of a feedback loop and the disturbance variances and covariances of the endogenous variables involved in the feedback loop. If the default start values of a SEM computer program do not lead to a converged solution, then it is up to the researcher to provide better initial estimates—see Bollen (1989, pp. 137–138) or Kline (2004, p. 121) for guidelines. Even if estimation converges, the solution may be inadmissible, that is, it contains Heywood cases such as a negative variance estimate or other kinds of illogical results such as an estimated standard error that is so large that no interpretation is reasonable. Computer programs for SEM do not always give error messages when the output contains illogical estimates, so one must carefully inspect the solution.

Estimation of nonrecursive models can also be susceptible to *empirical underidentification* (Kenny, 1979) even if the model is actually identified. This happens when estimates of certain key direct effects are close to zero, which effectively drops those paths from the structural model. This can alter a nonrecursive structural model in such a way that the rank condition is violated because of paths with close-to-zero coefficients—see Kline (2004, p. 250) for an example. Perhaps due to these difficulties just described, one sees relatively few models with feedback loops in the behavioral science literature. But in some disciplines, especially economics, they are much more common, which suggests that the challenges of estimating such models are not insurmountable.

EXAMPLE

This example concerns the estimation of feedback effects between stress and depression with data from a stratified random sample of 983 native-born Chinese Americans and immigrants of Chinese descent collected by Shen and Takeuchi (2001). These authors also administered measures of the degree of acculturation and SES to the same cases plus measures of additional constructs not mentioned here. The data analyzed for this example are summarized in Table 3.2.

Presented in Figure 3.5 is a latent variable path model with two exogenous factors, acculturation and SES, an endogenous stress factor, and an endogenous single indicator of depression symptoms. The structural model contains a direct feedback loop between the stress and depression variables plus a disturbance correlation. Note that the SES factor in Figure 3.5 has effect indicators as per the original authors' model. However, it may be impossible to estimate this model with the specification that the SES factor has cause indicators only because of identification problems. The structural part of the model in the figure is identified because it satisfies the necessary order condition and the sufficient rank condition. For example, with two endogenous variables in the structural model, there should be at least one variable excluded from the equation of each, which is here true: the exogenous SES factor is excluded from the equation of the endogenous stress factor, and the exogenous acculturation factor is excluded from the equation of the endogenous depression variable. The order condition is thus satisfied. Evaluation that the rank condition for the structural model of Figure 3.5 is not demonstrated here. The mea-

Table 3.2. Input Data (Correlations, Standard Deviations) for Analysis of a Model with a Feedback Loop Between Stress and Depression (Figure 3.5)

Variable	*1*	*2*	*3*	*4*	*5*	*6*	*7*	*8*
Acculturation indicators								
1. Acculturation Scale	1.00							
2. Generation Status	.44	1.00						
3. Percent Life in U.S.	.69	.54	1.00					
Socioeconomic status indicators								
4. Education	.37	.08	.24	1.00				
5. Income	.23	.05	.26	.29	1.00			
Stress indicators								
6. Interpersonal	.12	.08	.08	.08	−.03	1.00		
7. Job	.09	.06	.04	.01	−.02	.38	1.00	
Single indicator								
8. Depression	.03	.02	−.02	−.07	−.11	.37	.46	1.00
SD	.78	.41	.24	3.27	3.44	.37	.45	.32

Note: Data from Shen and Takeuchi (2001); N = 983.

surement part of the model is also identified; therefore, the whole model of Figure 3.5 is identified.

Presented in Appendices A through C is syntax for EQS 6 (Bentler, 2003), LISREL 8 (SIMPLIS; Jöreskog & Sörbom, 2003), and Mplus 3 (Muthén & Muthén, 2003), which specifies the model in Figure 3.5. Some problems were encountered in the analysis with EQS. Specifically, several user-supplied start values were necessary in order to obtain a converged solution. However, EQS was unable to estimate the standard error of the disturbance covariance. Analyses with LISREL and Mplus were unremarkable, and the final solutions are quite similar across all three programs. Values of selected fit indexes reported by LISREL indicate reasonable overall model fit: χ^2 (14) = 58.667, CFI = .976, SRMR = .031, RMSEA = .057 with the 90% confidence interval .044-.073. Also, values of all absolute correlation residuals are all less than .10. For interpretive ease, only the completely standardized solution from LISREL is reported in Figure 3.5. A brief summary of the other results follows.

The model in Figure 3.5 explains 31.7% and 47.0% of the variance, respectively, of the endogenous stress and depression variables. These values are the squared multiple correlations calculated in LISREL (and other SEM computer programs) as 1.0 minus the ratio of the estimated disturbance variance over the model-implied variance for each endoge-

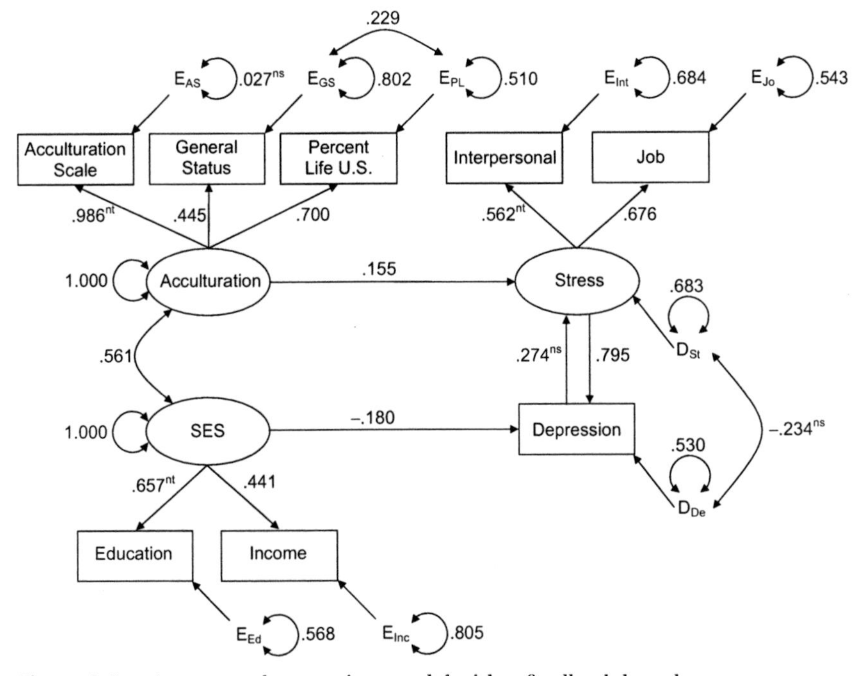

Figure 3.5. A structural regression model with a feedback loop between stress and depression, and the completely standardized solution. Estimates for the residual terms are proportions of unexplained variance. The unstandardized counterparts of the above estimates are all statistically significant at the .05 level except for those designated "ns," which means not significant, and "nt," which means not tested because this parameter is fixed to 1.0 in the unstandardized solution.

nous variable. However, Bentler and Raykov (2000) noted that squared multiple correlations computed as just described may be inappropriate for endogenous variables involved in feedback loops. This is because there may be a model-implied correlation between the disturbance of such a variable and one of its presumed causes, which is here true: For example, the model in Figure 3.5 implies a nonzero correlation between the disturbance of the stress factor and one of its presumed causes, depression. We can use the tracing rule from path analysis to see this predicted correlation: $D_{St} \rightarrow$ Stress $\rightarrow$ Depression. This predicted correlation violates the requirement of least squares estimation that the residuals (i.e., disturbances) are uncorrelated with the predictors (i.e., presumed direct causes). Bentler and Raykov described a general approach to estimating explained variance in either observed or latent endogenous variables of nonrecursive structural models that avoids this problem. This method generates the *Bentler-Raykov corrected* R^2, which takes account of nonzero model-implied correlations between disturbances and predictors. Values of the Bentler-Raykov corrected R^2 reported by EQS 6 for the endogenous stress and depression variables are, respectively, .432 and .482. The corrected R^2 for stress (.432) is quite a bit higher than the uncorrected R^2 for the same variable calculated by LISREL (.317), but the former may be more accurate.

The magnitude of the standardized direct effect of stress on depression (.795) is roughly three times the magnitude of the direct effect in the opposite direction (.274). The latter is also not statistically significant at the .05 level. Overall, it is not unexpected that greater stress predicts more depression symptoms and vice versa. It is more surprising, however, that greater acculturation is associated with higher levels of stress (standardized direct effect = .155), controlling for SES. Higher SES is also associated with fewer symptoms of depression (standardized direct effect = –.180).

Variables in feedback loops have indirect effects—and thus total effects—on themselves and each other, which is apparent in effects decompositions calculated by SEM computer programs for nonrecursive structural models. For example, an indirect standardized effect of stress on depression for the present example corresponds to the path

$$\text{Stress} \rightarrow \text{Depression} \rightarrow \text{Stress} \rightarrow \text{Depression}$$

and can be estimated with the tracing rule as the product $.795 \times .274 \times .795$, or .173. However, there are additional indirect effects of stress on depression because cycles of mutual influence in feedback loops are theoretically infinite. Mathematically, these product terms head fairly quickly to zero, but the total indirect effect of stress and depression is the sum of all possible cycles. It is tedious and error-prone to use the tracing rule to calculate by hand total indirect and total effects for even relatively simple models with feedback loops. Fortunately, many SEM computer programs including LISREL offer as an option the automatic calculation of all total indirect and total effects. For this example, the standardized total indirect effect of stress on depression reported by LISREL is .221, and the standardized total indirect effect of depression on stress is .076. The total effects of depression and stress on each other are calculated as the totals of their respective direct and indirect effects. For the standardized total effect of stress on depression, this sum equals $.795 + .211$, or 1.016, and the standardized total effect of depression on stress equals $.274 + .076$, or .350. Note that standardized total effect of stress on depression (1.016) exceeds 1.0, but this result is not illogical for the model of Figure 3.5. This is because this standardized total effect is not a correlation and thus its range is not from –1.0 to +1.0.

SUMMARY, PRACTICAL SUGGESTIONS, AND EXEMPLARS

The analysis of latent composites in measurement models or feedback loops in structural models are not without potential difficulties. However, there are times when the researcher's hypotheses about measurement or presumed causal effects are better expressed in models where some arrows are "backwards" compared with standard models. The representation of formative measurement through the specification of MIMIC factors with both cause and effect indicators avoids some of the identification problems of models where some factors are measured by cause indicators only. If the latter is required in a measurement model, then that factor should emit at least two direct effects onto factors each measured with effect indicators only. See Diamantopoulos and Winklhofer (2001) for an example of the analysis of a MIMIC model with cause and effect indicators of organizational resource commitment.

Problems to look out for in the analysis of models with feedback loops include nonconvergence of iterative estimation and empirical underidentification. The former can be avoided by providing accurate start values for direct effects in the feedback loop and associated disturbance variances and covariances. In other words, modelers should not rely exclusively on default start values of SEM computer programs. It also helps to "build" the model up in a series of steps, first without all direct effects in the feedback loop, gradually adding paths until the whole model is "assembled." For additional examples of the analysis of structural models with feedback loops, see Ahmed and Mosley (2002), who estimated reciprocal effects between use of mother–child health care and contraceptives in developing countries, and Sharkey (2002), who estimated reciprocal effects between nutritional risk and disability among homebound older adults.

APPENDIX A: EQS SYNTAX

```
/title
 Figure 5
/specifications
 cases = 983; variables = 8; data_file = 'Shen.ess ';
/labels
 V1 = ACC_SCL; V2 = GEN_STAT; V3 = PER_LIFE; V4 = EDUC;
 V5 = INC; V6 = INT_STR; V7 = JOB_STR; V8 = DEP;
 F1 = Accultur; F2 = Ses; F3 = Stress;
/equations
 V1 = F1 + E1; V2 = *F1 + E2; V3 = *F1 + E3;
 V4 = F2 + E4; V5 = *F2 + E5;
 V6 = F3 + E6; V7 = *F3 + E7;
```

```
F3 = .05*F1 - .6*V8 + D3; V8 = -.05*F2 + *F3 + E8;
/variances
F1 = .4*; F2 = *5; E1-E7 = *; E8 = .05*; D3 = .03*;
/covariances
F1,F2 = .6*; D3,E8 = -.01*; E2,E3 = .3*;
/print
fit=all; covariance = yes; effect = yes;
/end
```

APPENDIX B: LISREL (SIMPLIS) SYNTAX

```
Figure 5
Observed Variables
 ACC_SCL GEN_STAT PER_LIFE EDUC INC INT_STR JOB_STR DEP
Latent Variables: Accultur Ses Stress Depressi
Covariance matrix from file Shen.cov
Sample Size is 983
Equations
 ACC_SCL = 1*Accultur
 GEN_STAT PER_LIFE = Accultur
 EDUC = 1*Ses
 INC = Ses
 INT_STR = 1*Stress
 JOB_STR = Stress
 DEP = 1*Depressi
 Stress = Accultur Depressi
 Depressi = Ses Stress
Set the error variance of DEP to 0
Let the errors of GEN_STAT and PER_LIFE correlate
Let the errors of Stress and Depressi correlate
Path Diagram
LISREL Output: SC ND = 3 EF
End of Program
```

APPENDIX C: MPLUS SYNTAX

```
TITLE: Figure 5
DATA:
 FILE IS "Shen.dat"; TYPE IS STDEVIATIONS CORRELATION;
 NGROUPS = 1; NOBSERVATIONS = 983;
VARIABLE: NAMES ARE ACC_SCL GEN_STAT PER_LIFE EDUC INC INT_STR
JOB_STR DEP
ANALYSIS:
 TYPE IS GENERAL; ESTIMATOR IS ML; ITERATIONS = 1000;
 CONVERGENCE = 0.00005;
MODEL:
 Accultur BY ACC_SCL GEN_STAT PER_LIFE; Ses BY EDUC INC;
 Stress BY INT_STR JOB_STR; Stress ON Accultur DEP;
 DEP ON Stress Ses; DEP WITH Stress; GEN_STAT WITH PER_LIFE;
OUTPUT: SAMPSTAT RESIDUAL STANDARDIZED;
```

NOTES

1. Note that F1 is represented in Figure 3.1(a) as exogenous. This is consistent with confirmatory factor analysis (CFA) measurement models where all factors are exogenous. However, in a latent variable path model, some factors are specified as endogenous. In this case, F1 would have a disturbance.
2. See Kline (2004, pp. 229–231) for a method to take account of measurement error in observed exogenous variables.
3. In general, single-factor MIMIC models with one cause indicator are equivalent versions of single-factor reflective measurement models with all effect indicators, assuming that each indicator depends on only one factor and independent residuals (measurement errors; e.g., Kline, 2004, pp. 192–194).
4. The correlation residuals are also just as consistent with the possibility that the direct effect in the other direction, Adjustment → Achievement, is not zero. In the view of the author, however, the direct effect Achievement → Adjustment is more plausible.
5. Because the original model (Figure 3.3) and the respecified model with a MIMIC factor and the path Achievement Adjustment are not hierarchical, the chi-square difference test cannot be performed.

REFERENCES

Ahmed, S., & Mosley, W. H. (2002). Simultaneity in the use of maternal-child health care and contraceptives: Evidence from developing countries. *Demography, 39*, 75–93.

Arbuckle, J. L. (2003). *Amos 5* [Computer software]. Chicago: Smallwaters.

Beckie, T. M., Beckstead, J. W., & Webb, M. S. (2001). Modeling women's quality of life after cardiac events. *Western Journal of Nursing Research, 23*, 179–194.

Bentler, P. M. (2003). *EQS 6.1 for Windows* [Computer software]. Encino, CA: Multivariate Software.

Bentler, P. M., & Freeman, E. H. (1983). Tests for stability in linear structural equation systems. *Psychometrika, 48*, 143–145.

Bentler, P. M., & Raykov, T. (2000). On measures of explained variance in nonrecursive structural equation models. *Journal of Applied Psychology, 85*, 125–131.

Berry, W. D. (1984). *Nonrecursive causal models.* Beverly Hills, CA: Sage.

Bollen, K. A. (1989). *Structural equations with latent variables*. New York: Wiley.

Bollen, K., A., & Lennox, R. (1991). Conventional wisdom on measurement: A structural equation perspective. *Psychological Bulletin, 110*, 305–314.

Brito, C., & Pearl, J. (2003). A new identification condition for recursive models with correlated errors. *Structural Equation Modeling, 9*, 459–474.

Cudek, R. (1989). Analysis of correlation matrices using covariance structure models. *Psychological Bulletin, 105*, 317–327.

Diamantopoulos, A., & Winklhofer, H. M. (2001). Index construction with formative indicators: An alternative to scale development. *Journal of Marketing Research, 38*, 269–277.

Edwards, J. R., & Bagozzi, R. P. (2000). On the nature and direction of relationships between constructs and measures. *Psychological Methods*, *5*, 155–174.

Finkel, S. E. (1995). *Causal analysis with panel data*. Thousand Oaks, CA: Sage.

Heise, D. R. (1975). *Causal analysis*. New York: Wiley.

Hershberger, S. L. (1994). The specification of equivalent models before the collection of data. In A. von Eye & C. C. Clogg (Eds.), *Latent variables analysis* (pp. 68–105). Thousand Oaks, CA: Sage.

Jöreskog, K. G., & Sörbom, D. (2003). *LISREL 8.54 for Windows* [Computer software]. Lincolnwood, IL: Scientific Software International.

Kano, Y. (2001). Structural equation modeling with experimental data. In R. Cudeck, S. Du Toit, & D. Sörbom (Eds.), *Structural equation modeling: Present and future* (pp. 381–402). Lincolnwood, IL: Scientific Software International.

Kaplan, D. (2000). *Structural equation modeling*. Thousand Oaks, CA: Sage.

Kaplan, D., Harik, P., & Hotchkiss, L. (2001) Cross-sectional estimation of dynamic structural equation models in disequilibrium. In R. Cudeck, S. Du Toit, & D. Sörbom (Eds.), *Structural equation modeling: Present and future* (pp. 315–339). Lincolnwood, IL: Scientific Software International.

Kenny, D. A. (1979). *Correlation and causality*. New York: Wiley.

Kline, R. B. (2004). *Principles and practice of structural equation modeling* (2nd ed.). New York: Guilford Press.

MacCallum, R. C., & Browne, M. W. (1993). The use of causal indicators in covariance structure models: Some practical issues. *Psychological Bulletin*, *114*, 533–541.

MacCallum, R. C., Wegener, D. T., Uchino, B. N., & Fabrigar, L. R. (1993). The problem of equivalent models in applications of covariance structure analysis. *Psychological Bulletin*, *114*, 185–199.

Maruyama, G. M. (1998). *Basics of structural equation modeling*. Thousand Oaks, CA: Sage.

Muthén, L., & Muthén, B. (2003). *Mplus (Version 3)* [Computer software]. Los Angeles: Authors.

Nunnally, J. C., & Bernstein, I. H. (1994). *Psychometric theory* (3rd ed.). New York: McGraw-Hill.

Sharkey, J. R. (2002). The interrelationship of nutritional risk factors, indicators of nutritional risk, and severity of disability among home-delivered mean participants. *The Gerontologist*, *42*, 373–380.

Shen, B.-J., & Takeuchi, D. T. (2001). A structural model of acculturation and mental health status among Chinese Americans. *American Journal of Community Psychology*, *29*, 387–418.

Steiger, J. H. (2002). When constraints interact: A caution about reference variables, identification constraints, and scale dependencies in structural equation modeling. *Psychological Methods*, 7, 210–227.

Worland, J., Weeks, G. G., Janes, C. L. & Strock, B. D. (1984). Intelligence, classroom behavior, and academic achievement in children at high and low risk for psychopathology: A structural equation analysis. *Journal of Abnormal Child Psychology*, *12*, 437-454.

CHAPTER 4

POWER ANALYSIS IN COVARIANCE STRUCTURE MODELING

Gregory R. Hancock

Within the structural equation modeling (SEM) literature, methodological examinations of what constitutes adequate sample size, and the determinants thereof, are plentiful (see, e.g., Gagné & Hancock, in press; Jackson, 2003; MacCallum, Widaman, Zhang, & Hong, 1999; Marsh, Hau, Balla, & Grayson, 1998). Predictably, practitioners gravitate toward those sources providing evidence and/or recommendations supporting the acceptability of smaller sample sizes. Such recommendations, however, tend to be based on issues such as model convergence and parameter bias; they do not address statistical power. Thus, while there might be relatively large models for which, say, $n = 50$ yields reliable rates of model convergence with reasonable parameter estimates, such a sample size might be quite inadequate in terms of power for statistical tests of interest.

Within SEM, two contexts for power analysis are relevant: that in which tests of specific parameters within a given model are of interest, and that in which the test of data–model fit for an entire model is of interest. For each context, a researcher might have a need for *post hoc power analysis* or

Structural Equation Modeling: A Second Course, 69–115

a priori power analysis. For the former, sample data have been gathered and the statistical test has already been conducted; the researcher now wishes to estimate the power associated with that particular statistical test. For the latter, more directly termed *sample size determination*, the researcher's goal is to determine the necessary sample size to achieve sufficient power for all statistical tests of interest. Doing so prior to entering into a complex research endeavor makes sound practical sense, and indeed is becoming increasingly necessary for researchers seeking funding for their applied research.

In the current chapter, power analysis will be addressed as it relates to testing parameters within a model, as well as to testing data–model fit for an entire model. In each context both post hoc and a priori power analysis issues and procedures will be presented, making the overarching framework for this chapter represented by the diagram in Figure 4.1. Because of the wide array of models to which this framework applies, the focus of the presentation will be on the most common scenario. This is the single-sample covariance structure model, where maximum likelihood (ML) estimation is used with data meeting the standard distributional assumption of conditional multivariate normality. As such, models involving multiple samples, models with mean structures, and models under non-normal data conditions will not be addressed extensively. However, after offering a didactically oriented treatment of power analysis for the common SEM scenario, extensions of the core principles presented to these alternative scenarios will be touched upon briefly at the chapter's end.

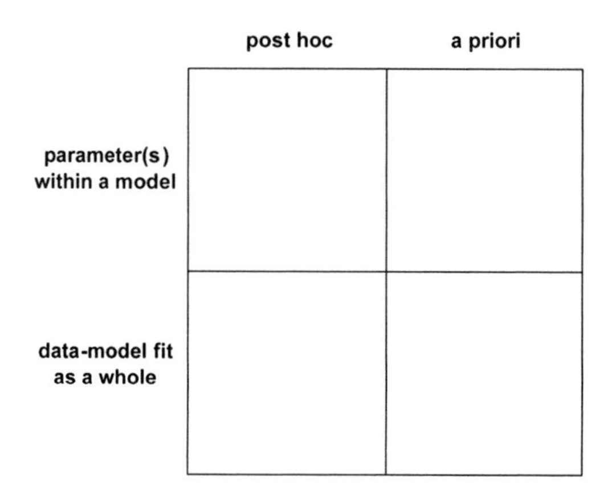

Figure 4.1. Framework for power analysis topics addressed in this chapter.

POWER ANALYSIS FOR TESTING PARAMETERS WITHIN A MODEL

Power analysis in SEM is built on the same basic ideas as power analysis in other settings (e.g., in analysis of variance). These involve (1) null and alternative hypotheses, (2) test statistics to assess the null hypothesis, and (3) central and noncentral distributions. Each of these is described below in the specific context of parameter testing in SEM, after which post hoc and a priori power analysis are developed and illustrated with examples.

Null and Alternative Hypotheses

For any research question involving specific parameters in a population (e.g., means, variances), there are null and alternative hypotheses regarding those parameters. Within the context of SEM, consider the example of a two-tailed test of an unstandardized path in a latent variable path model, where that path is from factor F1 to factor F2. In this case the null hypothesis would commonly be H_0: $b_{F2F1} = 0$, while the corresponding alternative would be H_1: $b_{F2F1} = 0$. More generally, for any single parameter θ, one can imagine a null hypothesis H_0: $\theta = \theta_0$ and its corresponding alternative as H_1: $\theta \neq \theta_0$, where the most frequent choice of θ_0 is 0. That is, researchers most commonly care about detecting any degree of nonzero contribution of a given parameter to the model; the current chapter focuses on power analysis associated with this purpose.

The above expression regarding a single parameter extends to any number of parameters as contained in a parameter vector $\boldsymbol{\theta}$. Consider a model with a "full" set of parameters in vector $\boldsymbol{\theta}_F$, and a model with a "reduced" set of parameters in vector $\boldsymbol{\theta}_R$, where elements in $\boldsymbol{\theta}_R$ constitute a subset of elements in $\boldsymbol{\theta}_F$; the models therefore have a hierarchical (nested) relation. The reduced model contains what we might term the *peripheral* parameters, which define a context for the parameters of key interest; the full model contains the peripheral parameters as well as the *focal* parameters, the latter being those for which power is of interest. As an example, $\boldsymbol{\theta}_F$ might contain all the parameters contained in an oblique confirmatory factor analysis (CFA) model (i.e., interfactor covariances, loadings, and error variances), while $\boldsymbol{\theta}_R$ could contain the parameters in a competing orthogonal CFA model (i.e., loadings and error variances); the focal parameters are thus the interfactor covariances. The null hypothesis in this case could be expressed directly as pertaining to the interfactor covariances contained in $\boldsymbol{\theta}_F$, specifically that they are all zero; correspondingly, the alternative hypothesis would be that one or more are nonzero. More generally, and more useful for the purposes of this chapter, the null and alternative hypotheses may be framed in terms of the overall badness of fit of the population data to the model with $\boldsymbol{\theta}_F$ and to the model with $\boldsymbol{\theta}_R$. Expressed generically, we may write the null hypothesis for the population as

H_0: Badness of data–model fit with $\boldsymbol{\theta}_R$ = Badness of data–model fit with $\boldsymbol{\theta}_F$.

This would occur in the above CFA case, for example, when the interfactor covariances are all zero (i.e., orthogonality exists). The corresponding alternative hypothesis for the population is thus

H_1: Badness of data–model fit with $\boldsymbol{\theta}_R$ > Badness of data–model fit with $\boldsymbol{\theta}_F$,

where a false null hypothesis implies that the badness of fit associated with the reduced model (with relatively fewer parameters in $\boldsymbol{\theta}_R$) is necessarily greater than that associated with the full model (with relatively more parameters in $\boldsymbol{\theta}_F$). This would occur in the above CFA case, for example, when at least one of the interfactor covariances is nonzero (i.e., some obliqueness exists). Thus, the null hypothesis is that the reduced and full models have equivalent data–model fit in the population, while the alternative hypothesis is that the reduced model has worse data–model fit in the population.

To be more concrete, when assessing the population-level data–model fit for models with full and reduced sets of parameters, one starts with the fit function associated with the desired method of estimation. For maximum likelihood (ML) estimation, which is the focus in this chapter, the relevant degree of data–model fit at the population level for a covariance structure model involving p measured variables is the familiar F_{ML} (sometimes denoted more completely as $F_{ML}(\boldsymbol{\theta})$):

$$F_{ML} = \ln|\hat{\boldsymbol{\Sigma}}| + tr(\boldsymbol{\Sigma}\hat{\boldsymbol{\Sigma}}^{-1}) - \ln|\boldsymbol{\Sigma}| - p \tag{1}$$

where $\boldsymbol{\Sigma}$ is the population covariance matrix for the observed data and $\hat{\boldsymbol{\Sigma}}$ is the population covariance matrix implied by the model with parameters in $\boldsymbol{\theta}$ (sometimes denoted more completely as $\hat{\boldsymbol{\Sigma}}(\boldsymbol{\theta})$).

Fitting population data $\boldsymbol{\Sigma}$ to both reduced and full models would yield population fit function values $F_{ML}(\boldsymbol{\theta}_R)$ and $F_{ML}(\boldsymbol{\theta}_F)$, respectively, where $F_{ML}(\boldsymbol{\theta}_R) - F_{ML}(\boldsymbol{\theta}_F) \geq 0$ as the full model has additional parameters and hence cannot have more badness of data–model fit in the population. That is, for testing one or more parameters in a covariance structure model, the null hypothesis may be expressed as H_0: $F_{ML}(\boldsymbol{\theta}_R) = F_{ML}(\boldsymbol{\theta}_F)$, or equivalently

$$H_0: F_{ML}(\boldsymbol{\theta}_R) - F_{ML}(\boldsymbol{\theta}_F) = 0, \quad (2)$$

indicating that the constraint of parameters in the reduced model has no effect on the data–model fit in the population. In the orthogonal and oblique CFA example, the null hypothesis states that the factors do not covary in the population, and therefore restricting them from doing so has no effect on the population-level data–model fit. The corresponding alternative hypothesis may be expressed as H_1: $F_{ML}(\boldsymbol{\theta}_R) > F_{ML}(\boldsymbol{\theta}_F)$, or equivalently

$$H_1: F_{ML}(\boldsymbol{\theta}_R) - F_{ML}(\boldsymbol{\theta}_F) > 0, \quad (3)$$

indicating that the additional parameters characterizing the difference between the full and reduced models do have at least some nonzero degree of impact on population-level data–model fit. For the CFA example, the alternative hypothesis states that there is at least some degree of covariation among the factors, and therefore restricting them from covarying entirely would have a deleterious effect on the population-level data–model fit. This framing of the null and alternative hypotheses, in terms of relative data–model fit of full and reduced models in the population, will prove particularly useful for power analysis.

Test Statistics to Assess Null Hypotheses

The test of the null hypothesis regarding any population parameter of interest (e.g., a difference between population means, $\mu_1 - \mu_2$) is facilitated by the sample statistic estimating the parameter (e.g., $\bar{Y}_1 - \bar{Y}_2$), which in turn is usually transformed to a more general test statistic (e.g., an observed t). For example, to test H_0: $b_{F2F1} = 0$ regarding the unstandardized path from factor F1 to factor F2 in a latent variable path model, the population parameter b_{F2F1} is estimated by that value derived from sample data, $\hat{b}_{F2F1}$; this is in turn typically transformed to an observed z-statistic, where $z = \dfrac{\hat{b}_{F2F1}}{SE(\hat{b}_{F2F1})}$ (assuming ML estimation). In fact, for any parameter θ, to test H_0: $\theta = 0$ the parameter estimate derived from sample data, $\hat{\theta}$, is customarily transformed to $z = \dfrac{\hat{\theta}}{SE(\hat{\theta})}$ automatically within the SEM software.

To be more general, let us return to parameter assessment framed in terms of comparative data–model fit for full and reduced models in the population. For any model with parameters in vector $\boldsymbol{\theta}$, the amount of badness of data–model fit in the population, F_{ML} (see Equation 1), can be estimated from observed sample data by

$$\hat{F}_{\text{ML}} = \ln|\hat{\boldsymbol{\Sigma}}| + tr(\mathbf{S}\hat{\boldsymbol{\Sigma}}^{-1}) - \ln|\mathbf{S}| - p\,, \tag{4}$$

where $\mathbf{S}$ is the sample covariance matrix. The sample data yielding $\mathbf{S}$ are used to derive estimates for model parameters in $\boldsymbol{\theta}$ that imply a covariance matrix $\hat{\boldsymbol{\Sigma}}$ so as to minimize the estimated badness of data–model fit in $\hat{F}_{\text{ML}}$. As values of the relevant second-order moment data in $\mathbf{S}$ fluctuate randomly from sample to sample, in turn yielding fluctuating parameter estimates in $\hat{\boldsymbol{\theta}}$ with implied covariance matrix $\hat{\boldsymbol{\Sigma}}$, the value of $\hat{F}_{\text{ML}}$ for a model with *df* degrees of freedom will vary from sample to sample with its own sampling distribution. Under specific data and model conditions, discussed below, the $\hat{F}_{\text{ML}}$ sampling distribution converts to a more familiar χ^2 distribution such that

$$(n-1)\hat{F}_{\text{ML}} = \chi^2_{df}. \tag{5}$$

And certainly there are other ways to assess overall data–model fit as well (e.g., CFI, SRMR); this χ^2 test statistic, however, allows one to conduct a *statistical* test of hypotheses regarding the overall data–model fit, and it is this statistical assessment upon which power analysis will be grounded.

So now let us use this information to focus specifically on the statistical comparison of models with reduced and full sets of parameters ($\boldsymbol{\theta}_{\text{R}}$ and $\boldsymbol{\theta}_{\text{F}}$, respectively). For testing the null hypothesis in Equation 2, the test statistic $\hat{F}_{\text{ML}}(\hat{\boldsymbol{\theta}}_{\text{R}}) - \hat{F}_{\text{ML}}(\hat{\boldsymbol{\theta}}_{\text{F}})$ may be determined from sample data that have been fit to both reduced and full models. This estimated difference in data–model fit, which has $df_{\text{diff}} = df_{\text{R}} - df_{\text{F}}$, will fluctuate from sample to sample with its own sampling distribution; fortunately, under specific data and model conditions (discussed below), this statistic converts to a χ^2 distribution (with test statistic χ^2_{diff}) as

$$(n-1)[\hat{F}_{\text{ML}}(\hat{\boldsymbol{\theta}}_{\text{R}}) - \hat{F}_{\text{ML}}(\hat{\boldsymbol{\theta}}_{\text{F}})] = \chi^2_{\text{diff}}\,. \tag{6}$$

The χ^2_{diff} statistic allows one to evaluate hypotheses regarding differences in data–model fit for models with and without focal parameters (or, more generally speaking, with and without constraints on those parameters). Note again that one might also compare models using data–model fit indices such as the popular information criteria (e.g., AIC); however, for the purpose of providing a *statistical* comparison, upon which power analysis will be grounded, this χ^2_{diff} test statistic is essential.

Central Distributions (for H_0), Noncentral Distributions (for H_1), and Power

H_0 and central distributions. When a null hypothesis is true, the statistic used to test H_0 follows a sampling distribution described as a *central* distribution, often with familiar distributional characteristics when certain underlying (and rather ideal) assumptions hold. Within the context of SEM, the test statistic (z) for evaluating H_0: $b_{F2F1} = 0$, for example, will asymptotically approach a standard normal distribution when population data are (conditionally) multivariate normal. Correspondingly, the square of this test statistic, z^2, will asymptotically approach a χ^2 distribution (with 1 *df*) under the same distributional conditions, thereby approximating the difference in data–model fit with and without that particular parameter. More generally, the test statistic facilitating the comparison of data–model fit for reduced and full models (as per Equation 6), or assessing overall data–model fit (as per Equation 5), will asymptotically approach a χ^2 distribution under specific data and model conditions (as discussed later).

Given the ubiquity of the χ^2 distribution in SEM, let us consider it in a bit more detail. Figure 4.2 depicts sampling distribution behavior for a test statistic labeled as T, which happens to correspond to a χ^2 distribution with 5 *df*. Shown on the left side of the figure is the case where the null hypothesis H_0 is true, in which observed test statistic values labeled as T_0 follow the central sampling distribution and have expected value $E(T_0)$. For central χ^2 distributions the expected value equals the number of degrees of freedom, $E[\chi^2] = df$; the shape of these distributions around that expectation, as the reader may also be aware, tends to be quite skewed for smaller numbers of degrees of freedom, and becomes more normal as *df* approaches 30 or so.

As relates to SEM, then, when using a χ^2 statistic to assess overall data–model fit for a correct model (i.e., $F_{ML} = 0$), the sample χ^2 value should randomly fluctuate around the model's number of degrees of freedom (under conditional multivariate normality). With respect to comparing reduced and full models, most relevant to the current section of this chapter, recall that the null condition is that the set of additional parameters contained in $\boldsymbol{\theta}_F$ is unnecessary; that is, $F_{ML}(\boldsymbol{\theta}_R) = F_{ML}(\boldsymbol{\theta}_F)$. In such cases one possibility is that the reduced and full models both have exact

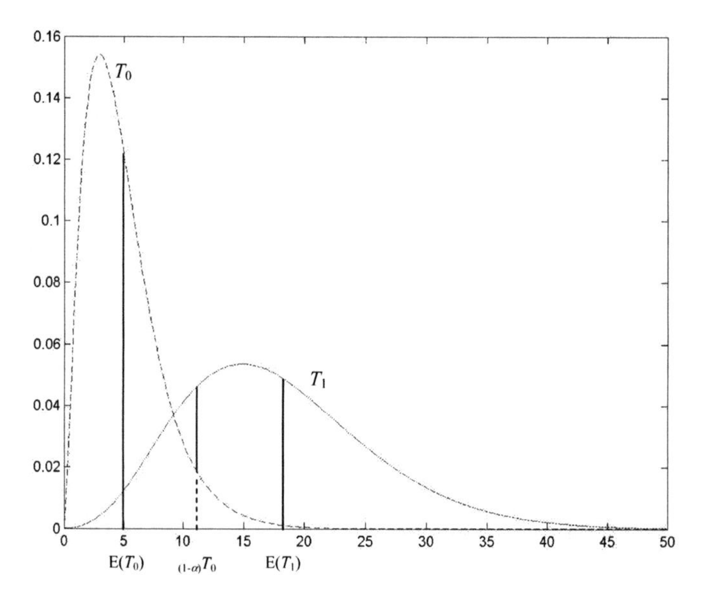

Figure 4.2. Sampling distribution for test statistic T under null (T_0) and alternative (T_1) conditions.

fit in the population (i.e., $F_{\text{ML}}(\boldsymbol{\theta}_{\text{R}}) = F_{\text{ML}}(\boldsymbol{\theta}_{\text{F}}) = 0$), and as such their corresponding χ^2 sampling distributions are centrally located at df_{R} and df_{F}, respectively. When fitting sample data to both models, the χ^2 difference statistic $\chi^2_{\text{diff}} = \chi^2_{\text{R}} - \chi^2_{\text{F}}$ follows a central χ^2 sampling distribution located at df_{diff}, where $df_{\text{diff}} = df_{\text{R}} - df_{\text{F}}$. A second possibility is that, although $F_{\text{ML}}(\boldsymbol{\theta}_{\text{R}}) = F_{\text{ML}}(\boldsymbol{\theta}_{\text{F}})$, neither the reduced model nor the full model is correct in the population (i.e., $F_{\text{ML}}(\boldsymbol{\theta}_{\text{R}}) - F_{\text{ML}}(\boldsymbol{\theta}_{\text{F}}) \neq 0$). In such cases the reduced and full model test statistics do not follow central χ^2 distributions; and more importantly, recent simulation evidence by Yuan and Bentler (2004) indicated that for common sample sizes, χ^2_{diff} is not satisfactorily represented by a central χ^2 distribution (even under conditional multivariate normality). Thus, for the null case when $F_{\text{ML}}(\boldsymbol{\theta}_{\text{R}}) = F_{\text{ML}}(\boldsymbol{\theta}_{\text{F}})$, the full (and thus reduced) model must be correct, or, as Yuan and Bentler indicated, reasonably close to correct, in order for the χ^2_{diff} test statistic to be well approximated by a central χ^2 distribution, and hence for inference relying upon that test statistic (for data–model fit as a whole or for comparing full and reduced models) to be reasonably valid.

H_1 and noncentral distributions. In general, when H_0 is false and instead a specific alternative H_1 is true, the sampling distribution for the test statistic differs from that under the null condition. Such a sampling distribution is referred to as a *noncentral* distribution. Within the context of SEM,

when H_0: $b_{F2F1} = 0$ is false, for example, the test statistic (z) will ideally follow a noncentral distribution, which is also normal in shape but is centered at a point other than zero. More generally, when a null hypothesis H_0 is false and a particular alternative H_1 is true, the χ^2 test statistic facilitating the comparison of data–model fit for reduced and full models, or of the assessment of overall data–model fit, will follow a noncentral distribution. That distribution might or might not actually be χ^2 in nature, as expanded upon below.

In Figure 4.2 a particular noncentral sampling distribution of observed test statistic values (labeled as T_1) is depicted to the right of the central distribution, with expected value $E(T_1)$. Such distributions are often described in terms of the expected value's location with respect to that of the central distribution, expressed as a *noncentrality parameter* λ, where

$$\lambda = E(T_1) - E(T_0); \tag{7}$$

this quantity in turn plays a critical role in power analysis, as described later. For the noncentral distribution depicted in Figure 4.2, the noncentrality parameter is $\lambda = 12.83$. Given that $E(T_0) = 5$ for the central χ^2 distribution with 5 *df*, Equation 7 implies that $E(T_1)=17.83$ for the noncentral distribution shown. That is, $E(T_1) = E(T_0) + \lambda$, which for χ^2 distributions yields the general relation

$$E[\chi^2] = df + \lambda. \tag{8}$$

Thus, given that H_0 is false and that the alternative H_1 represents a specific degree of the null's falseness as reflected in λ, sample χ^2 values should randomly fluctuate around $df + \lambda$ according to that specific noncentral distribution.

The above information has implications for χ^2 values assessing overall data–model fit as well as for χ^2 values used in model comparisons. When using a χ^2 test statistic to assess overall data–model fit for a false model (i.e., $F_{ML} > 0$), then the corresponding model χ^2 value determined from the sample as per Equation 5 should indeed fluctuate around $df + \lambda$ (assuming conditional multivariate normality). In this case the true degree of noncentrality for the population, not surprisingly given Equation 5 for samples, is

$$\lambda = (n - 1)F_{ML}. \tag{9}$$

With respect to comparing models with and without specific parameters, the alternative condition is when one or more of the additional parameters contained in $\boldsymbol{\theta}_F$ are not zero, and hence $F_{ML}(\boldsymbol{\theta}_R) > F_{ML}(\boldsymbol{\theta}_F)$.

In such cases one possibility is that the full model has exact fit in the population ($F_{ML}(\boldsymbol{\theta}_F) = 0$) while the reduced model does not ($F_{ML}(\boldsymbol{\theta}_R) > 0$). Here the model χ^2 sampling distribution for the full model is centrally located at df_F, while that for the reduced model is noncentral with expected value $df_R + \lambda$. The noncentrality parameter λ results from the improperly omitted (or, more generally, improperly constrained) parameters in the reduced model, and may be determined as

$$\begin{aligned}\lambda = \lambda_R - \lambda_F &= (n-1)F_{ML}(\boldsymbol{\theta}_R) - (n-1)F_{ML}(\boldsymbol{\theta}_F) \\ &= (n-1)[F_{ML}(\boldsymbol{\theta}_R) - F_{ML}(\boldsymbol{\theta}_F)],\end{aligned} \tag{10}$$

which equals

$$\lambda = (n-1)[F_{ML}(\boldsymbol{\theta}_R)] \tag{11}$$

when the full model is correct. When fitting sample data to both models, the χ^2 difference statistic $\chi^2_{diff} = \chi^2_R - \chi^2_F$ asymptotically approaches a noncentral χ^2 sampling distribution located at $df_{diff} + \lambda = (df_R - df_F) + \lambda$. When comparing models under a false null hypothesis, where the full model has better data–model fit in the population than the reduced model, the second possibility is that neither model is actually correct (i.e., $F_{ML}(\boldsymbol{\theta}_F) > 0$ and $F_{ML}(\boldsymbol{\theta}_R) > 0$). For this non-null scenario, recent simulation evidence by Yuan and Bentler (2004) indicated that χ^2 difference tests do not follow known noncentral χ^2 distributions with sufficient fidelity to yield reliable tests when using common sample sizes.

To recap, then, tests of models as a whole, or of parameters within those models, start with null and alternative hypotheses. The statistical choice between null and alterative is facilitated by specific test statistics whose behavior is governed by central and noncentral distributions. Those distributions will be of a familiar χ^2 form when data have favorable distributional properties, and when certain model conditions hold. Specifically, under conditional multivariate normality, the test statistic assessing overall data–model fit will follow a central χ^2 given the null condition that the model is correct in the population, and it will follow a noncentral χ^2 given alternative non-null conditions in which the model is not correct in the population. As for statistical comparisons of models with and without focal parameters (or, more generally, constraints on those parameters), the test statistic comparing the full and reduced models will follow a central χ^2 under the null condition that $F_{ML}(\boldsymbol{\theta}_R) = F_{ML}(\boldsymbol{\theta}_F)$ and, as Yuan and Bentler (2004) showed, when the full model is correct (or reasonably close to correct) in the population. The test statistic will follow a noncentral χ^2 under the alternative condition that $F_{ML}(\boldsymbol{\theta}_R) > F_{ML}(\boldsymbol{\theta}_F)$ and, following Yuan and Bentler, that the full model is correct (or reasonably so) in the population. It is under this key assumed condition, that the full model is reasonably well specified, that power analysis proceeds.

Power. The above basic principles regarding null and alternative hypotheses, test statistics, and distributions that are central and noncentral are fundamental to all of power analysis, whether in SEM other otherwise. Returning to Figure 4.2, under conditions when a null hypothesis H_0 is true, a point of particular interest within the central distribution is the upper tail critical value at the $(1 - \alpha) \times 100$th percentile employed in hypothesis testing with χ^2 distributions. This point is labeled as $_{(1-\alpha)}T_0$ in Figure 4.2, and within the central χ^2 distribution with 5 *df* it corresponds to 11.07 for the commonly used $\alpha = .05$ level. Relative to the central distribution, the noncentral distribution is shifted to the right considerably. In fact, because of its relatively large degree of noncentrality ($\lambda = 12.83$), most of the noncentral distribution exceeds the point defined by the α-level critical value $_{(1-\alpha)}T_0$ in the central distribution. The proportion of the noncentral distribution exceeding the point $_{(1-\alpha)}T_0$ is, by definition, the power π of the statistical test of H_0 given the condition that this specific alternative hypothesis H_1 is true. Symbolically,

$$\pi = \text{pr}[T_1 > {}_{(1\text{-}\alpha)}T_0 \mid H_1] \,. \tag{12}$$

For the noncentral distribution in Figure 4.2, the power is $\pi = .80$, implying that if one repeatedly conducted the α-level statistical test of interest under the specific non-null (and hence noncentral) condition shown, 80% of those tests would be expected to lead to a rejection of H_0 (assuming conditional multivariate normality). Thus, if we know the distributional form of the test statistic T under both the null (central) and the specific alternative (noncentral) conditions, then knowledge of the noncentrality parameter λ will allow for the determination of power. This holds true whether dealing with z, t, F, or χ^2 distributions, with the latter holding particular interest for power analysis within SEM. These principles are drawn upon below to facilitate post hoc and a priori power analysis for testing parameters within a model.

POST HOC POWER ANALYSIS FOR TESTING PARAMETERS WITHIN A MODEL

In post hoc power analysis, the power of a particular statistical test is estimated after sample data have been gathered and the statistical test has been conducted. Following from the above discussion, in order to estimate the power π of the test comparing full and reduced models, we must have an estimate of the degree of noncentrality in Equation 10. Following from Equation 8, a noncentrality parameter should be generally estimable as

$$\hat{\lambda} = \chi^2 - df, \tag{13}$$

or more specifically in this case,

$$\hat{\lambda} = \chi^2_{\text{diff}} - df_{\text{diff}}. \tag{14}$$

Indeed, such an approach should provide an estimate of the desired noncentrality parameter, $\hat{\lambda}$, and is easily computable using the results of fitting sample data to both the full and reduced models. Seminal work by Satorra and Saris (1985; see also Kaplan, 1995; Saris & Satorra, 1993), however, showed that a better estimate of noncentrality (hence yielding a better estimate of power) could be obtained by following a fairly simple modeling process. The steps in this strategy for determining post hoc power associated with the α-level test of the difference between a full model with parameters $\boldsymbol{\theta}_F$ and a reduced model with parameters $\boldsymbol{\theta}_R$ are detailed below. Note that a researcher must initially decide whether power is of interest as it pertains to α-level testing of the whole set of focal parameters simultaneously, or to testing each focal parameter individually. In this latter and more common case, there will actually exist pairs of full and reduced models each differing by a single focal parameter and where the remaining focal parameters are temporarily treated as peripheral. In this case the process below would thus be repeated for each focal parameter.

Step 1. Fit the sample data to the full model using ML estimation, obtaining the focal and peripheral parameter estimates in $\hat{\boldsymbol{\theta}}_F$ and the resulting model-implied covariance matrix $\hat{\boldsymbol{\Sigma}}(\hat{\boldsymbol{\theta}}_F)$. Assess the data–model fit utilizing accepted indices and cut-off criteria (see, e.g., Hu & Bentler, 1999); proceed with this post hoc power analysis process only if satisfactory data–model fit exists. Note that if the full model is indeed correct in the population, then the values in $\hat{\boldsymbol{\Sigma}}(\hat{\boldsymbol{\theta}}_F)$ can provide more efficient estimates of the corresponding population moments in $\boldsymbol{\Sigma}$ than the sample data in **S**.

Step 2. Having obtained the relevant model-implied moment information from the full model ($\hat{\boldsymbol{\Sigma}}(\hat{\boldsymbol{\theta}}_F)$), use those model-implied moment data as input to estimate the data–model fit for a reduced model (with peripheral parameters in $\boldsymbol{\theta}_R$). In doing so, fix the values for all peripheral parameters to the estimated values in $\hat{\boldsymbol{\theta}}_F$; fix the remaining focal parameter(s) in the reduced model to the hypothesized value(s), typically zero. Equivalently, all start values can be set as described above, with the modeling software set to have zero iterations; in this way the model should be frozen at the desired values from the full model, but with the constraints of the reduced model. Under these conditions the parameter values from the full model should be perfectly explaining portions of the model-implied moments used as input data, because those moments actually came from the full model. In fact, the only badness of fit associated with this reduced model comes directly from its constrained (in this case, to zero) focal parameter(s). As such, the reduced model will yield a model χ^2 test statistic that, by itself, represents an estimate of the noncentrality associated specifically with the constraint(s) of the reduced model (i.e., a direct estimate of λ).

Step 3. Using the value of $\hat{\lambda}$ from Step 2 and the number of degrees of freedom associated with the parameter(s) being tested, $df_{\text{diff}} = df_{\text{R}} - df_{\text{F}}$, determine the power estimate $\hat{\pi}$ for the χ^2 test at the desired α-level. Appendix A is provided to facilitate estimation of the level of power, assuming the common α-level of .05. Although more complete resources obviously exist, such as tables by Haynam, Govindarajulu, and Leone (1973), or values derivable from functions within statistical software (e.g., SAS, GAUSS), the tabled values in Appendix A should be quite adequate for df_{diff} up to 50.

For example, consider the power associated with the detection of non-zero relations among three factors in a CFA model. For the comparison of the oblique and orthogonal forms, which yields $df_{\text{diff}} = 3$ when testing all interfactor covariances simultaneously, imagine an estimated noncentrality parameter of $\hat{\lambda} = 2.60$. Based on Appendix A, the estimated power $\hat{\pi}$ is between .20 and .25. For more precision, linear interpolation between the listed noncentrality parameters of 2.10 and 2.71 may be used, which would yield a power of .241 (which is accurate, in this case, to three decimal places). Linear interpolation may similarly be used for df_{diff} values between those tabled, yielding interpolated noncentrality parameters within which a linearly interpolated power estimate may be derived. In practice, however, it seems most likely that the $df = 1$ column will be of most use, facilitating the derivation of power estimates associated with a .05-level test of individual focal parameters within a model.

Example of Post Hoc Power Analysis for Testing Parameters Within a Model

Imagine that a researcher wished to assess the differential impact of reading curriculum on young children's reading proficiency, after controlling for their initial reading readiness. At the start of kindergarten (i.e., prior to receiving reading instruction) each teacher used six-point rating scales to form ratings for each of 1,000 children on phonemic awareness (*phon*, V1), the alphabetic principle (*alpha*, V2), and print concepts and conventions (*print*, V3). These three measures were all believed to be observable manifestations of the underlying construct of Reading Readiness. Half of these kindergartners were assigned to receive a traditional phonics-based curriculum for 2 years, while the other half received a whole-language curriculum for 2 years (assume, for simplicity, no classroom-level effects). Then, at the end of first grade, three reading assessments were made, focusing on vocabulary (*voc*, V4), comprehension (*comp*, V5), and language (*lang*, V6). These were all believed to be observable manifestations of an underlying construct of Reading Proficiency. Finally, the two treatments defined a treatment dummy variable, V7, where V7=1 indicated phonics and V7=0 indicated whole-language. A covariance matrix **S** based on data fabricated for this example appears below for V1 through V7, respectively; data are for the total sample of 1,000 children:

$$\mathbf{S} = \begin{bmatrix} 1.801 & & & & & & \\ .975 & 1.858 & & & & & \\ .817 & 1.201 & 1.782 & & & & \\ 1.524 & 2.403 & 2.106 & 22.582 & & & \\ 1.486 & 2.179 & 1.948 & 14.587 & 15.976 & & \\ 1.113 & 1.720 & 1.620 & 12.523 & 10.517 & 16.040 & \\ .125 & .206 & .184 & .444 & .390 & .390 & .250 \end{bmatrix}.$$

The overall covariance structure model of interest in this example is a variation on a multiple-indicator multiple-cause (MIMIC) model, with a latent covariate (see, e.g., Hancock, 2004), and is depicted in Figure 4.3. The primary question of interest in such a model is whether treatment populations differ in Reading Proficiency after controlling for initial Reading Readiness. Within this example, each of the above steps in post hoc power analysis will be applied, focusing individually on the three structural parameters b_{F2F1}, c_{F1V7}, and b_{F2V7}; note that it is the latter of these parameters that directly addresses the primary research question of interest.

Step 1. ML estimation was used to fit the observed covariance matrix shown above to the model in Figure 4.3. As readers may verify with the modeling software of their choosing, the data–model fit was excellent: $\chi^2(12, N = 1{,}000) = 10.427$, CFI = 1.000, SRMR = .012, and RMSEA = .000 with 90% C.I. = (.000, .029). In addition, loading parameter estimates were $\hat{b}_{V2V1} = 1.463$, $\hat{b}_{V3F1} = 1.251$, $\hat{b}_{V5F2} = 0.853$, and $\hat{b}_{V6F2} = 0.726$; and, error variance estimates within the measurement model were $\hat{c}_{E1} = 1.143$, $\hat{c}_{E2} = .449$, $\hat{c}_{E3} = .751$, $\hat{c}_{E4} = 5.477$, $\hat{c}_{E5} = 3.528$, and $\hat{c}_{E6} = 7.030$. As for the structure, salient variance parameter esti-

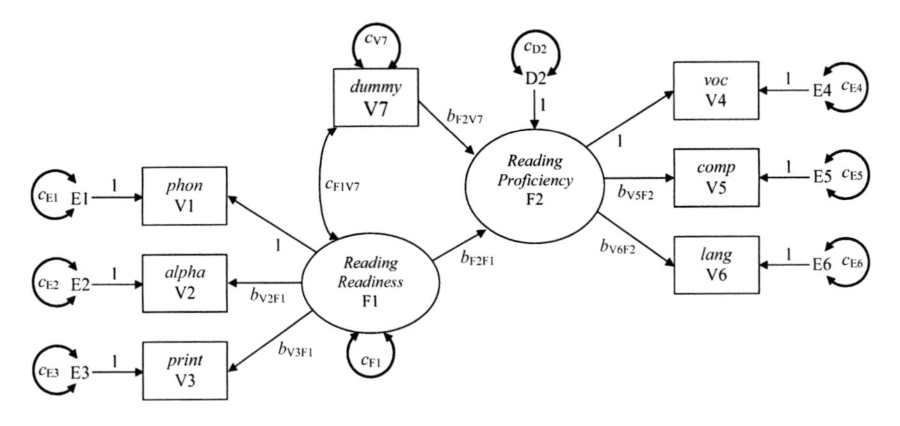

Figure 4.3. MIMIC model with latent covariate for example of post hoc power analysis for testing parameters within a model.

mates were $\hat{c}_{V7}$ = .250, $\hat{c}_{D1}$ = .579, and $\hat{c}_{D2}$ = 12.674, while the key structural parameters (with z statistics) were estimated as $\hat{b}_{F2F1}$ =2.481 (z = 11.448), $\hat{c}_{F1V7}$ = .563 (z = 9.524), and $\hat{b}_{F2V7}$ = .463 (z = 1.707). In the context of the model, the implication of these latter parameter estimates is that the phonics population (V7 = 1) was actually higher on initial Reading Readiness than the whole-language population (V7 = 0), that early Reading Readiness has a positive impact on later Reading Proficiency, and that a population difference in later Reading Proficiency was not detected between the phonics and whole-language populations after controlling for initial Reading Readiness.

Step 2. The above parameter estimates yield the following model-implied covariance matrix for V1 through V7, respectively:

$$\hat{\boldsymbol{\Sigma}} = \begin{bmatrix} 1.801 & & & & & & \\ .963 & 1.858 & & & & & \\ .824 & 1.205 & 1.782 & & & & \\ 1.699 & 2.485 & 2.126 & 22.582 & & & \\ 1.449 & 2.120 & 1.814 & 14.592 & 15.976 & & \\ 1.233 & 1.803 & 1.543 & 12.414 & 10.591 & 16.040 & \\ .141 & .206 & .176 & .465 & .397 & .338 & .250 \end{bmatrix}.$$

This matrix was used as input in subsequent post hoc power analyses; all peripheral parameters were now fixed to their full model parameter estimates, with the exception of a single structural parameter of interest, which was fixed to zero. Repeating this process for each focal structural parameter individually, resulting noncentrality parameter estimates (with subscripts specific to each parameter) were $\hat{\lambda}_{F2F1} = 230.125$, $\hat{\lambda}_{F1V7} = 112.876$, and $\hat{\lambda}_{F2V7} = 3.473$.

Step 3. Each of the estimated noncentrality parameters from Step 2 is associated with a noncentral χ^2 distribution with $df = 1$. Following from Appendix A, the estimated power of the .05-level test of the path from F1 (Reading Readiness) to F2 (Reading Proficiency) is $\hat{\pi} > .95$, while for the path from V7 (dummy) to F1 (Reading Readiness) the power estimate is also $\hat{\pi} > .95$. Obviously these tests had sufficient power to detect nonzero relations that truly existed as both structural parameter estimates' associated *z*-statistics were statistically significant (at any reasonable α level). For the path from V7 (dummy) to F2 (Reading Proficiency), which was not statistically significant at the .05 level, the power is estimated to be between .45 and .50. Using linear interpolation, the estimate is $\hat{\pi} = .46$. This implies that under conditional multivariate normality, if the true value of the unstandardized path from V7 to F2 is .463 (the estimated value) and all other parameter estimates hold for the population as well, then only 46% (i.e., $\hat{\pi} = .46$) of studies designed like this one (e.g., with $n = 1{,}000$) would detect a nonzero relation between the treatment dummy variable (V7) and the Reading Proficiency factor (F2) using a .05-level test.

A PRIORI POWER ANALYSIS (SAMPLE SIZE DETERMINATION) FOR TESTING PARAMETERS WITHIN A MODEL

Whereas in post hoc power analysis the test for which power is assessed has already been conducted using the sample data in hand, in a priori power analysis the necessary sample size is determined prior to gathering data in order to be able to conduct the α-level statistical test(s) of interest with a desired level of power π. For a given number of degrees of freedom, corresponding to the number of focal parameters to be assessed in a single test, an α-level χ^2 test has power as per Equation 12 that is governed by the degree of noncentrality associated with the specific alternative hypothesis H_1. To achieve a target level of power, customarily $\pi = .80$, there must be a specific degree of noncentrality for the alternative distribution; such necessary noncentrality parameters appear in Appendix A for many different numbers of degrees of freedom for $\alpha = .05$-level tests. For the common case of $df = 1$, for example, the alternative distribution must have noncentrality of at least $\lambda = 7.85$ to yield a minimum of $\pi = .80$ power for a .05-level test (assuming conditional multivariate normality).

The key in a priori power analysis, then, is to determine what conditions lead to the required noncentrality. Following from the previous presentation, and from the reader's prior experience with power analysis, one expects the degree of falseness of the null hypothesis H_0 (i.e., $F_{ML}(\boldsymbol{\theta}_R) - F_{ML}(\boldsymbol{\theta}_F)$) and the sample size n to play key roles in noncentrality, and hence power. With regard to H_0, the difference between the focal parameter values in the full model's parameter vector $\boldsymbol{\theta}_F$ and those in the reduced model's parameter vector $\boldsymbol{\theta}_R$ (usually values of zero) yields a difference in fit functions, $F_{ML}(\boldsymbol{\theta}_R) - F_{ML}(\boldsymbol{\theta}_F)$, whose magnitude directly relates to noncentrality and power. To elaborate briefly, the parameters of the full model in $\boldsymbol{\theta}_F$ yield model-implied population moment information in $\hat{\Sigma}(\theta_F)$. Using this population moment information as input for the full and reduced models yields $F_{ML}(\boldsymbol{\theta}_R)$ and $F_{ML}(\boldsymbol{\theta}_F)$, respectively. Given that the moment information was based on $\boldsymbol{\theta}_F$, we expect $F_{ML}(\boldsymbol{\theta}_F) = 0$; however, the data–model fit associated with the reduced model, $F_{ML}(\boldsymbol{\theta}_R)$, should be nonzero under a false null hypothesis. The magnitude of $F_{ML}(\boldsymbol{\theta}_R)$, then, becomes critical. As per Equation 11, multiplying this fit function difference by the quantity $n - 1$ yields the noncentrality parameter λ; this relation will prove key in a priori power analysis.

Let us now take the information in the above paragraph and rearrange it as needed. A researcher knows how many focal parameters are being tested using an α-level test, and thus df_{diff} is known; from this information the necessary noncentrality parameter λ may be determined (from, for example, Appendix A) in order to yield the desired level of power. Next, the numerical values of the parameters in $\boldsymbol{\theta}_R$ and $\boldsymbol{\theta}_F$ must be specified; this includes the focal parameters being tested and the peripheral parameters not being tested. For the focal parameters, typically (although not necessarily) the null values are 0; that is, the researcher is interested in testing whether the focal parameters are nonzero. As for the alternative values of those focal parameters, the researcher must posit specific numerical values. Just as in sample size determination in other areas of statistical data analysis, researchers must make educated guesses as to the nature of the effect(s) they wish to have sufficient power to detect. These guesses might be made based on theory or prior research, or might be selected to be conservative so as to ensure sufficient power to detect effects of at least that magnitude.

As stated above, numerical values must be specified not just for the focal parameters, but for *all* remaining peripheral parameters as well. These peripheral parameters might include factor loadings, error variances, interfactor covariances, and so forth, and they define the context in which the tests of the focal parameters are being conducted. The numerical values of these parameters can have a bearing on the noncentrality associated with the focal parameter tests, and hence must be chosen with care as well (see below). Once designated, these values are part of both the reduced and full model parameter vectors, $\boldsymbol{\theta}_R$ and $\boldsymbol{\theta}_F$, respectively, thus being held constant while determining $\hat{\boldsymbol{\Sigma}}(\boldsymbol{\theta}_F)$ and hence $F_{ML}(\boldsymbol{\theta}_R)$.

Finally, following from Equation 11, the necessary sample size may be determined as

$$n = 1 + \left[\frac{\lambda}{F_{ML}(\boldsymbol{\theta}_R)} \right]. \tag{15}$$

That is, the sample size is determined that magnifies the badness of fit of the reduced model so as to provide sufficient noncentrality to achieve the desired level of power for the α-level test of the focal parameter(s). The practical steps involved in this process are detailed more explicitly below, followed by an example using a latent variable path model. Note that the researcher must initially decide whether α-level testing of the whole set of focal parameters simultaneously is of interest, or if testing each focal parameter individually is the objective for the sample size determination. In this latter and more common case, there will exist pairs of full and reduced models each differing by a single focal parameter and where the remaining focal parameters are temporarily treated as peripheral. In this case the process below would thus be repeated for each focal parameter.

Step 1. This step involves critical preliminary decisions. First, decide upon the desired level of power π and the α level for the focal parameter test(s), which are customarily .80 and .05, respectively. Second, determine the corresponding target noncentrality parameter(s); for .05-level tests, the values in Appendix A are useful. For tests involving a single parameter, setting power to π = .80 leads to the now familiar noncentrality parameter λ = 7.85.

Third, and most critical, researchers must select numerical values for all parameters of the full model. This is probably the most difficult part of any a priori power analysis, and even more so in SEM as values for both focal and peripheral parameters must be selected. To help, for many models researchers might find it easier to select parameter values in a standardized rather than unstandardized metric (e.g., standardized path coefficients); exceptions to this will certainly exist, however, such as for latent growth models where the units are an essential part of the model. Even if researchers do find it more convenient to think in terms of standardized units, they might find it challenging to think within the context of the model under examination. For example, while one might have a theory-based sense of the magnitude of the relation between V1 and V4, estimating that magnitude after controlling for the roles of V2 and V3 might be more challenging. Prior research might not inform this choice as V2 and V3 might not have been included in prior analyses. A further challenge is that the values researchers posit for focal and/or peripheral parameters might wind up being logically impossible, yielding local problems such as variable variances in excess of 1 (when attempting to choose parameter values on a standardized metric), or more global problems such as a nonpositive definite model-implied population covariance matrix $\hat{\Sigma}(\boldsymbol{\theta}_F)$. Given that the challenges of parameter designation for the full model have been surmounted, the values of the focal parameters for the reduced model must be assigned. This task is easier, given that values of zero are most commonly chosen.

Step 2. From the values assigned in $\boldsymbol{\theta}_F$, the relevant model-implied moment information must be determined for the population. For covariance structure models this is $\hat{\Sigma}(\boldsymbol{\theta}_F)$, which, if standardized path values were chosen, will resemble a correlation matrix with values of 1 along the diagonal. These moment values can be determined by path tracing, the algebra of expectations using the structural equations, or in some cases from the modeling software itself. With regard to the latter, one might fix all parameters in the full model to the values in $\boldsymbol{\theta}_F$ and request the model-implied covariance matrix; or, one might use the values in $\boldsymbol{\theta}_F$ as start values, set the number of iterations to zero, and again request the model-implied covariance matrix. In either of the software strategies, input information must be supplied (i.e., a covariance matrix for covariance structure models), but this input is essentially irrelevant as the program will not iterate to fit that information. By whatever means, then, the researcher has determined the necessary moment information in $\hat{\Sigma}(\boldsymbol{\theta}_F)$.

Step 3. Having obtained the relevant model-implied population moment information from the full model, that information must now be used as input data to estimate the population data–model fit for the reduced model (with parameters in $\boldsymbol{\theta}_R$). In doing so, fix the values for all parameters to the values in $\boldsymbol{\theta}_R$ (i.e., with the remaining parameters in $\boldsymbol{\theta}_F$ to zero). Equivalently, all start values may be set to the values in $\boldsymbol{\theta}_R$, with the modeling software set to have zero iterations. Note, because the necessary sample size is not yet known, an arbitrary sample size will generally have to be assigned in order for the modeling software to analyze the supplied population moment information. Now, the only badness of data–model fit associated with this reduced model comes directly from the focal parameters that have been constrained (usually to zero); as such, the reduced model will yield a model fit function value $F_{ML}(\boldsymbol{\theta}_R)$. From this value, along with the target noncentrality parameter, the necessary sample size may now be computed using Equation 15. For the common case of examining each parameter separately, necessary sample size will be derived for each. Choosing the largest of these sample sizes, and rounding up to the nearest integer, will ensure sufficient power for tests of all focal parameters (under the assumed distributional and model conditions).

Example of A Priori Power Analysis for Testing Parameters Within a Model

Imagine that a researcher is planning to assess the impact of latent Reading Self-Concept and latent Math Self-Concept on ninth graders' latent Math Proficiency. In the past, three questionnaire items have been used as measured indicators of each of the self-concept constructs (V1 through V3 for reading, and V4 through V6 for math), although additional questionnaire items are available. As for Math Proficiency, two well-established standardized tests are available for use with the students of interest. The full model the researcher plans to use is depicted in Figure 4.4. The focal parameters will be those relating the latent variables: c_{F1F2}, b_{F3F1}, and b_{F3F2}. Each of these will be of individual interest; that is, necessary sample size should be determined for the test of each parameter separately.

Step 1. The level of power desired for these single-parameter (1 *df*) tests will be set at the customary π = .80 for .05-level tests. The corresponding target noncentrality parameter to be used for all parameters, as shown in Appendix A, is λ = 7.85. For the peripheral parameters in the measurement model, imagine that past research has shown standardized loadings for Reading Self-Concept indicators to be approximately b_{V1F1} = .5, b_{V2F1} = –.7, and b_{V1F1} = .6; corresponding error variances for a standardized model would be c_{E1} = .75, c_{E2} = .51, and c_{E3} = .64. For the Math Self-Concept indicators, prior evidence has shown standardized loadings to be

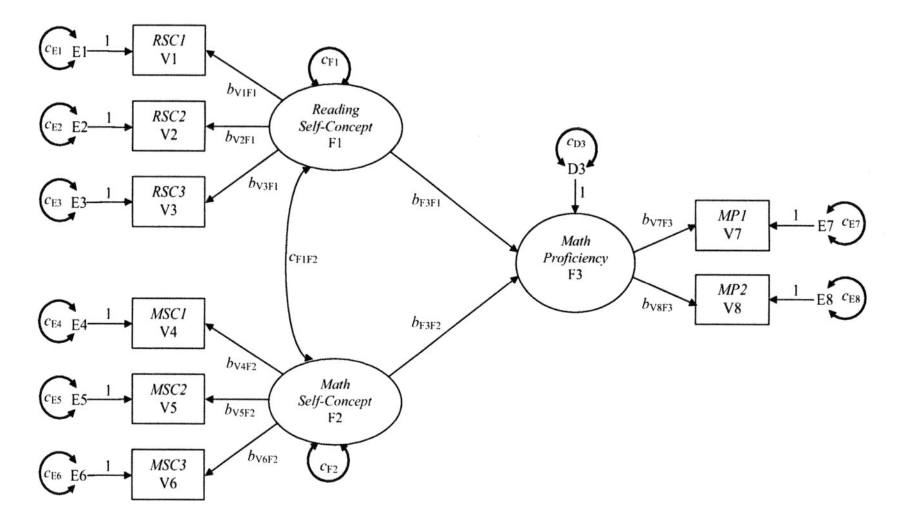

Figure 4.4. Latent variable path model for example of a priori power analysis for testing parameters within a model.

approximately $b_{V4F2} = .8$, $b_{V5F2} = .9$, and $b_{V6F2} = -.6$, thus having corresponding error variances for a standardized model of $c_{E4} = .36$, $c_{E5} = .19$, and $c_{E6} = .64$. The two Math Proficiency tests have reported reliability estimates of .88 and .93, thus suggesting standardized loading values of $b_{V7F3} = .938$ and $b_{V8F3} = .964$ (the square roots of the reliabilities), respectively; corresponding error variances would in turn be $c_{E7} = .12$ and $c_{E8} = .07$.

For the focal parameters, the researcher expects a negative relation between the Reading and Math Self-Concept factors of standardized magnitude $c_{F1F2} = -.30$. The expected direct effects on Math Proficiency, in a standardized metric, are set at $b_{F3F1} = .20$ for Reading Self-Concept and .50 for b_{F3F2}=Math Self-Concept. Although the researcher suspects the true relations might be a bit higher, slightly conservative values are chosen to ensure sufficient power. To preserve the standardized framework, the remaining peripheral parameters in the latent portion of the model are $c_{F1} = 1$, $c_{F2} = 1$, and $c_{D3} = .77$ (which the reader may verify using, for example, path tracing).

Step 2. Having articulated theoretical values for all parameters in the full model, the model-implied covariance matrix $\hat{\boldsymbol{\Sigma}}(\boldsymbol{\theta}_F)$ may be determined. In this example, as the reader may verify, the matrix is

$$\hat{\boldsymbol{\Sigma}}(\boldsymbol{\theta}_F) = \begin{bmatrix} 1.000 & & & & & & & \\ -.350 & 1.000 & & & & & & \\ .300 & -.420 & 1.000 & & & & & \\ -.120 & .168 & -.144 & 1.000 & & & & \\ -.135 & .189 & -.162 & .720 & 1.000 & & & \\ .090 & -.126 & .108 & -.480 & -.540 & 1.000 & & \\ .023 & -.033 & .028 & .330 & .371 & -.248 & 1.000 & \\ .024 & -.034 & .029 & .339 & .382 & -.254 & .904 & 1.000 \end{bmatrix}.$$

Step 3. Now, using the above matrix as input, and fixing all peripheral parameters to their hypothesized values, three reduced models are run: one with c_{F1F2} constrained to zero, one with b_{F3F1} constrained to zero, and one with b_{F3F2} constrained to zero. In doing so, the resulting maximum likelihood fit function values reflecting the badness of data–model fit associated with each single constraint are $F_{F1F2} = .053$, $F_{F3F1} = .031$, and $F_{F3F2} = .262$ (rounded).

Finally, following from Equation 16, the necessary sample sizes are determined as $n_{F1F2} = 149.11$, $n_{F3F1} = 254.23$, and $n_{F3F2} = 30.96$. Thus, the sample size needed to ensure at least .80 power for tests of all focal parameters is $n = 255$. This makes the standard distributional assumption of conditional multivariate normality, assumes that the model at hand is correct, and assumes that the parameter values used for the peripheral and focal parameters are also correct (if, in fact, any focal parameters are actually larger than planned, power will exceed .80 at this sample size).

A Simplification for Models with Latent Variables

As should be clear from the above presentation, a considerable amount of information must be assumed in order to determine sample size in an a priori power analysis. For models with latent variables, where the relations among latent variables will almost always contain the focal parameters, assumptions must be made about the nature of the supporting measurement model both in terms of each factor's number of indicator variables and the magnitude of those indicators' loadings. In practice, however, such information might not be well known. In fact, a researcher might actually wonder what quality of measurement model is necessary to achieve the power desired for tests within the structural portion of the model. While the quality of a factor's measurement model has been operationalized in a variety of ways (e.g., Fornell & Larcker, 1981), Hancock and Kroopnick (2005) recently demonstrated that one particular index is directly tied to noncentrality and power analysis in covariance structure models. This measure is referred to as *maximal reliability* in the context of scale construction (e.g., Raykov, 2004), and as the measure of construct reliability *coefficient H* in the context of latent variable models (see, e.g., Hancock & Mueller, 2001). Specifically, Hancock and Kroopnick illustrated that the precise nature of the measurement model is actually irrelevant; instead, the researcher may simply specify a single number as a place-holder for the measurement model of each factor in the latent variable path model, where that number reflects an anticipated (or minimum achievable) construct reliability associated with each factor's measurement model.

For a given factor with k indicators, the value of coefficient H may be computed using standardized loadings l_i as follows:

$$H = \frac{\sum_{i=1}^{k}\left(\frac{l_i^2}{1 - l_i^2}\right)}{1 + \sum_{i=1}^{k}\left(\frac{l_i^2}{1 - l_i^2}\right)}. \quad (17)$$

There are a number of interpretations for this index, detailed by Hancock and Mueller (2001). Most simply, it represents the proportion of variance in the construct that is theoretically explainable by its indicator variables. As such it is bounded by 0 and 1, is unaffected by loadings' sign, can never decrease with additional indicators, and is always greater than (or equal to) reliability of strongest indicator (i.e., max (l_i^2)). Thus, the assessment of construct reliability draws information from the measured indicators only inasmuch as that information is germane to the construct itself. For Reading Self-Concept (F1), Math Self-Concept (F2), and Math Proficiency (F3) from the previous example, these values may be computed to be H_{F1} = .650, H_{F2} = .868, and H_{F3} = .953, respectively.

Following from Hancock and Kroopnick (2005), as far as power is concerned, how one achieves the particular level of construct reliability for a given factor is irrelevant. A construct reliability of H = .650 could be achieved with three indicators of standardized loadings of .5, –.7, and .6, as in the previous example. It could also be achieved, for example, with 10 indicators each having a standardized loading of .396, or with a single indicator of standardized loading .806 (= $.650^{1/2}$). Assessments of power, or in the current case, of necessary sample size to achieve a desired level of power, are precisely the same. For simplicity, then, the model in the previous example could have been depicted as shown in Figure 4.5, where factors have single indicators each with a standardized loading equal to the square root of the factor's construct reliability. As these single indicators are place-holders for the respective measurement models, they are designated as "V" terms.

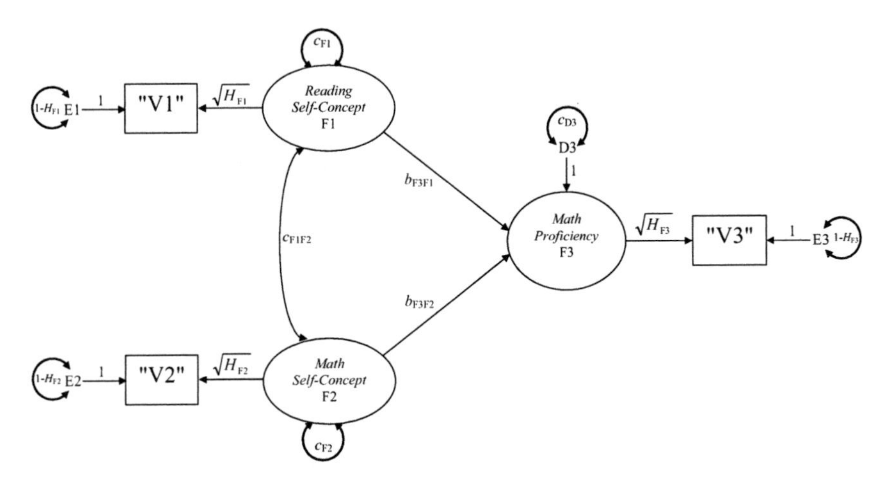

Figure 4.5. Simplified latent variable path model for example of a priori power analysis for testing parameters within a model.

To be explicit, then, in this case the peripheral measurement model parameters are set to $b_{V1F1} = .806$ for F1 (with $c_{E1} = .350$), $b_{V2F2} = .932$ for F2 (with $c_{E2} = .132$), and $b_{V3F3} = .976$ for F3 (with $c_{E3} = .047$). These values, along with the focal parameters and peripheral structural parameters of the full model from the previous example, yield the following considerably simpler model-implied covariance matrix (shown here to five decimal places):

$$\hat{\boldsymbol{\Sigma}}(\boldsymbol{\theta}_F) = \begin{bmatrix} 1.00000 & & \\ -.22536 & 1.00000 & \\ .03937 & .40045 & 1.00000 \end{bmatrix}.$$

As the reader may now verify, in going through Step 3 in the outlined process for a priori power analysis for testing parameters within a model, the same values for F_{ML} will arise for the focal parameters, and hence the same required sample sizes will be derived.

While this simplification is somewhat intriguing in and of itself, its practical value is the reason it is presented in this chapter. Specifically, it means that for models with latent variables where the focal parameters are contained in the structural portion of the model, the researcher does not need to specify a measurement model a priori in order to conduct a sample size determination. The researcher merely needs to estimate the level of construct reliability achievable for each factor, or perhaps articulate a commitment to achieving a minimum quality of measurement model for each individual factor.

This simplification will also allow a preliminary sensitivity analysis, of sorts, whereby the researcher can determine the degree of sensitivity of the sample size estimate to the quality of the measurement models. In the current example, we can easily determine what would happen if, for example, the construct reliability associated with Reading Self-Concept (F1) were to be increased to, say, .80. As the reader may verify, the new fit function values for the focal parameters would become $F_{F1F2} = .065$, $F_{F3F1} = .038$, and $F_{F3F2} = .264$, leading to required sample sizes of $n_{F1F2} = 121.77$, $n_{F3F1} = 207.58$, and $n_{F3F2} = 30.73$. The sample size now needed to ensure at least .80 power for tests of all focal parameters is $n = 208$. That means that by improving the quality of the measurement model for Reading Self-Concept from $H_{F1} = .650$ to $H_{F1} = .800$, the desired level of power may be ensured with 47 fewer subjects. This increase in construct reliability might be accomplished in practice, for example, by replacing RSC1 in the original with an indicator of standardized loading .85, or by adding to the original set a fourth indicator of standardized loading .83, or by adding four indicators of standardized loadings .60. In fact, for purposes of power analysis, how it is done is irrelevant.

In the end, then, taking advantage of this simplification requires some familiarity with the coefficient H measure, but gaining such familiarity may indeed be a worthwhile investment in this case (see Hancock & Mueller, 2001). Doing so will afford the researcher less specificity in terms of number and quality of measured indicator variables, making a priori power analysis for testing parameters within a model a simpler process.

POWER ANALYSIS FOR TESTING DATA–MODEL FIT AS A WHOLE

The previous section addressed post hoc power analysis and a priori power analysis (sample size determination) for testing parameters within a model. The results of those methods are built upon critical assumptions, and as such in practice they should be considered to yield approximations to true power or necessary sample size. In particular, in addition to distributional assumptions, the entire endeavor rests on the assumed reasonableness of the model within which the focal parameters are embedded. The current section treats power associated with testing data–model fit as a whole.

As presented at the beginning of the previous section on power analysis for testing parameters within a model, there are three common elements to consider prior to addressing post hoc or a priori power analysis: null and alternative hypotheses, relevant test statistics, and appropriate central/noncentral distributions. Each of these will be addressed specifically as they bear on testing data–model fit as a whole in SEM, setting the stage for methods of post hoc and a priori power analysis.

Null and Alternative Hypotheses

A possible, albeit rather optimistic, null hypothesis regarding one's covariance structure model is that the model is correct in the population. If this is indeed the case, then the covariance matrix implied by the model ($\hat{\mathbf{\Sigma}}$) and the corresponding information observed in the population data ($\mathbf{\Sigma}$) are identical; that is, there is exact fit. Unfortunately, there is not a reciprocal relation between model truth and model fit. That is, as discussed by Hershberger (Chapter 2, this volume), many different models can give rise to the same data moments, and thus while a correct model implies equal model-implied and observed moments, the converse need not be true: equal model-implied and observed moments need not imply a correct model. In the end then, while a researcher may wish to express null and alternative hypotheses in terms of population model–reality correspondence, the null and alternative models tested are only about population data–model fit. In this framework a rejection of the null hypothesis of exact data–model fit leads to the reasonable inference of an incorrect model (i.e., no model–reality correspondence); a failure to reject the null hypothesis leads merely to a tentative retention of that model as one in a set of models that could exactly represent the population processes at work.

Even though we can express null and alternative hypotheses in terms of exact data–model fit, as described above, in fact there are two primary drawbacks of doing so. The first is the backwards nature of the hypothesis testing to which it leads. Specifically, whereas rejecting a null hypothesis in favor of an alternative is typically the desired outcome in an analytical endeavor, in testing exact fit in SEM one would wish to retain the null hypothesis as it represents consonance between data and the theoretical model of interest. Given that statistical power refers to the probability of rejecting an hypothesized model rather than retaining it, relative to more traditional analytical scenarios the power analysis for testing data–model fit as a whole in SEM must be reframed (as will be done below).

The second drawback of testing exact data–model fit, as alluded to above, is its unrealistic optimism. In general, researchers do not expect of their model a complete fidelity to the underlying population mechanisms, but rather that the model is a reasonable approximation to reality. Theoretically trivial internal misspecifications such as minute unmodeled relations (e.g., error covariances, cross loadings), while indicative of a false theoretical model, may be deemed quite acceptable at a practical level. For this reason, rather than discussing models as true or false, practicality suggests discussing models as having *acceptable* or *unacceptable* data–model fit. In fact, as MacCallum, Browne, and Sugawara (1996) suggested, for testing purposes it is useful to define the null hypothesis for a given model as containing population data–model fit at the threshold between acceptable and unacceptable fit (termed "close fit" and "not close fit," respectively, by MacCallum et al.). The researcher's goal, then, becomes the rejection of this null in favor of an alternative in which the model of theoretical interest indeed has acceptable data–model fit, and in executing the study itself to amass sufficient power for such a favorable rejection. (Such acceptable data–model fit is also, as discussed previously, necessary to ensure validity of the tests of parameters within the model.)

So, in order to differentiate between acceptable and unacceptable fit, a metric for data–model fit needs to be established. One could characterize data–model fit using the fit function F_{ML} directly, or perhaps of the noncentrality parameter λ, where relatively smaller values are indicative of less discrepancy between observed and model-implied moments and hence more acceptable fit, and relatively larger values indicate a greater discrepancy in the fit between data and model and thus less acceptable fit. Unfortunately, the metric for fit functions and noncentrality parameters are not particularly intuitive, in part because their values are tied to the complexity of the models themselves. As a result, Steiger and Lind (1980) recommended the root mean square error of approximation (RMSEA) ε as an index to characterize the degree of discrepancy between model-implied and observed moments, and hence a metric for assessing degree of acceptability of data–model fit. The RMSEA represents a dispersal of data–model discrepancy (F_{ML}) across degrees of freedom:

$$\varepsilon = \sqrt{\frac{F_{ML}}{df}}. \tag{18}$$

Given the relation between F_{ML} and noncentrality, specifically $\lambda = (n - 1)F_{ML}$ from Equation 9, the RMSEA may also be expressed using the noncentrality parameter:

$$\varepsilon = \sqrt{\frac{\lambda}{df(n-1)}}. \tag{19}$$

Using the RMSEA as a metric for data–model fit, we may define the value of ε_0 as representing the boundary between acceptable and unacceptable data–model fit, which in turn will help to articulate the null and alternative hypotheses. One could use, for example, the null hypothesis H_0: $\varepsilon = \varepsilon_0$, that the population data–model fit is at the threshold between acceptability and unacceptability. The alternative hypothesis in this case, H_1: $\varepsilon \neq \varepsilon_0$, contains two possibilities: H_1: $\varepsilon < \varepsilon_0$, that the data–model fit is acceptable, and H_1: $\varepsilon > \varepsilon_0$, that the data–model fit is unacceptable. For the purposes of power analysis, however, interest tends to be in being able to reject data–model fit at the threshold in favor of acceptable data–model fit. As such, we will hereafter define the alternative hypothesis as H_1: $\varepsilon < \varepsilon_0$, and hence the corresponding null hypothesis as H_0: $\varepsilon \geq \varepsilon_0$.

As for the numerical choice of ε_0, some authors have suggested .05 as representing a reasonable threshold between acceptable (<.05) and unacceptable (>.05) data–model fit (Browne & Cudeck, 1993; Browne & Mels, 1990; Steiger, 1989). This value will be adopted for power analysis in the current chapter. That is, the null and alternative hypotheses regarding population data–model fit will be H_0: $\varepsilon \geq .05$ and H_1: $\varepsilon < .05$, respectively. Thus, sample information must be gathered in a general attempt to reject the null hypothesis containing unacceptability in favor of the alternative of acceptability. Test statistics for such hypothesis testing are discussed next.

Test Statistics to Assess Null Hypotheses

If one held as the null hypothesis of choice that there was exact data–model fit in the population, this could be represented in a variety of ways: $F_{\text{ML}} = 0$, $\lambda = 0$, and $\varepsilon = 0$, among others. To test these one could conceive of using sample estimates $\hat{F}_{\text{ML}}$, $\hat{\lambda}$, or $\hat{\varepsilon}$, respectively. The formula for sample $\hat{F}_{\text{ML}}$ values was presented previously (Equation 4), as was that for the estimated noncentrality $\hat{\lambda}$ (Equation 13). Following from Equations 18 and 19 above, and Equation 13 previously, an RMSEA estimate $\hat{\varepsilon}$ may be determined as

$$\hat{\varepsilon} = \sqrt{\frac{\hat{F}_{\text{ML}}}{df}} = \sqrt{\frac{\hat{\lambda}}{df(n-1)}} = \sqrt{\frac{\chi^2 - df}{df(n-1)}}. \tag{20}$$

Under the null hypothesis of exact fit, sample values of $\hat{F}_{\text{ML}}$, $\hat{\lambda}$, and $\hat{\varepsilon}$ will deviate randomly from zero for covariance structure models as **S** fluctuates randomly from $\mathbf{\Sigma}$ (and from $\hat{\mathbf{\Sigma}}$ in overidentified models), following their own sampling distributions. Note, however, that all of these test statistics are really just translations of the model χ^2; as such, sample values convert to model χ^2 statistics that follow a central distribution with expected value equal to the model *df*. In this case, a model χ^2 test statistic exceeding the α-level critical value would lead to a rejection of a null hypothesis of exact fit, whereas a model χ^2 test statistic failing to exceed the α-level critical value would lead to a retention of a null hypothesis of exact fit.

As mentioned above, because a retention of the null hypothesis is actually the desired outcome in the exact data–model fit testing paradigm, a shift is required—quite literally a distributional shift to a tolerable and rejectable level of data–model misfit. As this was framed using the RMSEA, specifically $\varepsilon_0 \geq .05$, $\hat{\varepsilon}$ is the appropriate test statistic of choice. That is, if $\hat{\varepsilon}$ is statistically significantly less than .05 at some acceptable α level, then we will reject the null hypothesis containing unacceptable population data–model fit in favor of the alternative of acceptable population data–model fit, $\varepsilon_0 < .05$. Otherwise, the null hypothesis containing unacceptable data–model fit will be retained. The last components necessary for power analysis, then, are the central and noncentral distributions associated with $\hat{\varepsilon}$.

Central Distributions (for H_0) and Noncentral Distributions (for H_1)

H_0 and central $\hat{\varepsilon}$ distributions. Under a null condition that $\varepsilon_0 = .05$ (ignoring, for the moment, $\varepsilon_0 > .05$), we expect $\hat{\varepsilon}$ to fluctuate randomly around .05. This sampling distribution is thus a central distribution with respect to this null condition. Because of the relation between the sample RMSEA and χ^2 as contained in Equation 20, we can translate this null RMSEA sampling distribution into a more familiar χ^2 distribution. Following Equation 19, which relates ε to a χ^2 noncentrality parameter λ, we see that the central distribution describing the variation of $\hat{\varepsilon}$ around ε_0 can be translated to a more familiar noncentral χ^2 distribution with noncentrality parameter λ_0. Specifically, from Equation 19 the following relation can be derived:

$$df(n-1)\varepsilon_0^2 = \lambda_0 . \tag{21}$$

So, for example, if one has designated the customary threshold $\varepsilon_0 = .05$ for a model with $df = 5$ that is to be evaluated using data from $n = 1{,}027$ subjects, the corresponding noncentral χ^2 distribution with 5 df has noncentrality parameter $\lambda_0 = (5)(1{,}027 - 1)(.05^2) = 12.83$. This is the noncentral distribution pictured in Figure 4.2. Thus, the central fluctuation of $\hat{\varepsilon}$ statistics around .05 can be represented by the noncentral fluctuation of model χ^2 statistics around 17.83, their expected value in a distribution with $df = 5$ and $\lambda = 12.83$.

H_1 and noncentral $\hat{\varepsilon}$ distributions. Although the alternative hypothesis of $H_1{:}\varepsilon < .05$ contains many possibilities for ε, power analysis requires the designation of a specific alternative numerical value (either through post hoc estimation or a priori selection). That is, one must state how much better (smaller) than $\varepsilon = .05$ one is willing to consider the data–model fit in the population. Most optimistic would be the designation of an alternative value $\varepsilon_1 = 0$, implying the belief (or hope) that the model has exact data–model fit in the population. Somewhat less optimistic, but more realistic, are values between 0 and .05. Sampling distributions of $\hat{\varepsilon}$ under conditions with ε_1 closer to 0 are more distinct from those under the null $\varepsilon_0 = .05$ condition, while sampling distributions of $\hat{\varepsilon}$ under conditions with ε_1 closer to .05 are considerably less distinct from those under the null $\varepsilon_0 = .05$ condition.

Just as in the null case where the central sampling distribution of $\hat{\varepsilon}$ values translates to a χ^2 distribution, so too does the noncentral distribution of $\hat{\varepsilon}$ values in the alternative case. Following from Equation 21, the corresponding χ^2 distribution has noncentrality parameter λ_1, where

$$\lambda_1 = df(n-1)\varepsilon_1^2. \quad (22)$$

Thus, for a study with data from $n = 1{,}027$ subjects, the optimistic value of $\varepsilon_1 = 0$ for a model with $df = 5$ would correspond to the designation of $\lambda_1 = 0$, meaning that the noncentral sampling distribution of $\hat{\varepsilon}$ statistics centered at $\varepsilon_1 = 0$ transforms to a central χ^2 distribution (with $df = 5$, in this case). On the other hand, the same study with an alternative value of, say, $\varepsilon_1 = .02$, would have its noncentral sampling distribution of $\hat{\varepsilon}$ statistics (centered at $\varepsilon_1 = .02$) transform to a 5-df noncentral χ^2 distribution with $\lambda_1 = 5(1{,}027 - 1)(.02)^2 = 2.05$. Thus, the central fluctuation of $\hat{\varepsilon}$ statistics around .02 under these study conditions can be represented by the noncentral fluctuation of model χ^2 statistics around 7.05, their expected value in a distribution with $df = 5$ and $\lambda = 2.05$. This distribution sits somewhere between those pictured in Figure 2, and thus has a greater degree of overlap with the $\varepsilon = .05$ distribution to the right than does the $\varepsilon = 0$ distribution to the left. The designation of ε_1, therefore, will have obvious implications for power, as described next.

POST HOC POWER ANALYSIS FOR TESTING DATA–MODEL FIT AS A WHOLE

Post hoc power analysis for testing data–model fit as a whole addresses the question of how much power a particular study appears to have had to reject the null hypothesis containing unacceptable data–model fit. Following from the above prefactory information, post hoc power analysis involving the RMSEA may be conducted using the more familiar family of χ^2 distributions, making it more attractive for power analysis than methods using other data–model fit indices (e.g., GFI and AGFI; MacCallum & Hong, 1997). Encouraging though this may sound, accessibility to the all important noncentral χ^2 distributions is somewhat limited, requiring availability of published tables (e.g, Haynam et al., 1973) or familiarity with functions for generating relevant information in such programs as SAS and GAUSS. The current chapter, however, will not require access to either, and instead will present tables useful for conducting post hoc power analysis of data–model fit as a whole.

Following Hancock and Freeman's (2001) adaptation of the SAS programs offered by MacCallum and colleagues (1996), which draw from noncentral χ^2 distributions, power tables for an $\alpha = .05$-level one-tailed test were derived for $\varepsilon_1 = .00$, .02, and .04, where ε_0 was set to the recommended value of .05. For each level of ε_1, model degrees of freedom were varied in increments of 5 from $df = 5$ to 250, while sample size in each df

condition was varied in increments of 50 from $n = 50$ up to the first sample size to achieve power (π) in excess of .99. For $df = 5$ to 100, π values appear for $\varepsilon = .00$, .02, and .04 in Appendix B; for $df = 110$ to 250, π values appear in Appendix C.

Before using appendix values for post hoc power analysis of overall data–model fit, some initial patterns in the values are worth noting. First, power increases as values of ε_1 get closer to 0, which is expected because this puts a greater separation between the alternative sampling distribution around ε_1 and that of the null sampling distribution around $\varepsilon_0 = .05$. Power values also increase with increases in sample size n, as one would generally expect and as is explicit in the magnitude of the noncentrality parameter in Equation 22. Imagine, for example, $\varepsilon_1 = .02$ and $\varepsilon_0 = .05$, where the model has $df = 50$. When $n = 101$, $\lambda_1 = 2.00$ (for ε_1) and $\lambda_0 = 12.50$ (for ε_0); when sample size increases to $n = 201$, $\lambda_1 = 4.00$ (for ε_1) and $\lambda_0 = 25.00$ (for ε_0). Thus, although λ_1 and λ_0 maintain the same ratio, it is the magnitude of their difference that determines the separation of their sampling distributions, and hence their power. When $n = 101$, the separation is 10.50 (= 12.50 – 2.00), while when $n = 201$, the separation is 21.00 (= 25.00 – 4.00). Finally, as for degrees of freedom, for a given value of ε_1 the power increases as df increases. While perhaps not as obvious as the impact of sample size, in fact the explanation follows precisely as it above did for sample size, using the relation from Equation 22 but this time holding sample size constant. An examination of such relations is left to the reader.

Now consider the following generic examples of post hoc power analysis, helping to familiarize the reader with the use of the relevant appendices. Imagine that a researcher had conducted a study with $n = 400$ subjects to test a model with $df = 55$. The data–model fit obtained in the study, as assessed by RMSEA, was $\hat{\varepsilon}_1 = .02$. Following from Appendix B, for a one-tailed test (at the $\alpha = .05$ level) the estimated power to reject $\varepsilon_0 = .05$ in favor of acceptable data–model fit (estimated to be $\varepsilon_1 = .02$) is $\hat{\pi} = .92$.

As a second example, consider a study with $n = 250$ subjects used to test a model with $df = 120$, where the sample RMSEA was $\hat{\varepsilon}_1 = .04$. Following from Appendix C, for a one-tailed test (at the $\alpha = .05$ level) the estimated power to reject $\varepsilon_0 = .05$ in favor of acceptable data–model fit (estimated to be $\varepsilon_1 = .04$) is $\hat{\pi} = .33$.

As for values of n, df, or ε_1 not addressed in Appendices B and C, forms of interpolation may be used to approximate the power π. With regard to sample size, although the relation between n and π (for any given df and ε_1) is not precisely linear, simple linear interpolation provides a reasonable approximation. For example, given a study using $n = 180$ subjects that yielded an RMSEA estimate of $\hat{\varepsilon}_1 = .02$ for a model with $df = 20$, the power will be between .19 (for n = 150) and .26 (for n = 200). Thus, the linearly interpolated power estimate is $\hat{\pi}$ = .19 + [(180 – 150)/(200 – 150)](.26 – .19) ≈ .23.

Similarly, with regard to degrees of freedom, the relation between df and π (for any given n and ε_1) is satisfactorily approximated between tabled values by linear interpolation. For example, given $\hat{\varepsilon}_1$ = .02 for a model with df = 38, a sample size of n = 250 yields estimated power between .50 (for df = 35) and .55 (for df = 40), or an interpolated estimate of $\hat{\pi}$ = .50 + [(38 – 35)/(40 – 35)](.55 – .50) ≈ .53.

Regarding ε_1, π may be approximated (although slightly underestimated) by linear interpolation for ε_1 values between .00 and .04. As an example, for a model with df = 50 that is fit using data from n = 350 subjects, an RMSEA value of $\hat{\varepsilon}_1$ = .025 corresponds to power between .26 (for ε_1 = .04) and .84 (for ε_1 = .02). The corresponding linearly interpolated estimate is thus of $\hat{\pi}$ = .26 + [(.04. – 025)/(.04 – .02)](.84. – 26) ≈ .68.

Finally, for combinations in which more than one element (n, df, or ε_1) is not the tabled value, linear interpolation may occur using each variable in a stepwise fashion. With untabled n and df values for a given ε_1 value, for example, a quartet of numbers (in adjacent rows and columns) define the power estimate's domain. Interpolation between the two left values and interpolation between the two right values yields two values between which a final interpolation may be conducted to yield the desired power estimate.

A PRIORI POWER ANALYSIS (SAMPLE SIZE DETERMINATION) FOR TESTING DATA–MODEL FIT AS A WHOLE

A priori power analysis for testing data–model fit as a whole addresses the question of how many subjects one needs in order to achieve a desired level of power to reject the null hypothesis containing unacceptable data–model fit. As was done above for post hoc power analysis, the current chapter presents tables useful for conducting sample size determination for tests of data–model fit as a whole. Following Hancock and Freeman's (2001) adaptation of the SAS programs offered by MacCallum and colleagues (1996), which draw from noncentral χ^2 distributions, a sample size table for an α = .05-level one-tailed test was created for ε_1 = .00, .02, and .04, where ε_0 was again set to the recommended value of .05. For each level, model degrees of freedom were varied in increments of 5 from df = 5 to 250, while tabled power levels included π = .50 to .95 in increments of .05, plus .99. The reader is reminded that .80 is generally the recommended minimum level of power used for planning purposes;

lower levels are shown merely for reference. Sample sizes for $\pi = .00$, .02, and .04 appear in Appendix D.

As an example, imagine that a researcher was planning to evaluate data–model fit of a model with $df = 15$, and the desired power is .80 for an $\alpha = .05$-level test associated with the RMSEA. As presented previously, the sample size required would depend on the degree to which the model's level of data–model fit, and hence the reference distribution's degree of noncentrality, deviate from the null $\varepsilon_0 = .05$. For a perfectly specified model ($\varepsilon_1 = .00$), Appendix D shows that a minimum of $n = 530$ is required. As perfect fit is rather optimistic, one might more reasonably consider $\varepsilon_1 = .02$, for which a minimum sample size of $n = 741$ is necessary. On the most conservative side tabled, true data–model fit of $\varepsilon_1 = .04$ would require a considerably larger $n = 4349$ in order to achieve power of .80. For researchers interested in planning to have sufficient power to reject $\varepsilon_0 \leq .05$ in favor of acceptable data–model fit, selecting $\varepsilon_1 = .02$ seems like a desirable balance between the generally unrealistic optimism of $\varepsilon_1 = .00$ and the frequent impracticality associated with the recommended sample size for $\varepsilon_1 = .04$.

For conducting a priori power analysis, the levels of ε_1 covered by Appendix D should allow for an adequate picture of sample size necessary in planning a study. However, some interpolation might be needed for df values not directly represented. Given the nature of the nonlinear relation between df and n for a given true data–model fit ε_1 and desired power π, simple linear interpolation often provides what may be an unacceptable overestimate of the sample size necessary to achieve a desired level of power. Thus, for all power levels within Appendix D, exponential functions of df were empirically derived by Hancock and Freeman (2001) to facilitate optimum linear interpolation in df exponentiated to the bth power. Appearing in Table 4.1 are these b exponents, which may be used to facilitate interpolation linearly in df^b as follows. To get an estimate of the sample size necessary at the target df, compute:

$$n_{df_\mathrm{T}} = n_{df_\mathrm{L}} - \left(\frac{df_\mathrm{L}^b - df_\mathrm{T}^b}{df_\mathrm{L}^b - df_\mathrm{H}^b}\right)(n_{df_\mathrm{L}} - n_{df_\mathrm{H}}), \tag{23}$$

where n is sample size, df_T refers to the target (T) number of degrees of freedom, df_L is the nearest tabled number of degrees of freedom lower (L) than the target value, and df_H is the nearest tabled number of degrees of freedom higher (H) than the target value.

Consider a researcher interested in a model with $df = 8$, who chooses to plan a study using $\varepsilon_1 = .02$. Furthermore, a level of power of $\pi = .80$ is

Table 4.1. Exponents for *df* to Interpolate Sample Size

	ε_1		
π	.00	.02	.04
.50	−0.70	−0.76	−0.99
.55	−0.71	−0.78	−0.99
.60	−0.70	−0.79	−0.99
.65	−0.71	−0.81	−0.99
.70	−0.72	−0.82	−1.00
.75	−0.71	−0.84	−1.00
.80	−0.72	−0.86	−1.00
.85	−0.73	−0.87	−1.00
.90	−0.74	−0.89	−1.00
.95	−0.75	−0.92	−1.00
.99	−0.78	−0.95	−1.00

desired for the α = .05-level test associated with the RMSEA. Using Appendix D, entries for df = 5 and df = 10 indicate the desired sample size to be between 1,707 and 997. Using the exponent indicated by Table 4.1, $b = -.86$, linear interpolation in df^b suggests $1{,}707[(5^{-.86} - 8^{-.86})/(5^{-.86} - 10^{-.86})](1{,}707 - 997) \approx 1{,}182$. That is, to have a .80 probability of rejecting $\varepsilon_0 \geq .05$ in favor of acceptable data–model fit (using an α = .05-level test), given that the true level of data model fit in the population is $\varepsilon_1 = .02$, the study would require a minimum of 1,182 subjects. As the reader may also verify, had the researcher been interested in assuming exact data–model fit in the population, $\varepsilon_1 = .00$, the minimum required sample size would have been n = 800.

CONCLUSIONS REGARDING POWER ANALYSIS IN COVARIANCE STRUCTURE MODELING

The purpose of the current chapter was to acquaint the reader with power analysis in structural equation modeling, both at a parameter level as well as at a model level. Although presented in terms of single-sample covariance structure models, the principles extend to multisample models and models with mean structures as well. Utilizing those models' corresponding F_{ML} fit functions (see, e.g., Bollen, 1989), the methods herein may be applied fairly straightforwardly to those additional scenarios, whether addressing power of tests of overall data–model fit or of specific parameters (see Hancock, 2001, for power analysis within mean structure models).

Extending to cases where data do not reasonably subscribe to the distributional assumptions made herein (i.e., conditional multivariate normality), methods of power analysis might become more challenging. For post hoc power analysis, where the data have already been gathered, bootstrap resampling methods could be an option to approximate empirically the distribution of the statistics of interest under null and non-null conditions. For a priori power analysis, if one can specify the nature of the population data ahead of time, then a type of parametric bootstrap technique might be feasible as has been suggested previously with normal data (Muthén & Muthén, 2002). Such a priori specification of the distribution does not seem likely in practice, however. More reasonable might simply be to use the principles behind the normal theory techniques described here and plan on utilizing a robust model χ^2 test statistic (see Satorra & Bentler, 1994) and its corresponding adjusted RMSEA (see, e.g., Nevitt & Hancock, 2000). In this case, relevant χ^2 difference tests would themselves have to be adjusted, as described by Satorra and Bentler (2001). Still, caution is warranted in this case as the accuracy of resulting power estimates and sample sizes recommended using such adjusted methods remains to be investigated thoroughly.

In conclusion, allow me to offer some personal perspectives regarding the methods detailed in the current chapter, in particular concerning their relative importance and practicality. To start, for power analysis in SEM as well as across other statistical methods, I believe a priori methods to be *far* more important than post hoc methods. As pertains to the latter, conducting a post hoc power analysis after a null hypothesis has been rejected (and is believed false) just seems to be establishing what is already apparent, that there was sufficient power. Similarly, conducting a post hoc power analysis after the null hypothesis has been retained (but is still believed false) is also merely establishing what is already apparent, in this case that there was insufficient power. Put least optimistically, then, post hoc power analysis seems either self-congratulatory or self-pitying. Somewhat more optimistically, post hoc power analysis hopes to inform the next generation of investigations involving the model(s) at hand.

In preparation for that next generation, a priori power analysis can prove extremely useful in estimating the sample size necessary for testing specific parameters as well as for testing overall data–model fit. When $\varepsilon_1 = .02$ and models have $df \geq 60$, to achieve power of $\pi = .80$ sample sizes of $n = 300$ appear to suffice for testing overall data–model fit (as long as the number of estimated parameters is not so large as to precipitate convergence or estimation problems; see, e.g., Bentler & Chou, 1987). For models with $df \geq 30$ when $\varepsilon_1 = .02$, sample sizes of around $n = 500$ appear to be adequate. When the number of degrees of freedom gets smaller, the researcher will start to need an extremely large sample size for testing overall data–model fit. When df = 10, for example, to achieve π = .80 when ε_1 = .02, a model requires almost n = 1,000 cases. This might frustrate researchers who have come to expect smaller sample sizes to be adequate for smaller models. As mentioned at the beginning of this chapter, however, such recommendations in the literature have arisen based on achieving reliable rates of model convergence with reasonable parameter estimates, not on achieving power.

With regard to a priori power analysis for tests of specific parameters within structural models, sample size estimates are built upon the assumption that the model in which those parameters exist is correct (or reasonably so). This assumption includes not just the structure of the model, but also the assumed numerical values for focal parameters and the peripheral parameters in that model. If any of these are wrong, and most certainly they are, then the estimated sample sizes necessarily become inaccurate. That said, such inevitable inaccuracies should not deter us from conducting these preparatory analyses. With apologies to Alfred Lord Tennyson, 'tis better to have planned a study and failed than never to have planned at all. By choosing the values in models carefully and conservatively, researchers have the ability to ensure that they enter their studies hoping for the best and prepared for the worst. I hope that the current chapter will help researchers in these endeavors.

ACKNOWLEDGMENTS

This chapter benefited directly from the methodological and didactic insights of Jaehwa Choi, Jeffrey Greene, Marc Kroopnick, and Roy Levy; shortcomings of the chapter, however, are my responsibility alone.

Appendix A: Noncentrality Parameters for .05-Level χ^2 Tests

	df													
π	*1*	*2*	*3*	*4*	*5*	*6*	*7*	*8*	*9*	*10*	*20*	*30*	*40*	*50*
.10	.43	.62	.78	.91	1.03	1.13	1.23	1.32	1.40	1.49	2.14	2.65	3.08	3.46
.15	.84	1.19	1.46	1.69	1.89	2.07	2.23	2.39	2.53	2.67	3.77	4.63	5.35	5.98
.20	1.24	1.73	2.10	2.40	2.67	2.91	3.13	3.33	3.53	3.71	5.18	6.31	7.26	8.10
.25	1.65	2.26	2.71	3.08	3.40	3.70	3.96	4.21	4.45	4.67	6.45	7.82	8.97	9.99
.30	2.06	2.78	3.30	3.74	4.12	4.46	4.77	5.06	5.33	5.59	7.65	9.24	10.57	11.75
.35	2.48	3.30	3.90	4.39	4.82	5.21	5.56	5.89	6.19	6.48	8.81	10.60	12.10	13.42
.40	2.91	3.83	4.50	5.05	5.53	5.96	6.35	6.71	7.05	7.37	9.96	11.93	13.59	15.06
.45	3.36	4.38	5.12	5.72	6.25	6.72	7.15	7.55	7.92	8.27	11.10	13.26	15.08	16.68
.50	3.84	4.96	5.76	6.42	6.99	7.50	7.97	8.40	8.81	9.19	12.26	14.60	16.58	18.31
.55	4.35	5.56	6.43	7.15	7.77	8.32	8.83	9.30	9.73	10.15	13.46	15.99	18.11	19.98
.60	4.90	6.21	7.15	7.92	8.59	9.19	9.73	10.24	10.71	11.15	14.71	17.43	19.71	21.72
.65	5.50	6.92	7.93	8.76	9.48	10.12	10.70	11.25	11.75	12.23	16.05	18.96	21.40	23.55
.70	6.17	7.70	8.79	9.68	10.45	11.14	11.77	12.35	12.89	13.40	17.50	20.61	23.23	25.53
.75	6.94	8.59	9.76	10.72	11.55	12.29	12.96	13.59	14.17	14.72	19.11	22.44	25.25	27.71
.80	7.85	9.63	10.90	11.94	12.83	13.62	14.35	15.02	15.65	16.24	20.96	24.55	27.56	30.20
.85	8.98	10.92	12.30	13.42	14.39	15.25	16.04	16.77	17.45	18.09	23.20	27.08	30.33	33.19
.90	10.51	12.65	14.17	15.41	16.47	17.42	18.28	19.08	19.83	20.53	26.13	30.38	33.94	37.07
.95	12.99	15.44	17.17	18.57	19.78	20.86	21.84	22.74	23.59	24.39	30.72	35.52	39.54	43.07

Appendix B: Power for Models with *df* = 5 to 100

$\varepsilon_1 = .00$

	df																			
n	*5*	*10*	*15*	*20*	*25*	*30*	*35*	*40*	*45*	*50*	*55*	*60*	*65*	*70*	*75*	*80*	*85*	*90*	*95*	*100*
50	.06	.07	.08	.09	.10	.11	.11	.12	.13	.13	.14	.15	.15	.16	.16	.17	.18	.18	.19	.19
100	.08	.11	.13	.15	.18	.20	.22	.24	.26	.28	.30	.31	.33	.35	.37	.39	.40	.42	.44	.45
150	.10	.15	.19	.24	.28	.32	.35	.39	.43	.46	.49	.52	.55	.58	.61	.64	.66	.68	.70	.72
200	.13	.20	.27	.33	.40	.45	.51	.56	.60	.65	.68	.72	.75	.78	.81	.83	.85	.87	.88	.90
250	.16	.26	.36	.44	.52	.59	.65	.71	.76	.80	.83	.86	.88	.91	.92	.94	.95	.96	.97	.97
300	.19	.32	.44	.55	.64	.71	.78	.83	.87	.90	.92	.94	.96	.97	.97	.98	.99	.99	.99	.99
350	.23	.39	.53	.65	.74	.81	.87	.91	.93	.95	.97	.98	.99	.99	.99	.99	.99	.99		
400	.26	.46	.62	.74	.83	.88	.93	.95	.97	.98	.99	.99	.99	.99						
450	.31	.53	.70	.81	.89	.93	.96	.98	.99	.99	.99									
500	.35	.60	.77	.87	.93	.96	.98	.99	.99											
550	.39	.66	.82	.91	.96	.98	.99													
600	.44	.72	.87	.94	.98	.99														
650	.48	.77	.90	.96	.99															
700	.52	.81	.93	.98	.99															
750	.57	.85	.95	.99																
800	.61	.88	.97	.99																
850	.64	.90	.98																	
900	.68	.93	.99																	
950	.71	.94	.99																	
1000	.75	.96																		
1250	.87	.99																		
1500	.94																			
1750	.97																			
2000	.99																			

$\varepsilon_1 = .02$

	df																			
n	*5*	*10*	*15*	*20*	*25*	*30*	*35*	*40*	*45*	*50*	*55*	*60*	*65*	*70*	*75*	*80*	*85*	*90*	*95*	*100*
50	.06	.07	.08	.08	.09	.09	.10	.10	.11	.11	.12	.12	.13	.13	.14	.14	.15	.15	.15	.16
100	.08	.10	.11	.13	.15	.16	.18	.19	.20	.22	.23	.25	.26	.27	.29	.30	.31	.32	.34	.35
150	.09	.13	.16	.19	.22	.25	.27	.30	.33	.35	.38	.40	.42	.45	.47	.49	.51	.53	.55	.57
200	.11	.16	.21	.26	.30	.34	.38	.42	.46	.50	.53	.56	.59	.62	.65	.67	.70	.72	.74	.76
250	.13	.20	.27	.33	.39	.45	.50	.55	.59	.63	.67	.70	.74	.76	.79	.81	.84	.85	.87	.89
300	.15	.25	.33	.41	.48	.55	.61	.66	.71	.75	.78	.82	.84	.87	.89	.90	.92	.93	.94	.95
350	.18	.29	.40	.49	.57	.64	.70	.75	.80	.84	.87	.89	.91	.93	.94	.96	.96	.97	.98	.98
400	.20	.34	.46	.56	.65	.72	.78	.83	.87	.90	.92	.94	.96	.97	.97	.98	.99	.99	.99	.99
450	.23	.39	.52	.63	.72	.79	.84	.89	.92	.94	.96	.97	.98	.98	.99	.99	.99	.99		
500	.26	.44	.58	.69	.78	.85	.89	.93	.95	.97	.98	.98	.99	.99	.99					
550	.29	.48	.64	.75	.83	.89	.93	.95	.97	.98	.99	.99	.99							
600	.32	.53	.69	.80	.87	.92	.95	.97	.98	.99	.99									
650	.35	.57	.73	.84	.90	.94	.97	.98	.99											
700	.38	.61	.77	.87	.93	.96	.98	.99												
750	.40	.65	.81	.90	.95	.97	.99	.99												
800	.43	.69	.84	.92	.96	.98	.99													
850	.46	.72	.86	.94	.97	.99														
900	.49	.75	.89	.95	.98	.99														
950	.51	.78	.91	.96	.99															
1000	.54	.80	.92	.97	.99															
1250	.65	.89	.97	.99																
1500	.74	.95	.99																	
1750	.81	.97																		
2000	.86	.99																		
2500	.93																			
3000	.97																			
3500	.98																			
4000	.99																			

$\varepsilon_1 = .04$

	df																			
n	*5*	*10*	*15*	*20*	*25*	*30*	*35*	*40*	*45*	*50*	*55*	*60*	*65*	*70*	*75*	*80*	*85*	*90*	*95*	*100*
50	.05	.06	.06	.06	.06	.07	.07	.07	.07	.07	.07	.07	.08	.08	.08	.08	.08	.08	.08	.08
100	.06	.07	.07	.08	.08	.08	.09	.09	.09	.10	.10	.10	.11	.11	.11	.12	.12	.12	.13	.13
150	.07	.07	.08	.09	.10	.10	.11	.12	.12	.13	.13	.14	.14	.15	.15	.16	.16	.17	.17	.18
200	.07	.08	.10	.11	.11	.12	.13	.14	.15	.16	.17	.17	.18	.19	.20	.20	.21	.22	.23	.23
250	.08	.09	.11	.12	.13	.14	.16	.17	.18	.19	.20	.21	.22	.23	.24	.25	.26	.27	.28	.29
300	.08	.10	.12	.14	.15	.17	.18	.20	.21	.22	.24	.25	.26	.28	.29	.30	.31	.33	.34	.35
350	.09	.11	.13	.15	.17	.19	.21	.22	.24	.26	.27	.29	.30	.32	.34	.35	.36	.38	.39	.41
400	.09	.12	.14	.17	.19	.21	.23	.25	.27	.29	.31	.33	.35	.36	.38	.40	.42	.43	.45	.46
450	.10	.13	.16	.18	.21	.23	.26	.28	.30	.32	.34	.37	.39	.41	.43	.45	.46	.48	.50	.52
500	.10	.14	.17	.20	.23	.25	.28	.31	.33	.36	.38	.40	.43	.45	.47	.49	.51	.53	.55	.57
550	.11	.15	.18	.21	.25	.28	.31	.33	.36	.39	.41	.44	.46	.49	.51	.53	.55	.58	.60	.61
600	.11	.16	.19	.23	.26	.30	.33	.36	.39	.42	.45	.48	.50	.53	.55	.57	.60	.62	.64	.66
650	.12	.16	.21	.24	.28	.32	.35	.39	.42	.45	.48	.51	.54	.56	.59	.61	.63	.66	.68	.70
700	.12	.17	.22	.26	.30	.34	.38	.41	.45	.48	.51	.54	.57	.60	.62	.65	.67	.69	.71	.73
750	.13	.18	.23	.28	.32	.36	.40	.44	.47	.51	.54	.57	.60	.63	.66	.68	.70	.73	.75	.77
800	.13	.19	.24	.29	.34	.38	.42	.46	.50	.54	.57	.60	.63	.66	.69	.71	.73	.76	.78	.79
850	.14	.20	.25	.31	.35	.40	.45	.49	.53	.56	.60	.63	.66	.69	.72	.74	.76	.78	.80	.82
900	.14	.21	.27	.32	.37	.42	.47	.51	.55	.59	.62	.66	.69	.72	.74	.77	.79	.81	.83	.84
950	.15	.22	.28	.33	.39	.44	.49	.53	.57	.61	.65	.68	.71	.74	.77	.79	.81	.83	.85	.86
1000	.15	.22	.29	.35	.41	.46	.51	.56	.60	.64	.67	.71	.74	.76	.79	.81	.83	.85	.87	.88
1250	.18	.26	.34	.42	.49	.55	.60	.65	.70	.74	.77	.80	.83	.85	.88	.89	.91	.92	.93	.94
1500	.20	.30	.40	.48	.56	.62	.68	.73	.78	.81	.84	.87	.89	.91	.93	.94	.95	.96	.97	.97
1750	.22	.34	.45	.54	.62	.69	.75	.80	.84	.87	.90	.92	.93	.95	.96	.97	.98	.98	.99	.99
2000	.24	.38	.49	.59	.68	.75	.80	.85	.88	.91	.93	.95	.96	.97	.98	.98	.99	.99		
2500	.28	.45	.58	.69	.77	.83	.88	.92	.94	.96	.97	.98	.99	.99	.99	.99				
3000	.32	.51	.65	.76	.84	.89	.93	.95	.97	.98	.99	.99								

3500	.36	.57	.72	.82	.89	.93	.96	.98	.99	.99
4000	.39	.62	.77	.86	.92	.96	.98	.99		
4500	.43	.66	.81	.90	.95	.97	.99			
5000	.46	.71	.85	.93	.96	.98				
6000	.52	.78	.90	.96	.98	.99				
7000	.58	.83	.94	.98	.99					
8000	.63	.87	.96	.99						
9000	.67	.91	.98							
10000	.72	.93	.99							
12500	.80	.97								
15000	.86	.99								
17500	.90									
20000	.93									
25000	.97									
30000	.99									

Appendix C: Power for Models with *df* = 110 to 250

$\varepsilon_1 = .00$

	df											
n	*110*	*120*	*130*	*140*	*150*	*160*	*170*	*180*	*190*	*200*	*225*	*250*
50	.20	.21	.23	.24	.25	.26	.27	.28	.29	.30	.33	.35
100	.48	.50	.54	.57	.60	.62	.65	.67	.69	.71	.76	.80
150	.76	.78	.82	.85	.87	.89	.91	.92	.93	.94	.96	.98
200	.92	.93	.96	.97	.98	.98	.99	.99	.99	.99	.99	.99
250	.98	.99	.99	.99	.99	.99						
300	.99											

$\varepsilon_1 = .02$

	df											
n	*110*	*120*	*130*	*140*	*150*	*160*	*170*	*180*	*190*	*200*	*225*	*250*
50	.17	.18	.18	.19	.20	.21	.22	.22	.23	.24	.26	.28
100	.37	.40	.42	.44	.46	.49	.51	.53	.55	.57	.61	.65
150	.61	.64	.67	.70	.73	.75	.78	.80	.82	.84	.87	.90
200	.80	.83	.85	.88	.90	.91	.93	.94	.95	.96	.97	.98
250	.91	.93	.95	.96	.97	.98	.98	.99	.99	.99	.99	.99
300	.97	.98	.98	.99	.99	.99	.99					
350	.99	.99	.99									

$\varepsilon_1 = .04$

	df											
n	*110*	*120*	*130*	*140*	*150*	*160*	*170*	*180*	*190*	*200*	*225*	*250*
50	.09	.09	.09	.09	.09	.10	.10	.10	.10	.10	.11	.11
100	.13	.14	.14	.15	.16	.16	.17	.17	.18	.18	.19	.21
150	.19	.20	.21	.22	.23	.24	.25	.25	.26	.27	.29	.32
200	.25	.26	.28	.29	.30	.32	.33	.34	.36	.37	.40	.43
250	.31	.33	.35	.37	.38	.40	.42	.44	.45	.47	.51	.54
300	.37	.40	.42	.44	.46	.48	.50	.52	.54	.56	.60	.64
350	.44	.46	.49	.51	.54	.56	.58	.60	.62	.64	.69	.73
400	.49	.52	.55	.58	.60	.63	.65	.67	.69	.71	.76	.80
450	.55	.58	.61	.64	.67	.69	.71	.74	.76	.77	.82	.85
500	.60	.64	.67	.69	.72	.74	.77	.79	.81	.82	.86	.89
550	.65	.68	.71	.74	.77	.79	.81	.83	.85	.86	.90	.92
600	.69	.73	.76	.78	.81	.83	.85	.87	.88	.90	.93	.95
650	.73	.77	.79	.82	.84	.86	.88	.90	.91	.92	.95	.96
700	.77	.80	.83	.85	.87	.89	.91	.92	.93	.94	.96	.97
750	.80	.83	.85	.88	.90	.91	.93	.94	.95	.96	.97	.98
800	.83	.85	.88	.90	.92	.93	.94	.95	.96	.97	.98	.99
850	.85	.88	.90	.92	.93	.95	.96	.96	.97	.98	.99	.99
900	.87	.90	.92	.93	.95	.96	.97	.97	.98	.98	.99	
950	.89	.91	.93	.95	.96	.97	.97	.98	.98	.99		
1000	.91	.93	.94	.96	.97	.97	.98	.98	.99	.99		
1250	.96	.97	.98	.99	.99	.99	.99	.99	.99			
1500	.98	.99	.99	.99	.99							
1750	.99	.99										

Appendix D: Sample Size for Models with *df* = 5 to 250

ε_1 = .00

	π										
df	*.50*	*.55*	*.60*	*.65*	*.70*	*.75*	*.80*	*.85*	*.90*	*.95*	*.99*
5	673	732	793	858	929	1008	1099	1209	1354	1584	2062
10	428	464	503	543	586	635	690	756	842	977	1252
15	331	359	388	419	452	488	530	580	644	744	946
20	277	300	325	350	377	407	441	482	535	616	779
25	242	262	283	305	329	355	384	419	464	533	672
30	217	235	254	273	294	317	343	374	414	475	597
35	198	214	231	249	268	289	312	340	376	431	540
40	183	198	214	230	247	266	288	314	347	397	496
45	171	185	199	214	231	248	268	292	323	369	461
50	161	174	187	202	217	233	252	274	303	346	431
55	152	165	177	191	205	221	238	259	286	327	407
60	145	157	169	181	195	210	226	246	271	310	385
65	138	150	161	173	186	200	216	235	259	295	367
70	133	143	154	166	178	192	207	225	248	282	351
75	128	138	148	159	171	184	199	216	238	271	336
80	123	133	143	154	165	177	191	208	229	261	323
85	119	128	138	149	159	171	185	201	221	252	312
90	115	124	134	144	154	166	179	194	214	243	301
95	112	121	130	139	150	161	173	188	207	236	292
100	109	117	126	135	145	156	168	183	201	229	283
110	103	111	120	128	138	148	159	173	190	217	267
120	98	106	114	122	131	141	152	165	181	206	254
130	94	101	109	117	125	135	145	157	173	197	242
140	90	97	105	112	120	129	139	151	166	188	232
150	87	94	101	108	116	124	134	145	160	181	223
160	84	90	97	104	112	120	129	140	154	175	215
170	81	88	94	101	108	116	125	135	149	169	208
180	79	85	91	98	105	112	121	131	144	163	201
190	76	82	89	95	102	109	117	127	140	158	195
200	74	80	86	92	99	106	114	124	136	154	189
225	70	75	81	87	93	100	107	116	127	144	177
250	66	71	76	82	88	94	101	109	120	136	167

ε_1 = .02											
	π										
df	*.50*	*.55*	*.60*	*.65*	*.70*	*.75*	*.80*	*.85*	*.90*	*.95*	*.99*
5	926	1025	1131	1248	1379	1529	1707	1930	2234	2734	3829
10	569	625	685	750	821	902	997	1115	1273	1530	2086
15	433	474	518	565	616	674	741	823	934	1111	1491
20	358	392	427	464	505	551	604	669	756	893	1185
25	311	339	369	401	435	474	518	572	644	758	997
30	277	302	328	356	386	420	458	505	567	665	869
35	251	274	297	322	349	379	414	455	510	596	775
40	231	252	273	296	321	348	379	417	466	544	704
45	215	234	254	275	297	322	351	386	431	501	647
50	202	220	238	258	278	302	328	360	402	467	600
55	191	207	225	243	262	284	309	339	377	438	562
60	181	197	213	230	249	269	292	320	357	413	529
65	173	188	203	219	237	256	278	304	339	392	501
70	165	179	194	210	226	245	266	291	323	374	476
75	159	172	186	201	217	234	254	278	309	357	454
80	153	166	179	193	209	225	245	267	297	343	435
85	148	160	173	187	201	217	236	257	286	330	418
90	143	155	167	180	194	210	228	249	276	318	403
95	138	150	162	175	188	203	220	240	267	307	389
100	134	146	157	169	183	197	213	233	258	297	376
110	127	138	149	160	173	186	202	220	244	280	353
120	121	131	141	152	164	177	191	209	231	265	334
130	116	125	135	145	157	169	183	199	220	253	317
140	111	120	129	139	150	162	175	190	211	241	303
150	107	115	124	134	144	155	168	183	202	232	290
160	103	111	120	129	139	150	162	176	194	223	279
170	99	108	116	125	134	144	156	170	188	215	268
180	96	104	112	121	130	140	151	164	181	207	259
190	93	101	109	117	126	135	146	159	176	201	251
200	91	98	106	114	122	132	142	154	170	195	243
225	85	92	99	107	114	123	133	144	159	182	226
250	80	87	94	100	108	116	125	136	150	171	212

$\varepsilon_1 = .04$

	π										
df	*.50*	*.55*	*.60*	*.65*	*.70*	*.75*	*.80*	*.85*	*.90*	*.95*	*.99*
5	5621	6478	7413	8449	9616	10962	12568	14580	17328	21843	31738
10	2931	3360	3828	4346	4930	5603	6406	7412	8786	11043	15990
15	2028	2315	2628	2974	3364	3813	4349	5020	5936	7442	10740
20	1573	1790	2025	2286	2579	2916	3319	3822	4510	5640	8114
25	1298	1472	1661	1871	2106	2377	2699	3103	3653	4557	6537
30	1112	1259	1417	1592	1789	2015	2285	2622	3081	3835	5486
35	979	1105	1242	1393	1562	1757	1988	2277	2672	3319	4735
40	878	989	1109	1242	1391	1562	1765	2019	2364	2931	4171
45	798	898	1006	1124	1257	1410	1591	1817	2125	2629	3732
50	734	824	922	1029	1150	1287	1451	1655	1933	2387	3380
55	681	764	853	951	1061	1187	1336	1522	1775	2189	3093
60	637	713	796	886	987	1103	1240	1411	1643	2024	2853
65	599	670	746	830	924	1031	1158	1317	1532	1883	2650
70	566	632	704	782	870	970	1088	1236	1436	1763	2475
75	537	600	667	740	823	916	1027	1165	1353	1658	2324
80	512	571	634	704	781	869	974	1103	1279	1567	2191
85	489	545	605	671	744	828	926	1049	1215	1486	2074
90	469	522	579	642	711	790	884	1000	1157	1414	1970
95	451	501	556	615	682	757	846	956	1106	1349	1877
100	434	483	535	592	655	727	811	917	1059	1291	1793
110	405	450	498	550	608	674	752	848	978	1189	1647
120	381	422	467	515	569	630	702	790	910	1105	1526
130	360	399	440	485	535	592	659	741	853	1033	1423
140	342	378	417	460	506	560	622	699	803	971	1334
150	326	360	397	437	481	531	590	662	760	918	1257
160	311	344	379	417	459	506	561	630	722	870	1190
170	299	330	363	399	439	484	536	601	688	828	1130
180	287	317	349	383	421	464	514	575	658	791	1077
190	277	306	336	369	405	446	493	552	631	757	1029
200	268	295	324	356	390	429	475	531	606	727	986
225	247	272	299	328	359	394	435	486	554	663	895
250	231	254	278	305	333	366	403	450	511	610	821

REFERENCES

Bentler, P. M., & Chou, C. (1987). Practical issues in structural modeling. *Sociological Methods & Research*, *16*, 78–117.

Bollen, K. A. (1989). *Structural equations with latent variables*. New York: Wiley.

Browne, M. W., & Cudeck, R. (1993). Alternative ways of assessing model fit. In K. A. Bollen & J. S. Long (Eds.), *Testing structural equation models* (pp. 136–162). Newbury Park, CA: Sage.

Browne, M. W., & Mels, G. (1990). *RAMONA user's guide*. Unpublished report, Department of Psychology, Ohio State University.

Fornell, C., & Larcker, D. F. (1981). Evaluating structural equation models with unobservable variables and measurement error. *Journal of Marketing Research, 18*, 39–50.

Gagné, P., & Hancock, G. R. (in press). Measurement model quality, sample size, and solution propriety in confirmatory factor models. *Multivariate Behavioral Research*.

Hancock, G. R. (2001). Effect size, power, and sample size determination for structured means modeling and MIMIC approaches to between-groups hypothesis testing of means on a single latent construct. *Psychometrika, 66*, 373–388.

Hancock, G. R. (2004). Experimental, quasi-experimental, and nonexperimental design and analysis with latent variables. In D. Kaplan (Ed.), *The Sage handbook of quantitative methodology for the social sciences*. Thousand Oaks, CA: Sage.

Hancock, G. R., & Freeman, M. J. (2001). Power and sample size for the RMSEA test of not close fit in structural equation modeling. *Educational and Psychological Measurement, 61*, 741–758.

Hancock, G. R., & Kroopnick, M. H. (2005, April). *A simplified sample size determination for tests of latent covariance structure parameters*. Paper presented at the annual meeting of the American Educational Research Association, Montréal.

Hancock, G. R., & Mueller, R. O. (2001). Rethinking construct reliability within latent variable systems. In R. Cudeck, S. H. C. du Toit, & D. Sörbom (Eds.), *Structural equation modeling: Past and present. A Festschrift in honor of Karl G. Jöreskog* (pp. 195–261). Chicago: Scientific Software International.

Haynam, G. E., Govindarajulu, Z., & Leone, F. C. (1973). Tables of the cumulative non-central chi-square distribution. In H. L. Harter & D. B. Owen (Eds.), *Selected tables in mathematical statistics* (Vol. 1, pp. 1–78). Chicago: Markham.

Hu, L., & Bentler, P. M. (1999). Cutoff criteria for fit indexes in covariance structure analysis: Conventional criteria versus new alternatives. *Structural Equation Modeling: A Multidisciplinary Journal, 6*, 1–55.

Jackson, D. L. (2003). Revisiting sample size and number of parameter estimates: Some support for the *N:q* hypothesis. *Structural Equation Modeling: A Multidisciplinary Journal, 10*, 28–141.

Kaplan, D. (1995). Statistical power in structural equation modeling. In R. Hoyle (Ed.), *Structural equation modeling: Concepts, issues, and applications* (pp. 100–117). Thousand Oaks, CA: Sage.

MacCallum, R. C., Browne, M. W., & Sugawara, H. M. (1996). Power analysis and determination of sample size for covariance structure modeling. *Psychological Methods, 1*, 130–149.

MacCallum, R. C., & Hong, S. (1997). Power analysis in covariance structure modeling using GFI and AGFI. *Multivariate Behavioral Research, 32*, 193–210.

MacCallum, R. C., Widaman, K. F., Zhang, S., & Hong, S. (1999). Sample size in factor analysis. *Psychological Methods, 4*, 84–99.

Marsh, H. W., Hau, K-T., Balla, J. R., & Grayson, D. (1998). Is more ever too much? The number of indicators per factor in confirmatory factor analysis. *Multivariate Behavioral Research, 33*, 181–220.

Muthén, L., & Muthén, B. (2002). How to use a Monte Carlo study to decide on sample size and determine power. *Structural Equation Modeling: A Multidisciplinary Journal, 9*, 599–620.

Nevitt, J., & Hancock, G. R. (2000). Improving the root mean square error of approximation for nonnormal conditions in structural equation modeling. *Journal of Experimental Education, 68*, 251–268.

Raykov, T. (2004). Estimation of maximal reliability: A note on a covariance structure modelling approach. *British Journal of Mathematical and Statistical Psychology, 57*, 21–27.

Saris, W. E., & Satorra, A. (1993). Power evaluations in structural equation models. In K. A. Bollen & J. S. Long (Eds.), *Testing structural equation models* (pp. 181–204). Newbury Park, CA: Sage.

Satorra, A., & Bentler, P. M. (1994). Corrections to test statistics and standard errors in covariance structure analysis. In A. von Eye & C. C. Clogg (Eds.), *Latent variables analysis: Applications for developmental research* (pp. 399–419). Thousand Oaks, CA: Sage.

Satorra, A., & Bentler, P. M. (2001). A scaled difference chi-square test statistic for moment structure analysis. *Psychometrika, 66*, 507–514.

Satorra, A., & Saris, W. E. (1985). The power of the likelihood ratio test in covariance structure analysis. *Psychometrika, 50*, 83–90.

Steiger, J. H. (1989). *EzPATH: A supplemental module for SYSTAT and SYSGRAPH*. Evanston, IL: SYSTAT.

Steiger, J. H., & Lind, J. M. (1980, June). *Statistically based tests for the number of common factors*. Paper presented at the annual meeting of the Psychometric Society, Iowa City, IA.

Yuan, K., & Bentler, P. M. (2004). On chi-square difference and *z* tests in mean and covariance structure analysis when the base model is misspecified. *Educational and Psychological Measurement, 64*, 737–757.

PART II

EXTENSIONS

CHAPTER 5

EVALUATING BETWEEN-GROUP DIFFERENCES IN LATENT VARIABLE MEANS

Marilyn S. Thompson and Samuel B. Green

Researchers conducting experimental or quasi-experimental research often seek to compare groups hypothesized to differ on one or more outcomes. In the simplest form, groups may be compared on a single dependent variable that is measured directly and assumed to reliably represent the construct of interest. Often, however, research questions involve group differences on multiple outcomes that are assumed to represent one or more constructs. In such conditions, many researchers routinely perform a multivariate analysis of variance (MANOVA) to assess mean differences among groups on the set of multiple dependent variables. Although MANOVA persists as the most commonly applied multivariate method for analyzing mean differences, methodologists have argued that structural equation modeling (SEM) methods for comparing groups on latent variable means are a more appropriate approach in many multivariate designs (e.g., Aiken, Stein, & Bentler, 1994; Cole, Maxwell, Arvey, & Salas, 1993; Green & Thompson, 2003a; Hancock, 1997). In this chapter, we discuss the utility of SEM for evaluating between-group differences on

Structural Equation Modeling: A Second Course, 119–169

latent variable means and illustrate SEM approaches for doing so. Specifically, we address the following key points:

- SEM is more appropriate when the interest is in the comparison of latent variable means, while MANOVA is acceptable for an emergent variable system.
- SEM provides a flexible approach to comparison of latent variable means that accounts for unreliability of measures, allows for inclusion of latent covariates, and can be more powerful than MANOVA.
- Between-group differences in latent variable means can be conducted using a structured means modeling (SMM) approach or a multiple-indicator multiple-cause (MIMIC) modeling approach.
- The SMM approach is more flexible than the MIMIC approach in that it allows for partial measurement invariance across groups.

We illustrate the MIMIC and SMM approaches for the two-sample case, but note that these methods extend to any number of samples and can be applied in experimental, quasi-experimental, and nonexperimental designs. Although models may also be specified to assess differences in latent means for repeated measures problems, this chapter focuses on methods for between-subjects designs. We begin with a discussion of how a construct may be related to a set of measured variables, and then consider the implications of this relationship in terms of choosing an appropriate model and analytic approach.

EMERGENT VERSUS LATENT VARIABLE SYSTEMS

In both the SEM and MANOVA approaches to assessing mean differences between groups, multiple measured variables are hypothesized, in some fashion, to conceptually represent constructs. To avoid confusion with the differing meaning of the term "factor" across the SEM and ANOVA/MANOVA literatures, we use the term "grouping variable" in this chapter to refer to a variable that differentiates individuals into groups and reserve the term "factor" to represent a construct or dimension underlying or emerging from the dependent variables. In this section, we initially disregard the grouping variable and focus on how the construct can be related to the measured variables.

A critical step in choosing an appropriate model and analytic approach is determining the directionality of the relationship between the construct and the measured variables, which may generally be described in one of two ways (Bollen & Lennox, 1991; Cohen, Cohen, Teresi, Marchi, & Velez,

1990; Cole et al., 1993). One possibility is for the measured variables to be contributory agents of the constructs, implying an *emergent* variable system with *causal indicators*. Alternatively, the constructs may be causal agents of the dependent measures, resulting in a *latent* variable system with *effect indicators* (see Kline, chapter 3, this volume, for more detail than presented in this section).

Emergent Variable Systems

In an emergent variable system, causal indicators are combined additively to form a linear composite representing the construct. Assume, for example, a researcher is interested in evaluating a model that includes physical health. Because there is no single observable and reliable measure of physical health, individuals are assessed on body mass index, blood pressure, cholesterol level, and presence of chronic diseases. It is more logical to regard these physical measures as affecting overall health, rather than overall health affecting these measures. Therefore, in this case, it is conceptually meaningful to form a linear composite of these measures and consider this composite as representative of the construct of physical health. In an SEM framework, an emergent variable system is schematically represented by arrows directed from the causal indicators to the emergent factor. Factors in an emergent variable system are often included in models to explain the relationship between these indicators and other variables. Another distinguishing feature of emergent variable systems is that indicators may or may not be correlated because causal indicators are not a function of factors. For example, body mass index might be unrelated to the presence of chronic disease because overall physical health is not their underlying cause. Accordingly, the validity of factors in an emergent variable system should be established by their relationship with variables external to the system.

Latent Variable Systems

In contrast, in most applications of SEM the measured variables are effect indicators that are correlated because they share common causes, that is, the same underlying factors. In latent variable systems, the measured variables are hypothesized to be linear combinations of factors plus error, so arrows are directed from factors and errors to the measured variables. Researchers in the behavioral sciences commonly conceptualize human behaviors to be effect indicators of underlying factors, such that the observed indicators are affected by the latent trait. For example, social self-concept is a latent trait that may be hypothesized to affect measures of shyness, comfort with adults, and comfort with peers. The measurement model relating factors to their effect indicators can be validated by collecting data on the measured variables and conducting a confirmatory factor analysis (CFA) to assess whether the intended factor structure accounts for the covariances among the indicators.

A construct is not inherently defined as a factor in either an emergent or latent variable system. Rather, the construct may be treated as part of an emergent or latent variable system according to whether the construct is measured by its consequences or its causes (Cohen et al., 1990). We consider again the construct of physical health and how it might be meaningfully reconstructed as a latent rather than an emergent variable. If physical health were viewed as the causal agent, then measures of the consequences of having good or poor physical health would be appropriate effect indicators. These consequences might include indicators such as degree of mobility, level of chronic pain, and number of medications taken daily. Although it might be theoretically sound to hypothesize physical health as part of either an emergent variable system (with causal indicators) or a latent variable system (with consequences as effect indicators), other constructs and their associated measures may fit more readily into one or the other framework.

ADVANTAGES OF SEM FOR COMPARING GROUPS ON LATENT VARIABLE MEANS

While SEM and MANOVA can both be used to assess differences in means between levels of a grouping variable on a construct represented by multiple dependent variables, these approaches are not interchangeable and differ markedly with respect to the underlying model (see Cole et al., 1993, for a thorough discussion). Although SEM is a very general model that can accommodate either emergent or latent variable systems within structural models, standard SEM approaches for evaluating differences in construct means are designed for latent variable systems. In contrast, MANOVA is an appropriate method for evaluating differences in means in emergent variable systems. The emergent variables in MANOVA are the mathematically derived linear combinations of the measured variables on which groups are maximally separated (i.e., the discriminant functions). However, because they are not necessarily the combinations of interest to the researcher, the results of MANOVA may be uninterpretable with respect to a latent construct. We suspect SEM approaches should more frequently be utilized for analyzing group differences on a construct and present below several arguments for this perspective.

Matching the Model and the Method

The distinction between emergent and latent variable systems should be made based on theory and is fundamental in deciding on a proper model and analytic approach for evaluating mean differences on multiple dependent variables. Nonetheless, researchers often fail to properly model the variable system. Researchers eager to apply SEM may incorrectly assume a variable system as latent when in fact it is theoretically emergent (see Cohen et al., 1990, for examples). Likewise, probably due to MANOVA's greater familiarity and accessibility using popular statistical software, researchers often test differences in a set of measures that should constitute a latent variable system via MANOVA—a technique that assumes an emergent system. The failure to utilize an appropriate model is problematic given that the results of MANOVA and SEM approaches to assessing differences in construct means are likely to yield similar results only under very restrictive conditions (Cole et al., 1993; Green & Thompson, 2003b). Cole and colleagues (1993) showed that the substantive conclusions would be nearly identical across the two methods for a truly single factor model with no measurement bias between groups. However, they also found that applying MANOVA to latent variable systems that differ in realistic ways from this restrictive condition leads to seriously deficient findings. Conclusions may differ dramatically across the methods to the extent that some of the measures also share variance from a source not accounted for in the model (such as method variance that can be represented in SEM by correlated errors) or that measurement bias between groups is present in one or more measured variables (which implies unequal intercepts, as will be discussed with the SMM approach).

SEM and MANOVA approaches also differ with respect to how measurement error is treated. Each measured variable, regardless of how it is related to the construct, presumably has measurement error associated with it. Linear composites in emergent variable systems, such as those assessed in MANOVA, may be heavily influenced by the propagation of measurement errors of the indicators. Tests for population differences in construct means based on these linear composites may fall short of finding true differences because unreliability of the composites can obscure true differences between groups. In contrast, latent variables in SEM are theoretically error-free representations of constructs. Accordingly, SEM in many cases may have greater power than MANOVA approaches for assessing differences in latent variable means (Hancock, 2003; Hancock, Lawrence, & Nevitt, 2000; Kano, 2001).

In summary, MANOVA and discriminant analysis are appropriate methods for emergent variable systems. For latent variable systems, however, SEM approaches are preferable for comparing population means on constructs between groups because they are conceptually consistent with the theorized relationships between constructs and their effect indicators and will in many cases be more powerful.

Getting in the Mood for Modeling

Statistical software used to compute MANOVAs allows the analyst to be somewhat lazy in that the relation between the construct and the measures is implied by the choice of analysis; that is, there is no flexibility in how the direction of these relationships is specified. In contrast, regardless of which software package is used for conducting SEM, the researcher must specify the model fully through program syntax (although syntax generation may be assisted by choosing options in dialogue boxes or creating path diagrams). Furthermore, after fitting the data to the initially hypothesized model, the analyst can empirically explore which parts of the model contribute to lack of fit and may choose to relax certain constraints imposed on the initial model. Careful thought and decisions are required at every stage of the SEM process, including model specification, testing, and interpretation. Additionally, latent covariates can be included in SEM, and the user has flexibility in how relationships between covariates and the constructs or measures are specified. In summary, taking an SEM approach to assessing differences in latent mean structures puts you in the mood for modeling.

EVALUATING DIFFERENCES IN FACTOR MEANS UNDER FACTOR STRUCTURE INVARIANCE

Differences in factor means can be evaluated using two types of models: the multiple indicators and multiple causes (MIMIC) model or the structured means model (SMM). The MIMIC model is the simpler of the two and can be applied using the knowledge acquired in a first course of SEM. The data for MIMIC models are variances and covariances among measured variables; the model parameters include variances, covariances, and path coefficients; and the adequacy of the model is assessed using traditional fit indices. In contrast, for SMM a multiple-group model is conducted on means in addition to variances and covariances among measured variables; the model includes intercept parameters as well as variances, covariances, and path coefficients; and the definition of fit must be expanded to permit the reproduction of sample means.

In this section, we introduce the structured means and MIMIC models for evaluating differences in factor means under strict factorial invariance. By strict factorial invariance, we mean the parameters of the factor model are identical across groups except for factor means. As we discuss later, the SMM approach can be applied under much less restrictive conditions and, in practice, typically is. We initially present the restrictive form of SMM for three reasons. First, as applied, the SMM method is a complex, stepwise approach involving comparisons between models. It is advantageous to discuss a particular specification of the structured means model before evaluating differences between various structured means models. Second, the specification of the SMM approach under strict factorial invariance is relatively simple and, thus, easier to grasp. Third, the specification of SMM under strict factorial invariance makes very clear the set of restrictive conditions under which the MIMIC model can be applied.

We created a fictitious study and data to illustrate the MIMIC and SMM approaches. The study involves evaluating the relationship between day-care use and positive outcome variables for preschool children raised in two-parent families. The preschool children are divided into two groups according to a nonexperimental design: (a) those with two working parents who consistently use day-care facilities (day-care group) and (b) those with one stay-at-home parent and minimal use of day-care facilities (home-care group). Six outcome variables are included to assess academic and social school readiness. Indicators of academic readiness include scales measuring vocabulary level (V1), knowledge of letters and numbers (V2), and classification skills (V3), while indicators of social readiness include scales assessing self-control (V4), positive interaction with adults (V5), and positive interaction with peers (V6). Our developmental theory suggests that the readiness factors are latent and affect their respective measures. For example, the covariances among the academic measures can be viewed as a function of academic preparation, or, as we refer to it, academic readiness. Data for these indicators are presented in Table 5.1 for the day-care group (N_1 = 200) and home-care group (N_2 = 200) and for the combined sample (N = 400). The data were created to be consistent with the restriction that the factor structure be invariant across groups. Later in this chapter, we introduce different data for this scenario to illustrate analyses for factor structures that vary across groups.

Structured Means Modeling

We take a building blocks approach to explain structured means modeling. We begin with a standard confirmatory factor analysis model (one-group CFA), add intercepts to the model to account for the means of the measures (one-group CFA with means), and finally expand the model to assess differences between factor means for independent populations using multiple-group analyses (multiple-group CFA with means).

One-Group CFA

We first conduct a CFA on the covariance matrix for the day-care group in Table 5.1. We specify a model in which each of the first three measures (V1 to V3) is a function of the academic readiness factor (F1) plus measurement error (E1 to E3), and each of the last three measures (V4 to V6) is a function of the social readiness factor (F2) plus measurement error (E4 to E6). For example, the equation for the second measure is

$$V2 = b_{V2F1}F1 + E2 . \tag{1}$$

Table 5.1. Covariance Matrices and Means for Measured Variables for the Day-Care Group, Home-Care Group, and Total—Dataset 1

Measured Variables	*Covariance Matrix*						
	V1	V2	V3	V4	V5	V6	*Means*
	Group 1—Day-Care Group (N_1 = 200)						
V1	138.00						50.40
V2	45.58	80.49					79.60
V3	35.19	23.56	56.34				98.88
V4	45.13	32.00	16.64	232.17			74.06
V5	35.33	10.49	10.56	79.74	149.16		49.12
V6	73.34	28.73	33.21	117.30	79.90	324.36	120.10
	Group 2—Home-Care Group (N_2 = 200)						
V1	127.61						53.70
V2	58.49	76.81					81.32
V3	29.09	19.72	54.29				101.82
V4	45.84	28.94	31.82	223.09			77.69
V5	20.33	13.27	5.81	62.25	135.72		51.62
V6	50.64	55.45	30.15	126.74	62.16	337.37	123.37
	Total—Combined for Day-Care and Home-Care Groups (*N* = 400)						
	V1	V2	V3	V4	V5	V6	V7
V1	135.19						
V2	53.33	79.21					
V3	34.48	22.86	57.33				
V4	48.37	31.96	26.83	230.35			
V5	29.82	12.93	9.99	73.08	143.64		
V6	64.53	43.40	34.00	124.69	72.90	332.72	
V7	.83	.43	.73	.91	.62	.82	.25

Note: V1: Vocabulary; V2: Letters/Numbers; V3: Classification; V4: Self-Control; V5: Adult interaction; V6: Peer interaction; V7: Day (0) versus home (1) care.

The factor loadings on the first and sixth measures are fixed to 1 to define the metric of F1 and F2, respectively. The factors are allowed to be correlated. In total, we are estimating 13 model parameters: two factor variances (c_{F1} and c_{F2}), one factor covariance (c_{F1F2}), four factor loadings (b_{V2F1}, b_{V3F1}, b_{V4F2}, b_{V5F2}), and six error variances (c_{E1} to c_{E6}). These parameters are estimated so that the reproduced matrix of the variances and covariances among the measured variables based on the model parameters is as similar as possible to the sample matrix of variances and covariances among the measured variables. Lack of fit between these matrices is a function of the constraints imposed on the model (e.g., no factor loading between F1 and V4). We assess global lack of fit using three indices: chi-square statistic (χ^2), standardized root mean square residual (SRMR), and root mean square error of approximation (RMSEA). For our data, the one-group CFA model fits very well, $\chi^2(8) = 10.73$, $p = .22$ SRMR = .04, RMSEA = .04 (90% C.I. of .00 to .10). In Appendix A we provide a discussion of these global fit indices and suggest why the values we provide for these indices might differ from the ones generated by any specific SEM program.

In computing the variances and covariances among measures, the means of the variables are subtracted from their raw scores. Consequently, we implicitly ignore the means of the measures in the CFA modeling process and treat V1 through V6 as deviations about their means. The means on these deviations scores, V1 through V6, are equal to zero. To be consistent with means of zero on V1 through V6, the means on the latent variables—factors and errors—are considered equal to zero. For example, given the means on F1 and E2 are equal to zero, the mean of V2 is also equal to zero:

$$\mu_{V2} = \mu_{b_{V2F1}F1\,+\,E2} = b_{V2F1}\mu_{F1} + \mu_{E2} = b_{V2F1}\,(0) + 0 = 0, \qquad (2)$$

where μ indicates the mean of the quantity that is subscripted.

One-Group CFA With Means

The data for SEM may involve not only variances and covariances among the measures, but also their means. A more complex model is required if data include means, as illustrated in Figure 5.1 for our daycare data. In particular, the model must explicitly incorporate additional parameters to allow for nonzero means for factors and measures. We allow for nonzero means by including in the model a unit predictor, which is equal to 1 for all individuals and is denoted as a triangle in the path diagram. This predictor has no variance and therefore is sometimes called a pseudo-variable. The unit predictor may be included in equations for fac-

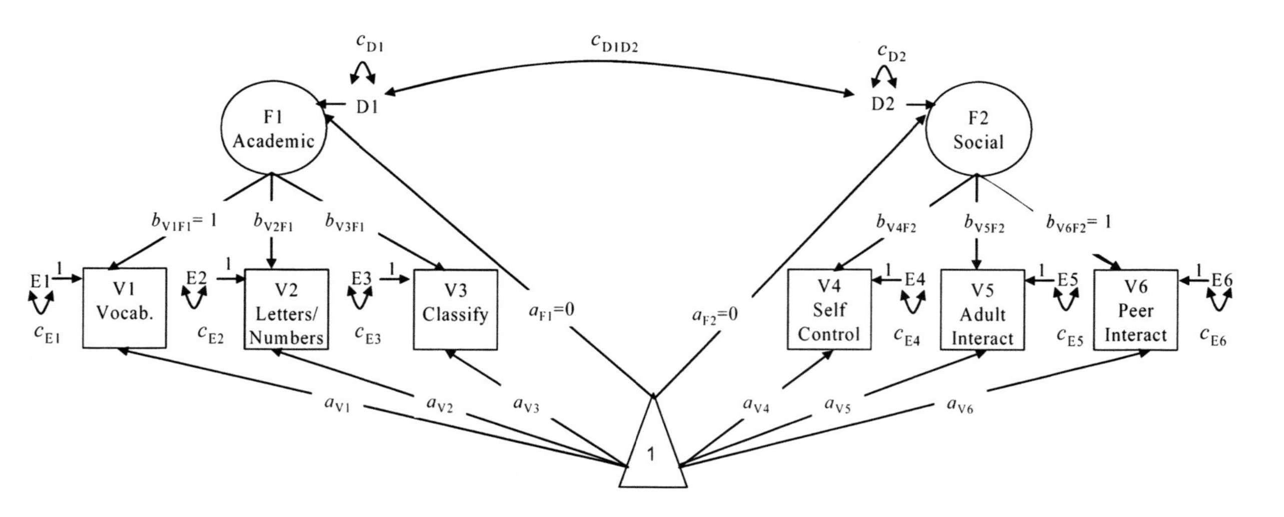

Parameter Estimates for Measured Variables

V#	$b_{V\#F1}$	$b_{V\#F2}$	$a_{V\#}$
V1	1		50.40
V2	.58		79.60
V3	.46		98.88
V4		.88	74.06
V5		.61	49.12
V6		1	120.10

Note. Error variances are not reported.

Parameter Estimates for Disturbances (D#) and Factors (#)

D# / F#	D1	D2	$a_{F\#}$
D1 / F1	79.24		0
D2 / F2	56.74	138.15	0

Figure 5.1. Results for one-group CFA with means for Dataset 1.

tors and measures, and the model parameters associated with this predictor are intercepts in that they are additive constants.

The addition of intercepts to the one-group CFA model allows us to incorporate the means of the measured variables. As we will see, mathematically these means must always be reproduced perfectly by the intercepts, and, thus, the intercepts part of the model is not open to empirical scrutiny (i.e., not falsifiable). In this sense, the addition of means to a one-group CFA adds no information over and above a one-group CFA without means. Our primary reason for presenting this model is pedagogical, as a building block to the multiple-group CFA with means.

Following the approach described by Bentler (1995), factors in a confirmatory factor model are treated as dependent variables rather than independent variables because they are a function of the unit predictor (as well as a disturbance term). For F1 in our example,

$$\text{F1} = a_{\text{F1}}1 + \text{D1}, \tag{3}$$

where 1 is the unit predictor, a_{F1} is the intercept for F1, and D1 is the disturbance term for F1. Given Equation 3 and assuming the mean of the disturbance is zero, the mean of the factor is a_{F1}:

$$\mu_{\text{F1}} = \mu_{a_{\text{F1}}1 + \text{D1}} = \mu_{a_{\text{F1}}1} + \mu_{\text{D1}} = a_{\text{F1}} + 0 = a_{\text{F1}}. \tag{4}$$

The factor intercept can be interpreted as the factor mean. As with the variance for a factor, the mean for a factor is arbitrary for a single-group analysis and must be defined by the researcher or the model will be underidentified. Analogous to fixing a factor variance to one, a mean for a factor may be defined by setting it to zero. As implied by Equation 4, a factor mean is set to zero by fixing the intercept for that factor to zero.

It should also be noted that the variances and covariances for factors are no longer model parameters because factors are now dependent variables (i.e., a function of a unit predictor and disturbances). Thus, factor intercepts, disturbance variances, and disturbance covariances are model parameters rather than factor means, variances, and covariances. However, factor intercepts are precisely the factor means, and the variances and covariances among the disturbances for the factors are equivalent to the variances and covariances among the factors (e.g., $c_{\text{D1}} = \sigma^2_{\text{F1}}$, $c_{\text{D2}} = \sigma^2_{\text{F2}}$, and $c_{\text{D1D2}} = \sigma_{\text{F1F2}}$). An alternative approach (see, e.g., Byrne, 1998) is not to make factors a function of the unit predictor and disturbances. Factors may then be treated as independent variables, and the means, variances, and covariances of factors are model parameters. Readers can easily translate our formulation of the structured means models to this alternative formulation by substituting the terms factor means, factor variances and factor covariances for factor intercepts, disturbance variances, and disturbance covariances, respectively.

So, in a one-group CFA with means, V1 through V6 are a function of not only factors and errors, but the unit predictor as well. For example, for the second measure,

$$V2 = a_{V2}1 + b_{V2F1}\,F1 + E2. \tag{5}$$

The intercept allows for the reproduction of the mean for V2. Recall that in order for the model to be identified, all factor intercepts are fixed at zero, and thus the factor means are set to zero. Given the mean of F1 is zero and the mean of E2 is assumed to be zero, the mean of V2 should be equal to its intercept:

$$\begin{aligned} \mu_{V2} &= \mu_{a_{V2}1 + b_{V2F1}F1 + E2} = \mu_{a_{V2}1} + b_{V2F1}\mu_{F1} + \mu_{E2} \\ &= a_{V2} + b_{V2F1}(a_{F1}) + 0 = a_{V2} + b_{V2F1}(0) = a_{V2} \end{aligned} \tag{6}$$

The means part of the model is just-identified. Mathematically, the intercepts for measures must be exactly equal to the means of the measures. These intercepts permit the model to reproduce perfectly the means of the measures, and consequently the means part of the model cannot produce lack of fit. An indication that the means part of the model is just-identified is that it contributes no degrees of freedom. For our example, the means part of the model contributes an additional six pieces of data—the means of V1 through V6—but requires an equal number of additional model parameters—the intercepts for V1 through V6. Because the means part of the model is just-identified, it adds no new information over and above a CFA excluding means of measures.

In Figure 5.1, we present the model parameters, including intercepts, for the day-care group data. As shown, we fixed the intercepts for F1 and F2 to be 0 and thus define the factor means to be 0. In addition, we defined the units for these factors as the units of V1 and V6 by fixing b_{V1F1} and b_{V6F2} to 1. Because the means portion of the model is just-identified, the estimated intercepts ($a_{V\#}$) for V1 through V6 must be identical to the means for V1 through V6 (as shown in Table 5.1). In addition, all parameter estimates and fit indices for the CFA without means are identical to those for the CFA with means. Although in practice there is little reason to include means in a one-group CFA, this step aids in understanding a multiple-group CFA with means.

Multiple-Group CFA With Means: The SMM Approach

We now consider how to evaluate differences in factor means using a multiple-group CFA approach. The data for a two-group CFA with means consists of means, variances, and covariances of measures for the two groups, as shown in Table 5.1 for our example. We assume that the form of the model is equivalent across groups, such that if analyzed separately each group would have the same model parameters and the same constraints imposed on these parameters (Bollen, 1989). For our example, the model specified in Figure 5.2 should fit for both the day-care and home-care groups. In addition, in this section, we also assume equivalence in the values of model parameters between groups, except for factor intercepts. The implication of these assumptions is that the covariance matrices among the measures are homogeneous. The factor intercepts are allowed to differ between groups so that the factor means—the focus of our analyses—can differ. EQS 6.1, SIMPLIS 8.5, and Mplus 2.14 computer programs for the SMM analyses are in Appendix B.

The model for the SMM approach. We could present the same path diagram showing the relationships between the factors and the measured variables for each group. However, because the structure of the model is the same for both groups, we present a single diagram for our example in Figure 5.2 and indicate notational differences in the parameters across groups. We allow for group differences in factor intercepts by using the presubscripts $G1$ and $G2$ for groups 1 and 2, respectively. The factor intercepts for groups 1 and 2 are ${}_{G1}a_{\mathrm{F1}}$ and ${}_{G2}a_{\mathrm{F1}}$ for F1 and ${}_{G1}a_{\mathrm{F2}}$ and for ${}_{G2}a_{\mathrm{F2}}$ for F2. We indicate that we constrain all other model parameters in the model to be equivalent across groups by not including presubscripts on those parameters.

A number of intercepts must be constrained in order to have an identified model. For our example, we have 12 means for the six measures in the two groups and therefore cannot estimate more than 12 intercept parameters for the means portion of the model. Only one of the intercepts for measures must be constrained to be equal between groups, but in this section we constrain all six of these intercepts. Similar to the one-group CFA with means, factor intercepts are factor means, and the factor intercepts in one group (arbitrarily defined as the first group) are constrained to zero so that the model is identified, regardless of what other constraints are imposed. However, the factor intercepts for the second group can be freely estimated. These factor intercepts are not only the factor means in the second group, but also differences in factor means between the second group and the first group because the factor means in the first group are equal to zero. In our example, the means portion of the model includes eight estimated parameters—six intercepts for the six

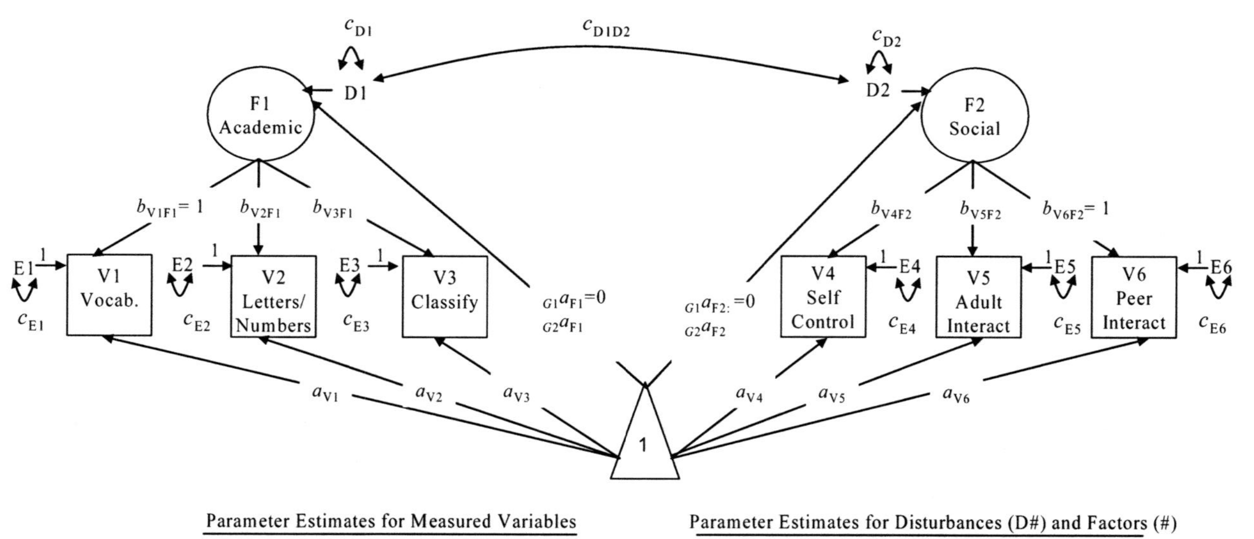

Parameter Estimates for Measured Variables

V#	$b_{V\#F1}$	$b_{V\#F2}$	$a_{V\#}$
V1	1		50.26
V2	.66		79.28
V3	.45		99.55
V4		.90	74.12
V5		.54	49.33
V6		1	119.80

Note. Error variances are not reported.

Parameter Estimates for Disturbances (D#) and Factors (#)

D# / F#	D1	D2	${}_{G1}a_{F\#}$	${}_{G2}a_{F\#}$
D1 / F1	76.59		0	3.59
D2 / F2	54.75	137.96	0	3.88

Figure 5.2. Results for structured means model for Dataset 1.

measures constrained equal across groups and two intercepts for the two factors in the second group. Given the 12 means for the six measures across the two groups, the means portion of the model contributes four degrees of freedom (= 12 means – 8 estimated parameters).

By now, we should have a good understanding of the intercept parameters. As in our example, the model must fit well in order to interpret the parameters. Although the intercepts for the measures are the same across groups, they represent the means for measures in the first group, that is, the group in which the factor intercepts are fixed to 0. Given the means of factors are constrained to zero in the first group and means of errors are assumed to be equal to zero, the intercept for a measure should be approximately equal to the mean for that measure in the first group. For V2 in the first group,

$$\begin{aligned} \mu_{G1}V2 &= \mu_{a_{V2}1 + (b_{V2F1})_{G1}F1 + E2} = \mu_{a_{V2}1} + b_{V2F1}\mu_{G1}F1 + \mu_{E2} \\ &= a_{V2} + b_{V2F1}({}_{G1}a_{F1}) + 0 = a_{V2} + b_{V2F1}(0) + 0 = a_{V2}. \end{aligned} \tag{7}$$

Because models generally do not fit perfectly, the intercepts for V1 through V6 are not exactly equal to means for these variables in the first group.

Given the intercepts for V1 through V6 represent the means for the measures in the first group, the only way to reproduce the means for the measures in the second group is through the factor intercepts (i.e., factor means) for the second group. For V2 in the second group,

$$\begin{aligned} \mu_{G2}V2 &= \mu_{a_{V2}1 + (b_{V2F1})_{G2}F1 + E2} = \mu_{a_{V2}1} + b_{V2F1}\mu_{G2}F1 + \mu_{E2} \\ &= a_{V2} + b_{V2F1}({}_{G2}a_{F1}) + 0 = a_{V2} + b_{V2F1}({}_{G2}a_{F1}), \end{aligned} \tag{8}$$

where a_{V2} is approximately equal to the mean for V2 in the first group. Therefore, given that the model is correct, any between-group difference in means on V2 is a function of ${}_{G2}a_{F1}$, the difference in means between groups for F1.

Assessment of factor mean differences. We want to interpret differences between factor means for our example, but can interpret these differences only if the model fits well. For our example, the model fits quite well, $\chi^2(33) = 41.78$, $p = .14$ SRMR = .05, RMSEA = .03 (90% C.I. .00 to .05).

If the model had not fit adequately, we would have needed to consider releasing some of the between-group constraints; we describe an approach for handling partial invariance in a later section. In this example, our model fits adequately and we can proceed to evaluating the factor mean differences. With two groups, the difference between group means for a factor is equal to the mean for this factor in the second group, given the mean for the same factor is fixed to zero in the first group. Consequently, for our example, the means are greater in the second group on both factors: 3.59 for F1 and 3.88 for F2. We can test whether the difference in population means for a factor is equal to zero by conducting a chi-square difference test for that factor. First, a second model must be fitted in which the intercept for the factor of interest is constrained to be equal to zero, the same as the intercept for the factor in the first group. Then the chi-square for the model without this constraint is subtracted from the chi-square for the model with this constraint. The difference is approximately distributed in large samples as a chi-square with degrees of freedom equal to the difference in the degrees of freedom for the two respective models. For our example, $\chi^2(34) = 53.44$, $p = .02$ for the model with equal intercepts for F1, and $\chi^2(34) = 49.04$, $p = .05$ for the model with equal intercepts for F2. The chi-square difference tests for the factor means are both significant at the .05 level: $\chi^2(34 - 33 = 1) = 53.44 - 41.78 = 11.66$, $p < .01$ for the first factor mean and $\chi^2(34 - 33 = 1) = 49.04 - 41.78 = 7.26$, $p < .01$ for the second factor mean. Alternatively, Wald z tests typically are reported by SEM packages to test the factor intercepts in the second group; the squared z statistics are estimates of the chi-square difference values. Although the Wald test is generally less accurate (Chou & Bentler, 1990), it is asymptotically equivalent to the chi-square difference test and is easier to obtain because it is not necessary to respecify the model. In this case, the Wald test yields very similar results: $z = 3.41$, $p < .01$ for the first factor mean and $z = 2.66$, $p < .01$ for the second factor mean (z^2 of 11.63 and 7.08 for F1 and F2, respectively).

Based on these results, we know the null hypothesis that factor means are equal across groups can be rejected for both factors. However, we do not know the strength of the effects of day-care versus home-care on school and academic readiness factors. Should we conclude the effect is greater on the social readiness factor because the mean difference was greater on this factor? To compute a standardized effect size for a factor, we divide the factor intercept for the second group by the square root of the variance of the disturbance for that same factor (Hancock, 2001). For our example, the standardized effect sizes are

$$\text{ES}_{\text{F1}} = \frac{G2a_{\text{F1}}}{\sqrt{c_{\text{D1}}}} = \frac{3.59}{\sqrt{76.59}} = .41 \text{ and } \text{ES}_{\text{F2}} = \frac{G2a_{\text{F2}}}{\sqrt{c_{\text{D2}}}} = \frac{3.88}{\sqrt{137.86}} = .33\,. \quad (9)$$

Based on these analyses, the standardized effect size is slightly greater on the academic readiness factor. The means for the home-care group were .41 factor standard deviations higher than the day-care group on academic readiness and .33 factor standard deviations higher on social readiness. Presumably, because factors are error free, the size of an effect should be larger than those found with measured variables. Accordingly, although these standardized effect sizes might be judged as small to moderate based on Cohen's (1988) guidelines, they might be considered small within a structured means analysis.

MIMIC Models

Comparable analyses can be conducted with the much simpler MIMIC model that does not require a multiple-group approach. With the MIMIC model, the measured variables include not only indicators of the factors, but also coded variables that differentiate among groups. The number of coded variables included in an analysis is equal to the number of groups minus one. With two groups, a single coded variable is included in the model that has direct effects on the factors, and those direct effects allow for differences in factor means. In our example, the coded variable is a 1-0 dummy-coded variable, although other types of coding could also be used. We arbitrarily coded 0 for all individuals in the day-care group and 1 for all individuals in the home-care group. As shown in Figure 5.3, the dummy-coded variable has direct effects on the readiness factors.

Before discussing the model further, we need to consider the format of the data. To conduct the analyses, scores on all variables from the multiple groups ($N_1 = 200$, $N_2 = 200$) are combined together in a single dataset ($N = 400$). For our example, the combined dataset includes scores on V1 through V6 (indicators of school readiness) as well as V7 (dummy-coded, child-care variable) for both the day-care and the home-care groups. A covariance matrix among the measured variables (e.g., V1 through V7) may then be computed, and, for our example, is shown at the bottom of Table 5.1.

It may seem confusing initially to be able to reach conclusions about factor mean differences by analyzing a covariance matrix. The covariance matrix contains two parts: (1) the variances and the covariances among the indicators of the factors and (2) the covariances between the dummy-coded variable and the indicators as well as the variance of dummy-coded variable. The covariance matrix among indicators is a total covariance matrix. It is a combination of the between-group covariance matrix among the means on the indicator variables and the within-group covariance matrix among indicator scores deviated about their means. No dis-

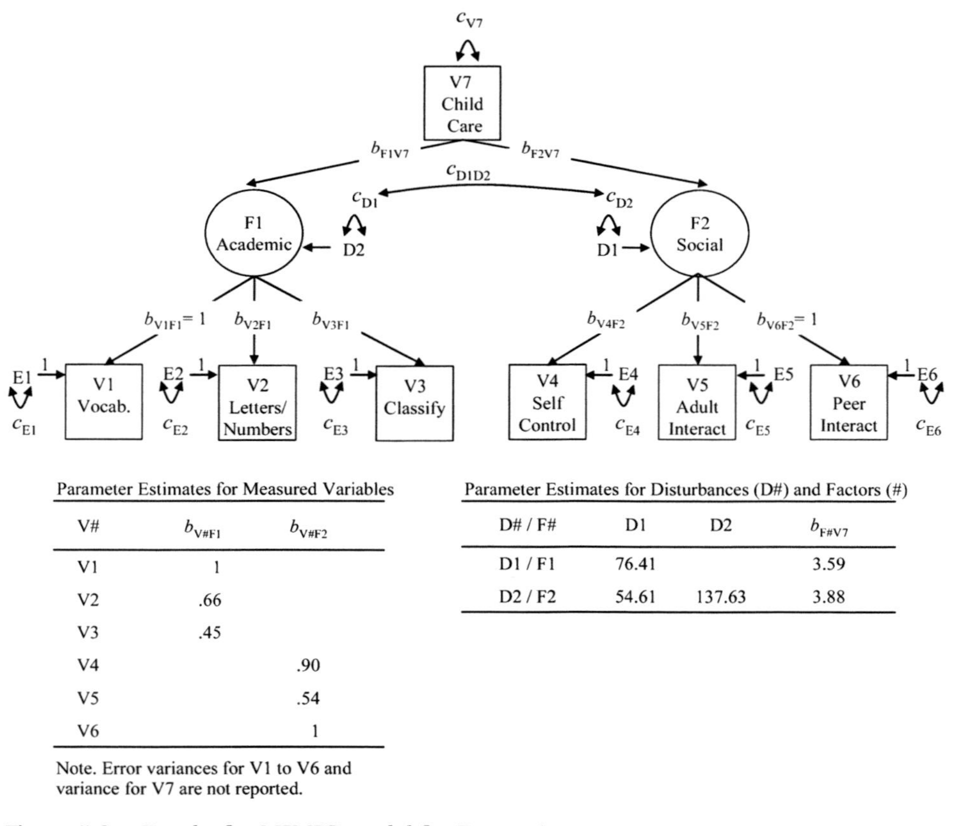

Parameter Estimates for Measured Variables

V#	$b_{V\#F1}$	$b_{V\#F2}$
V1	1	
V2	.66	
V3	.45	
V4		.90
V5		.54
V6		1

Note. Error variances for V1 to V6 and variance for V7 are not reported.

Parameter Estimates for Disturbances (D#) and Factors (#)

D# / F#	D1	D2	$b_{F\#V7}$
D1 / F1	76.41		3.59
D2 / F2	54.61	137.63	3.88

Figure 5.3. Results for MIMIC model for Dataset1.

tinction is necessary between the covariance matrices for the two groups (e.g., the within-day-care and within-home-care covariance matrices), as with the structured means model, because the MIMIC model assumes that the population covariance matrices are homogeneous and, therefore, can be pooled. Inclusion of the variance of the dummy-coded variable and the covariances between the dummy-coded variable and indicators allows the MIMIC analysis to differentiate between the between-group and within-group covariance matrices and, therefore, reach conclusions about differences in means.

The program for the MIMIC analysis is in Appendix B. The MIMIC model fits quite well, $\chi^2(12) = 15.24, p = .23$, SRMR = .03, RMSEA of .03 (90% C.I. of .00 to .06), and thus the parameter estimates may be interpreted. In addition, the standardized weights between the measured variables and the factors (not shown) are relatively high (i.e., greater than .50), allowing for a straightforward interpretation of the factors as academic and social readiness. Now to answer our question about differences in factor means, we interpret the unstandardized parameter estimates between V7 (dummy-coded variable) and F1 of 3.59 and between V7 and F2 of 3.88. Individuals in the home-care group (coded as 1) score, on average, 3.59 units higher on the academic readiness factor and 3.88 units higher on the social readiness factor in comparison with individuals in the day-care group (coded as 0). The Wald z tests are significant for both the mean difference on Factor 1, $z = 3.42, p < .01$ and on Factor 2, $z = 2.67, p < .01$. The asymptotically equivalent chi-square difference tests, with the path from the dummy-coded variable set to zero for each factor separately, yield similar results.

Based on the significance tests, the population factor means differ between the groups, but standardized effect size statistics are necessary to assess how strong the effects are. To compute a standardized effect size, we divide the unstandardized weight between the dummy-coded variables and the factor by the square root of the disturbance variance (Hancock, 2001). For our example, the standardized effect sizes are

$$ES_{F1} = \frac{b_{F1V7}}{\sqrt{c_{D1}}} = \frac{3.59}{\sqrt{76.41}} = .41 \text{ and } ES_{F2} = \frac{b_{F2V7}}{\sqrt{c_{D2}}} = \frac{3.88}{\sqrt{137.63}} = .33\,, \qquad (10)$$

and are interpreted as in the previous SMM analysis.

Although both the SMM and MIMIC models appear quite different on the surface, these models, as specified thus far, are very similar. They assume homogeneity of within-group population covariance matrices and include parameters to assess between-group mean differences. If the homogeneity assumption holds, the two approaches should yield similar results. For our example, both approaches yielded comparable parameter estimates, equivalent estimates of differences in factor means, and comparable tests of these differences.

EVALUATING DIFFERENCES IN FACTOR MEANS WHEN MODEL PARAMETERS VARY ACROSS GROUPS

We cannot reach valid conclusions about between-group differences in factor means without requiring some model parameters to be equal between groups (Meredith, 1993). As discussed previously, we have assumed to this point strict factorial equivalence, that is, between-group equivalence in factor loadings, intercepts for measures, error variances and covariances, and factor variances and covariances. However, with SMM models, it is unnecessary to require the variances and the covariances among the errors and the variances and the covariances among factors to be equivalent between groups. With SMM models, we would prefer strong factorial invariance, that is, between-group equivalence of factor loadings and intercepts for measures. Nonetheless, as illustrated by Byrne, Shavelson, and Muthén (1989), valid statistical inferences can be made under partial measurement invariance and partial intercept invariance. A minimum requirement for partial measurement invariance is that the loadings between each factor and one of its measures are equivalent between groups. Similarly, a minimum requirement for partial intercept invariance is that the intercepts for one measure for each factor are equivalent between groups. Although MIMIC models can accommodate differences in intercepts by including direct effects from the dummy-coded variable to measures, MIMIC models are more limited in their ability to handle partial invariance in contrast to the more flexible SMM models.

In this section, we consider methods for analyzing data when the set of model parameters is partially invariant. We again use the day-care versus home-care problem, but analyze a different dataset, as presented in Table 5.2. For these data, the numbers of observations for the day-care and the home-care groups are 250 and 150, respectively. Models requiring strict factorial invariance fail to fit these data and, thus, require respecification to allow for partial invariance.

Lack of Fit for Models Requiring Strict Factorial Invariance

For the dataset in Table 5.2, we begin by specifying a structured means model with all of its parameters constrained to be equal between groups except factor intercepts. The model fails to fit adequately, $\chi^2(33) = 263.03$, $p < .01$, SRMR = .12, and RMSEA = .13 (90% C.I. of .12 to .15).

Table 5.2. Covariance Matrices and Means for Measured Variables for the Day-Care Group, Home-Care Group, and Total—Dataset 2

Measured Variables	*Covariance Matrix*						
	V1	V2	V3	V4	V5	V6	*Means*
	Group 1—Day-Care Group ($N_1 = 250$)						
V1	154.54						49.14
V2	44.75	90.23					82.60
V3	40.98	22.77	78.76				104.95
V4	41.35	2.83	7.92	220.12			78.58
V5	23.28	9.12	.75	61.46	159.88		54.95
V6	47.08	22.88	16.05	125.08	84.31	332.26	119.91
	Group 2—Home-Care Group ($N_2 = 150$)						
V1	124.93						55.01
V2	52.19	80.67					81.22
V3	64.45	42.85	83.56				97.01
V4	59.95	33.10	38.34	290.65			72.67
V5	32.32	16.09	18.29	124.71	169.17		46.57
V6	87.15	39.70	51.82	174.20	108.39	355.22	124.28
	Total—Combined for Day-Care and Home-Care Groups ($N = 400$)						
	V1	V2	V3	V4	V5	V6	V7
V1	151.18						
V2	45.52	86.88					
V3	38.69	32.78	95.18				
V4	40.06	16.03	30.27	254.09			
V5	15.04	14.41	22.94	96.55	179.46		
V6	67.94	27.69	21.22	137.05	84.49	344.48	
V7	1.83	−.32	−1.87	−1.39	−1.97	1.03	.235

Note: V1: Vocabulary; V2: Letters/Numbers; V3: Classification; V4: Self-Control; V5: Adult interaction; V6: Peer interaction; V7: Day (0) versus home (1) care.

For the same dataset, the MIMIC model also fits poorly, $\chi^2(12) = 220.11$, $p < .01$, SRMR = .11, and RMSEA = .21 (90% C.I. of .18 to .23). Even though no constraints are directly imposed on the MIMIC model, the model implicitly assumes between-group constraints of model parameters and, thus, fails to fit our example data that are consistent with a model with partially invariant model parameters. Interestingly, both models yielded very similar parameter estimates, presumably because both assume strict factorial invariance.

Stepwise Method for Testing Differences in Factor Means Under Partial Invariance

Up to this point, we have considered multiple-group SMM analyses with a maximal number of between-group equalities. However, in practice researchers who use the SMM approach must make decisions only about between-group constraints on the factor loadings and intercepts for measures. In Table 5.3, we present a stepwise approach for assessing differences in factor means that incorporates decisions about these between-group constraints. We illustrate this stepwise approach in Table 5.4 using the second dataset.

Some overall comments about Tables 5.3 and 5.4 are useful before we discuss individual steps. First, Table 5.3 describes general procedures for analyzing between-group factor mean differences for two or more groups, although multiple comparison methods are not presented due to space limitations. Second, we previously defined the metric of a factor by fixing to 1 the loading between that factor and one of its measures in all groups. In Table 5.3, we define the metric of a factor by fixing the variance of that factor to 1 in one group and imposing between-group equality constraints of factor loadings (except in the first step where no between-group constraints are imposed). This latter procedure allows for the evaluation of between-group constraints on all factor loadings. Third, guidelines are presented to help researchers make data-driven decisions about what parameters should be constrained. Except for step 1, these decisions are based on the comparative fit of nested models. Just like the chi-square test for assessing fit of a model, the chi-square difference test for assessing differential fit of nested models is strongly influenced by sample size. Accordingly, differential fit should be assessed using not only the chi-square difference test, but also other fit indices, such as SRMR and RMSEA. Fourth, decisions about whether to constrain parameters based on fit indices are subjective, and different researchers can reach different decisions. Fifth, some confusion may occur in comparing fit of models in that sometimes models become more complex as analyses proceed from earlier to later steps, and sometimes models become less complex. Sixth, Lagrange multiplier (LM) tests, sometimes referred to as modification indices (MI), are used to assess what between-group constraints on factor loadings or intercepts of measures should be removed to maximize improvement in fit (Chou & Bentler, 1990). It is advantageous, when possible, to be less dependent on these empirical search methods and rely more on theory and past empirical findings to make these decisions. Next, we briefly describe the stepwise approach for assessing differences in factor means.

In step 1, we assess whether the same factor model fits adequately in all groups. If it does not, we cannot proceed. In addition, we assess whether

Table 5.3. Steps for Assessing Between-Group Differences in Factor Means

Steps	*Model Specification*	*Decision Rule*
Step 1. Evaluate same factor model in each of the groups and combined groups.	Factor variances and covariances • Fix variances of all factors to 1 in all groups. • Freely estimate all factor covariances in all groups. Factor loadings • Freely estimate loadings for all measures in all groups. Intercepts for factors and measures • For this step, means of measures are unnecessary and not analyzed. Thus, intercepts are not included in model.	Fit of model for step 1 within groups and cumulatively across groups is • Adequate. Factor model is appropriate for each group. Proceed to step 2a. • Inadequate. Factor model fits inadequately for at least one of the groups. Cannot proceed unless a model fits adequately in all groups.
Step 2a. Evaluate between-group equivalence of factor loadings.	Factor variances and covariances • Fix variances of all factors in one group to 1. • Freely estimate factor variances in all other groups. • Freely estimate factor covariances in all groups. Factor loadings ➢ Constrain all factor loadings to be equal between groups. Intercepts for factors and measures • Same as model for step 1.	Difference in fit between models for steps 1 and 2a is • Minimal. All factor loadings are considered essentially equivalent between groups and are constrained to be equal in all subsequent steps. Proceed to step 3. • Considerable. Presumably, factor loadings for one or more measures differ between groups. Proceed to step 2b.
Step 2b. Evaluate between-group equivalence of individual loadings. Multiple substeps involved in this step.	Factor variances and covariances • Same as model for step 2a. Factor loadings • Allow individual loadings to vary between groups if loadings were found in previous substeps of 2b to be nonequivalent between groups. ➢ Choose the between-group constraint on loadings between a particular factor and measure that would maximize an increase in fit if removed. This constraint—the focal constraint—is removed at this step. The focal constraint may be selected based on LM/MI tests of between-group constraints on loadings from the previous substep of 2b or step 2a. • Constrain all remaining factor loadings to be equal between groups. Intercepts for factors and measures • Same as model for step 1.	Difference in fit between model of current substep of 2b and model of previous substep of 2b (or model for step 2a if 1st substep of 2b) • Minimal. Loadings of focal constraint are considered essentially equivalent between groups. Therefore, this constraint is reimposed. All other constraints on loadings that have not been removed before this substep are maintained. Proceed to step 3a. • Considerable. Loadings of focal constraint differ between groups. Therefore, this between-group constraint is not imposed in future substeps or steps. Proceed to assess remaining between-group constraints not evaluated at previous substeps, unless only one constraint on loadings remains per factor. If only one constraint remains per factor, must assume the remaining constraint(s) on loadings are appropriate and proceed to step 3a.

(Table continues)

Table 5.3. (Continued)

Steps	*Model Specification*	*Decision Rule*
Step 3a. Evaluate between-group equivalence of intercepts for measures.	Disturbance variances and covariances[a] • Fix variances of all disturbances to 1 in one group. • Freely estimate disturbance variances in all other groups. • Freely estimate disturbance covariances in all groups. Factor loadings • Allow individual loadings to vary between groups if loadings were found to be nonequivalent between groups in step 2b. • Constrain all other factor loadings to be equal between groups. Intercepts for measures • Allow individual intercepts for measures to vary between groups if the loadings for these measures are allowed to vary for model of step 3a. ➢ Constrain all remaining intercepts for measures to be equal between groups. Intercepts for factors[b] • Fix intercepts for factors to 0 in one group. • Allow intercepts for factors in all other groups to be freely estimated.	Difference in fit between final model from step 2 and model from step 3a. • Minimal. All constrained intercepts are considered essentially equivalent between groups and are constrained to be equal in all subsequent steps. Proceed to step 4. • Considerable. Presumably, one or more of the constrained intercepts for measures differ between groups. Proceed to step 3b.
Step 3b. Evaluate between-group equivalence of individual intercepts for measures. Multiple substeps involved in this step.	Disturbance variances and covariances[a] • Same as model for step 3a. Factor loadings • Same as model for step 3a. Intercepts for measures • Allow individual intercepts for measures to vary between groups if the loadings for these measures are allowed to vary for model of step 3b. ➢ Choose a between-group constraint on intercepts for a measure that would maximize an increase in fit if removed. This constraint—the focal constraint—is removed at this step. The focal constraint may be selected based on LM/MI tests of between-group constraints on intercepts from the previous substep of 3b or step 3a. • Constrain all remaining factor intercepts to be equal between groups. Intercepts for factors[b] • Same as model for step 3a	Difference in fit between model of current substep of 3b and model of previous substep of 3b (or model of step 3a if 1st substep). • Minimal. Intercepts of focal constraint are considered essentially equivalent between groups. Therefore, this constraint is reimposed. All other constraints on intercepts for measures that have not been removed before this substep are maintained. Proceed to step 4. • Considerable. Intercepts of focal constraint differ between groups. Therefore, this between-group constraint is not imposed in future substeps or steps. Proceed to assess remaining between-group constraints on intercepts for measures that were not evaluated at previous substeps, unless only one constraint on intercepts remains per factor. If only one constraint remains per factor, must assume the remaining constraint(s) on intercepts are appropriate and proceed to step 4a.

Step 4a. Evaluate between-group equivalence of intercepts for factors (i.e., factor means).	Disturbance variances and covariances[a] • Same as model for step 3a. Factor loadings • Same as model for step 3a. Intercepts for measures • Allow individual intercepts for measures to vary between groups if the loadings for these measures are allowed to vary for model of step 4a. • Allow individual intercepts to vary between groups if intercepts were found to be nonequivalent between groups in step 3b. • Constrain all other intercepts to be equal between groups. Intercepts for factors[b] ➢ Fix intercepts for all factors to be equal to 0 in all groups.	Difference in fit between final model from step 3 and model from step 4a. • Minimal. All constrained intercepts for factors are considered essentially equivalent between groups. Accordingly, factor means are considered essentially equivalent. Need not conduct any further analyses. • Considerable. Presumably, one or more of the constrained intercepts for factors differ between groups. Proceed to step 4b.
Step 4b. Evaluate between-group equivalence of individual intercepts for factors (i.e., factor means). Multiple substeps involved in this step.	Disturbance variances and covariances[a] • Same as model for step 3a. Factor loadings • Same as model for step 4a. Intercepts for measures • Same as model for step 4a. Intercepts for factors[b] • Fix intercepts for factors to 0 in one group. ➢ At the *J*th substep, fix intercept(s) for the *J*th factor to 0 in all other groups. Allow intercepts for all other factors in all other groups to be freely estimated.	Difference in fit between final model from step 3 and model from current substep of 4b. • Minimal. Intercepts (i.e., means for factor) for the *J*th factor are considered essentially equivalent between groups. • Considerable. Intercepts for the *J*th factor (i.e., means for the *J*th factor) differ between groups. Proceed to the next substep, unless at the current step the intercepts for the last factor were evaluated.

Note: The description of the steps presumes that the factor model of interest has no within-group constraints imposed on the error variances, factor variances, and factor covariances other than those necessary to identify the metric of the factor and includes no error covariances. The steps can be easily modified if this supposition is incorrect. The symbol ➢ indicates the constraint(s) assessed in the step or substep.

[a]The variances and covariances among the disturbance terms are the variances and covariances among the factors. With some SEM programs, factor variances and covariances are directly specified rather than disturbance variances and covariances.

[b]The factor intercepts are the factor means. In some SEM programs, the factor means are directly specified.

Table 5.4. Results for Model Comparisons to Assess Between-Group Differences in Factor Means (Dataset 2)

Steps	*Results*[a]	*Decision Rule*
Step 1. Evaluate same factor model in each of the groups and combined groups.	*Model for step 1* Group 1: $\chi^2(8) = 8.02, p = .43$, SRMR = .031, RMSEA = .003 Group 2: $\chi^2(8) = 9.77, p = .28$, SRMR = .045, RMSEA = .039 Combined groups: $\chi^2(16) = 17.79, p = .34$, SRMR = .039, RMSEA = .017	Fit of model for step 1 is better in group 1 than group 2, but very good in both groups. Proceeded to step 2a.
Step 2a. Evaluate between-group equivalence of factor loadings.	*Model for step 2a* $\chi^2(20) = 26.07, p = .16$, SRMR = .048, RMSEA = .028 (LM/MI test: Largest χ^2 was for loadings for V3: $\chi^2(1) = 4.12$.) *Difference in fit between models for steps 1 and 2a* $\chi^2(4) = 8.28, p = .08$, SRMR = + .009, RMSEA = + .011	Difference in fit between models for steps 1 and 2a might be considered minimal by some and more than minimal by others. We considered the change in fit non-minimal and proceeded to step 2b to determine the constraints on the loadings that produced the lack of fit.
Step 2b. Evaluate between-group equivalence of individual loadings. Multiple substeps involved in this step.	*Substep 1* *Model for 1st substep of 2b—Focal constraint on loadings for V3* $\chi^2(19) = 22.04, p = .28$, SRMR = .043, RMSEA = .020 (LM/MI test: Largest χ^2 was for loadings for V6: $\chi^2(1) = 2.59$.) *Difference in fit between models for 1st substep of 2b and step 2a* $\chi^2(1) = 4.03, p = .04$, SRMR = + .005, RMSEA = + .008	*Substep 1* The LM/MI test from step 2a suggested that fit could be maximally improved if the constraint on the loadings for V3 was removed. In substep 1, we allowed the loadings for V3 to differ between groups. The improvement in fit was marginal. The loadings for V3 differed: 4.37 and 6.65 for the groups 1 and 2, respectively. To avoid misspecification, we decided to allow the loading for V3 to differ in all future analyses.

	Substep 2 *Model for 2nd substep of 2b—Focal constraint on loadings for V6* $\chi^2(18) = 19.16, p = .38$, SRMR = .041, RMSEA = .013 (LM/MI test: Largest χ^2 was for loadings for V1 or V2: $\chi^2(1) = 1.14$) *Difference in fit between models for substeps 1 and 2 of step 2b* $\chi^2(1) = 2.88, p = .09$, SRMR = + .002, RMSEA = + .007	*Substep 2* The LM/MI test from 1st substep of 2b suggested that fit could be maximally improved if the constraint on the loadings for V6 was removed. In substep 2, we allowed the loadings for V6 to differ between groups. The fit improved, but less than it did for substep 1. The loadings for V6 differed: 13.21 and 9.26 for groups 1 and 2, respectively. Again, to avoid misspecification, we decided to allow the loading for V6 to differ in all future analyses.
	Substep 3 *Model for 3rd substep of 2b—Focal constraint on loadings for V1* $\chi^2(17) = 17.86, p = .40$, SRMR = .039, RMSEA = .011 *Difference in fit between models for substeps 2 and 3 of step 2b* $\chi^2(1) = 1.30, p = .25$, SRMR = + .002, RMSEA = + .002	*Substep 3* The LM/MI test from 2nd substep of 2b suggested that fit could be maximally improved if the constraint on the loadings for V1 or V2 were removed. The choice was arbitrary. We report the results when we removed the constraint for V1. The fit improved minimally. The loadings for V1 differed: 9.96 and 7.24 for groups 1 and 2, respectively. Given these results, we decided to constrain the loadings for V1 to be equal between groups and continued to step 3a.
Step 3a. Evaluate between-group equivalence of intercepts for measures.	*Model for step 3a* $\chi^2(20) = 54.44, p < .01$, SRMR = .053, RMSEA = .066 (LM/MI test: Largest χ^2 was for intercept for V1 or V2: $\chi^2(1) = 23.41$.) *Difference in fit between models for substep 2 of 2b and 3a* $\chi^2(2) = 35.28, p < .01$, SRMR = + .012, RMSEA = + .053	Difference in fit between models for substep 2 of 2b and step 3a was substantial. We proceeded to step 3b to determine the constraints on the intercepts that produced the lack of fit.

(Table continues)

Table 5.4. (Continued)

Steps	*Results*[a]	*Decision Rule*
Step 3b. Evaluate between-group equivalence of individual intercepts for measures. Multiple substeps involved in this step.	*Substep 1* *Model for 1st substep of 3b—Focal constraint on intercepts for V1* $\chi^2(19) = 29.43, p = .19$, SRMR = .047, RMSEA = .037 (LM/MI test: Largest χ^2 was for intercepts for V4 or V5: $\chi^2(1) = 10.20$.) *Difference in fit between models for 1st substep of 3b and step 3a* $\chi^2(1) = 25.01, p < .01$, SRMR = + .006, RMSEA = + .029	*Substep 1* The LM/MI test from step 3a suggested that fit could be maximally improved if the constraint on the intercepts for V1 or V2 was removed. The choice was arbitrary. In substep 1, we allowed the loadings for V1 to differ between groups. The improvement in fit was considerable. The intercepts for V1 differed: 49.14 and 57.41 for groups 1 and 2, respectively. We allowed the intercepts for V1 to differ in all future analyses.
	Substep 2 *Model for 2nd substep of 2b—Focal constraint on loadings for V4* $\chi^2(18) = 19.16, p = .38$, SRMR = .041, RMSEA = .013 *Difference in fit between models for substeps 1 and 2 of step 3b* $\chi^2(1) = 10.27, p < .01$, SRMR = + .006, RMSEA = + .024	*Substep 2* The LM/MI test from 1st substep of 3b suggested that fit could be maximally improved if the constraint on the intercepts for V4 or V5 was removed. The choice was arbitrary. In substep 2, we allowed the intercepts for V4 to differ between groups. The fit improved. The intercepts for V4 differed: 78.58 and 85.62 for groups 1 and 2, respectively. We chose to allow the intercepts for V4 to differ in all future analyses. Only one between-group intercept constraint remained per factor. We had to assume that these constraints were appropriate and proceeded to step 4a.
Step 4a. Evaluate between-group equivalence of intercepts for factors (i.e., factor means).	*Model for step 4a* $\chi^2(20) = 56.39, p < .01$, SRMR = .061, RMSEA = .068 *Difference in fit between models for substep 2 of 3b and 4a* $\chi^2(2) = 37.23, p < .01$, SRMR = + .020, RMSEA = + .055	Difference in fit between models for substep 2 of 3b and step 4a was substantial. We proceeded to step 4b to determine the constraints on the factor intercepts in group 2 that produced the lack of fit.

Step 4b. Evaluate between-group equivalence of individual intercepts for factors (i.e., factor means). Multiple substeps involved in this step.	*Substep 1* *Model for 1st substep of 4b—Test of intercepts for F1* $\chi^2(19) = 21.28, p = .32$, SRMR = .041, RMSEA = .017 *Difference in fit between models for substep 1 of 4b and substep 2 of 3b* $\chi^2(1) = 2.12, p = .15$, SRMR = .000, RMSEA = + .004	*Substep 1* The improvement in fit was minimal. Based on the final model from step 3 (model for substep 2 of 3b), the intercepts for F1 differed: 0.00 and -.26 for groups 1 and 2, respectively. The standardized effect size was -.26; that is, the mean for the second group was .26 standard deviations lower than the mean for the first group. We cannot conclude that the means on the 1st factor differed in the population.
	Substep 2 *Model for 2nd substep of 4b—Test of intercepts for F2* $\chi^2(19) = 56.17, p < .01$, SRMR = .060, RMSEA = .070 *Difference in fit between models for substep 2 of 4b and substep 2 of 3b* $\chi^2(1) = 37.01, p < .01$, SRMR = + .019, RMSEA = + .057	*Substep 2* The improvement in fit was substantial. Based on the final model from step 3 (model for substep 2 of 3b), the intercepts for F2 differed: 0.00 and -1.35 for groups 1 and 2, respectively. The standardized effect size was -1.15; that is, the mean for the second group was 1.15 standard deviations lower than the mean for the first group. We can conclude that the means on the 2nd factor differed substantially in the population.

[a] Three global fit indices were computed: chi-square (χ^2), standardized root mean square residual (SRMR), and root mean square error of approximation (RMSEA). For any one of these indices, differences were computed by subtracting the global fit index for the more complex model from the global index for the less complex (nested) model. Greater positive values of differences indicate poorer fit of the less complex model.

the factor model fits adequately across all samples by conducting a multiple-group analysis with no between-group constraints on the factor model. The fit from the multiple-group analysis for this step is necessary to compare it with the model fit at the next step. Note that in this first step, as well as in step 2, only the covariance matrix is analyzed. Means are excluded from the analyses and intercepts are not included in the model.

In step 2a, we assess whether factor loadings are equivalent between groups by comparing the fit of models from steps 1 and 2a, including conducting a chi-square difference test. A model with between-group constraints for all loadings (step 2a) is compared with a model with no between-group constraints (step 1). However, a greater number of constraints are imposed on the factor variances for the model in step 1 than the model for step 2. Consequently, it is not obvious why a chi-square difference test can be applied in that the model for step 2a is not nested within the model for step 1. However, the models for steps 1 and 2a have equivalent models, as presented in Appendix C, and the equivalent model for step 2a is nested within the equivalent model for step 1.

If the model with between-group constraints for all loadings (step 2a) is rejected, a search is conducted in step 2b to evaluate which loadings between factors and measures can be freely estimated. It is advantageous to maintain at least two between-group constraints on loadings for each factor. For any one factor, a loading for one measure must be constrained to be equivalent between groups in order to have an identified model. In addition, as shown in substep 2 of 2b in our example (i.e., Table 5.4), when only two between-group constraints were imposed on loadings per factor, the LM/MI tests were not informative and produced the same chi-square value for both loadings.

Steps 3a and 3b are analogous to steps 2a and 2b, except in steps 3a and 3b between-group constraints of intercepts of measures are assessed. The final model of step 2 and the model of step 3a can be compared using a chi-square difference test because these two models have equivalent models that are nested within each other (see Appendix C). For any one factor, an intercept for one measure must be constrained to be equivalent between groups in order to have an identified model. Also as shown in step 3a and substep 1 of 3b in our example (i.e., Table 5.4), when two between-group constraints are imposed on intercepts per factor, the LM/MI tests were not informative and produced the same chi-square value for both intercepts.

We briefly consider how between-group differences in intercepts for measures affect between-group differences in measures under three conditions. First, if the factor loadings and intercepts for a measure are equivalent between groups, the difference in means for this measure reflects the difference in factor means and not intercepts. In this condition, a difference in means for a measure is unaffected by irrelevant sources (i.e., sources other than the underlying factor), except for sampling error. In our example, V2 meets this condition and ${}_{G2}\mu_{V2} - {}_{G1}\mu_{V2} = b_{V2F1}({}_{G2}\mu_{F1} - {}_{G1}\mu_{F1})$. Second, if factor loadings for a measure are equivalent between groups, but intercepts differ, the difference in means on that measure is a function of not only the difference in factor means, but also the difference in intercepts. In our example, V1 meets this condition and ${}_{G2}\mu_{V1} - {}_{G1}\mu_{V1} = b_{V1F1}({}_{G2}\mu_{F1} - {}_{G1}\mu_{F1}) + ({}_{G2}a_{V1} - {}_{G1}a_{V1})$. In this condition, some source other than the factor is causing differences in means for a measure, and, in this sense, the measure is biased. Third, if a measure has different factor loadings in different groups, the intercepts for that measure almost certainly would differ too. The differences in means for a measure are complexly determined and not reflective of the differences in factor means. In our example, V3 meets this condition and ${}_{G2}\mu_{V3} - {}_{G1}\mu_{V3} = [{}_{G2}b_{V3F1}({}_{G2}\mu_{F1}) - {}_{G1}b_{V3F1}({}_{G1}\mu_{F1})] + ({}_{G2}a_{V3} - {}_{G1}a_{V3})$.

In step 4, we evaluate the differences in factor means between groups. To help interpret the magnitude of the differences in factor means, standardized effect sizes are computed (Hancock, 2001). However, standardized effect sizes were calculated using a pooled estimate of the standard deviation of disturbances for a factor (equivalent to the factor standard deviation) because the disturbance variances were not constrained to be equal between groups. For the first factor, the standardized effect is

$$\mathrm{ES}_{F1} = \frac{{}_{G2}a_{F1}}{\sqrt{[({}_{G1}N){}_{G1}c_{D1} + ({}_{G2}N){}_{G2}c_{D1}]/({}_{G1}N + {}_{G2}N)}} \tag{11}$$

$$= \frac{-.26}{\sqrt{[(250)1 + (150)1.095]/(250 + 150)}} = -.255$$

The mean for factor 1 is .255 factor standard deviations (pooled) smaller for group 2 than for group 1.

The sign and magnitude of the differences in factor means are dependent on the decisions made in steps 2 and 3, and unfortunately these decisions can be ambiguous. As presented in step 3 in Table 5.4, it was unclear whether the intercepts of V1 or V2 should be constrained to be equal between groups for F1 and whether the intercepts of V4 or V5 should be constrained to be equal for F2. In Table 5.5, we show that the choice among these intercept constraints had an effect on the estimates of the differences in means for F1 and F2. In particular, if the constraints are imposed on V2 and V5 (model *A*), the means on the academic and the social factors (F1 and F2, respectively) were greater in the day-care group

Table 5.5. SMM Results for Models with Various Between-Group Constraints on the Intercepts for Measured Variables

	Group 1			*Group 2*			*Between-Group Difference*		
	Loadings		*Intercepts*	*Loadings*		*Intercepts*	*Means*	*p-value*	*Effect Size*
	F1	F2		F1	F2				
			Model *A* with intercept constraints for V2 and V5						
V1	9.06		49.14	9.06		57.41			
V2	5.20		82.60	5.20		82.60			
V3	4.39		104.95	6.66		98.78			
V4		9.62	78.58		9.62	85.62			
V5		6.23	54.95		6.23	54.95			
V6		13.21	119.91		9.26	136.74			
F1			0			–.26	–.26	.145	–.26
F2			0			–1.35	–1.35	<.001	–1.15
			Model *B* with intercept constraints for V1 and V4						
V1	9.06		49.14	9.06		49.14			
V2	5.20		82.60	5.20		77.85			
V3	4.39		104.95	6.66		92.70			
V4		9.62	78.58		9.62	78.58			
V5		6.23	54.95		6.23	50.40			
V6		13.21	119.91		9.26	129.97			
F1			0			+.65	+.65	<.001	+.64
F2			0			–.61	–.61	<.001	–.52
			Model *C* with intercept constraints for V1, V2, V4, and V5						
V1	10.06		49.64	10.06		49.64			
V2	4.41		81.08	4.41		81.08			
V3	4.14		104.95	6.76		93.70			
V4		9.03	79.43		9.03	79.43			
V5		6.82	53.96		6.82	53.96			
V6		13.27	119.91		9.68	132.92			
F1			0			+.49	+.49	<.001	+.49
F2			0			–.89	–.89	<.001	–.77

Note: Fits for models *A* and *B* were identical: $\chi^2(18) = 19.16, p = .38$, SRMR = .041, RMSEA = .013. For model *C*, $\chi^2(20) = 54.44, p < .01$, SRMR = .053, RMSEA = .066.

than in the home-care group. However, the effect on F1 was small (ES_{F1} = – .26) and nonsignificant at the .05 level, while the effect on F2 was large (ES_{F2} = – 1.35) and significant. In contrast, when constraints were imposed on V1 and V4 (model *B*), the mean on the academic factor was higher in the home-care than the day-care group, while the mean on the social factor was higher in the day-care group. For both factors, the differences were significant and had a moderate standardized effect size ($ES_{F1} = +.64$ and $ES_{F2} = -.52$). Although models *A* and *B* produced very different results in terms of differences in factor means, they are equivalent models and, therefore, have identical fit. Accordingly, the choice between models cannot be made based on model fit.

Estimates of factor mean differences are dissimilar for models *A* and *B* because the estimates for the two models rely on different measured variables. For simplicity, we focus on F1 to illustrate that the estimates of factor mean differences rely on the measures with constrained intercepts. For model *A*, the intercepts for V2 are constrained to be equal, and thus the estimate of the difference in factor means is a function of V2. The equations for the reproduced means for V2 are identical in groups 1 and 2 except for the factor means of the two groups:

$$_{G1}\overline{V2}_{\text{Reproduced}} = (5.20)\ _{G1}\overline{F1} + 82.60$$

$$_{G2}\overline{V2}_{\text{Reproduced}} = (5.20)\ _{G2}\overline{F1} + 82.60$$

Thus, the difference in reproduced means of V2 is

$$_{G2}\overline{V2}_{\text{Reproduced}} - {}_{G1}\overline{V2}_{\text{Reproduced}} = (5.20)(_{G2}\overline{F1} - {}_{G1}\overline{F1}).$$

The model accurately reproduced the difference in sample means for V2 of −1.38 (81.22–82.60) by estimating the difference in factor means to be −.265. In contrast, for model *B*, the intercepts for V1 are constrained to be equal, and thus the estimation of the difference in factor means is reliant on V1. Because the equations for the reproduced means for V1 are identical in groups 1 and 2 except for the factor means of the two groups, the difference in reproduced means for V1 is

$$_{G2}\overline{V2}_{\text{Reproduced}} - {}_{G1}\overline{V2}_{\text{Reproduced}} = (9.06)(_{G2}\overline{F1} - {}_{G1}\overline{F1})$$

Model *B* accurately reproduced the difference in sample means for V1 of 5.87 (55.01–49.14) by estimating the difference in factor means to be +.648. In summary, the factor mean was in favor of group 1 for model *A* because the mean on V2 was greater in that group, while the factor mean was in favor of group 2 for model *B* because the mean on V1 was greater in that group.

For models *A* and *B*, we rely on a single measure to estimate the difference in factor means. It seems risky to base an estimate on a single measure, given the possibility for misspecification and the variation in estimates of factor mean differences. An alternative might be to impose between-group constraints on intercepts for more than one measure. As shown in model *C*, we could impose between-group constraints on the intercepts of V1 and V2 for F1 and intercepts of V5 and V6 for F2. For our data, model *C* does not fit as well as models *A* and *B*, and on these grounds might be abandoned. On the other hand, the differences in factor means for model *C* are somewhere between those for models *A* and *B*.

It is crucial to have a correctly specified model so that estimates of mean differences on factors are unbiased. From this perspective, it makes sense to minimize the number of constraints. On the other hand, to the extent that a minimal number of constraints are imposed, we are essentially making decisions about mean differences for a factor based on a single measure. If the model is misspecified, however, the mean differences might be very biased. Researchers can avoid this paradox by carefully designing studies that have multiple measures with good construct validity for each factor. If the chosen measures are valid for all groups, they should demonstrate comparable loadings and intercepts in the stepwise SMM analysis, and the differences in factor means will be based on multiple measures.

ACCOMMODATING MORE COMPLEX RESEARCH DESIGNS

Designs With More Than Two Groups

Our examples have illustrated the MIMIC and SMM approaches for the two-group case with populations distinguished by a single grouping variable; traditionally, in ANOVA/MANOVA language, such a design might be called a one-way design with two groups. Extending the application of either the MIMIC or SMM approaches to three or more groups that differ according to a single dimension is relatively straightforward. With the MIMIC model, we specified group membership for the two-group case using a single dummy-coded variable and tested the coefficient relating the coded variable to a latent variable to infer whether the two populations differ on the latent means. More generally, we could use $J - 1$ dummy-coded variables simultaneously to evaluate differences among J groups in the population. The omnibus hypothesis that the population means for a factor are equal between groups may be tested by comparing the fit of a model in which the coefficients between all coded variables and the factor of interest are freely estimated with a model in which these parameters are constrained to zero. Tests of pairwise differences between factor means can be made by evaluating whether a particular coefficient between a coded variable and a factor is equal to zero or whether two coefficients are equal.

Methods used in MIMIC analyses with a single grouping variable can be extended to assess main and interaction effects in designs with multiple grouping variables (i.e., higher-way or factorial designs). Following analogous procedures used to conduct analysis of variance with regression methods, coded variables are included in the model for each grouping variable as well as cross-product terms between sets, and tests are conducted for particular hypotheses by constraining appropriate parameters. Of course, researchers should keep in mind the restrictive assumption implied by the MIMIC model; that is, parameters of models must be identical across all population groups.

The SMM approach described in Table 5.3 can be applied to evaluate differences in factor means among three or more groups. Evaluation of between-group constraints is more complex under these conditions than for the two-group case because parameters may vary with respect to some subset of the groups. However, the additional required steps are generalizations of the ones provided. For designs with multiple grouping variables, further research is necessary to determine how to specify models using the SMM approach to test main and interaction effects.

Inclusion of Manifest and Latent Covariates

The MIMIC and SMM approaches can readily be adapted to assess differences in factor means after controlling for one or more covariates. SEM can accommodate covariates specified either as measured variables or as factors based on multiple indicators. For example, we could examine differences between the home-care and day-care groups on the academic readiness factor, controlling for differences in social readiness—a latent covariate. As discussed earlier, a factor is a theoretically error-free representation of a construct so may provide a better way in which to operationalize a covariate if multiple indicators of the covariate are available. Inclusion of covariates can reduce the residual variance of the factors within groups and adjust factor means based on differences on the covariate.

Within a MIMIC approach, a manifest or latent covariate is included in the model along with the group-coded variable(s) as predictors of the factors. A covariance is included between the exogenous covariate and the group-coded variable. Evaluation of the difference in the population factor means, adjusted for the covariate, can then be made using the same parameters as before—the path from the group-coded variable to the factor. The same general approach can be adapted to accommodate multiple covariates. For example, using a MIMIC approach, Aiken and colleagues (1994) evaluated posttreatment mean differences in daily life factors across two treatment groups of drug addicts, after controlling for initial differences in daily life factors. Group differences were assessed by evaluating posttreatment factor means, adjusting for pretreatment differences specifically on the same factor. Adjustment for pretreatment differences was also made at the indicator level by specifying error covariances for corresponding indicators across pre- and posttreatment assessments. The flexibility of SEM for specifying covariates to control for any or all of the factors and indicators is an advantage over traditional methods such as MANCOVA, which does not allow selective pairings of covariates and outcomes.

Similarly, either manifest or latent covariates can be included in models as exogenous variables within an SMM approach. The covariates are added as predictors of the factors on which mean differences are to be assessed. The means of latent covariates, as well as those of the factors of interest, must be set to zero in the reference group. Additionally, while the MIMIC approach implies the homogeneity of slopes assumption by allowing a single parameter between the covariate and latent variable to be estimated, this invariance constraint is explicitly specified and can be tested in the SMM approach. Evaluation of the difference in latent population means, adjusted for the covariate, is made based on whether the intercept for the factor of interest differs from zero. See Hancock (2004) for further details and illustrated examples of MIMIC and SMM models with latent covariates.

SUMMARY AND CONCLUSIONS

We summarize the key points for this chapter on evaluating between-group differences in latent means by addressing the following simple question: "What steps can be taken that lead to the best chance for a successful analysis of differences in latent means?"

First and foremost, when planning the study, researchers should specify thoughtfully the models to be used. This includes the choice of whether a latent or emergent variable system is a more conceptually appropriate way to operationalize the constructs of interest. For an emergent variable system, MANOVA is a reasonable method. However, for latent variable systems, SEM should be used to allow the relationship between the constructs and measures to be properly specified. In addition, SEM is more powerful than MANOVA for latent variable systems (e.g., Kaplan & George, 1995). SEM methods also permit the use of theoretically error-free representations of constructs and flexibility in the specification of the model. However, it is extremely important to include valid construct measures in SEM analyses. The most unidimensional set of measures possible should be used to represent each construct to avoid reaching incorrect decisions about factor mean differences.

Second, researchers are confronted with the choice between SMM and MIMIC methods for conducting their analyses. Although this choice could be made using a variety of strategies, we have a preference for a data-driven approach that could incorporate both methods. Initial analyses would assess whether model parameters are invariant across groups. These analyses would potentially involve CFAs within groups, multiple-group CFA with between-group constraints, and structured means modeling with between-group constraints. If the results of these analyses suggest that the model parameters (except factor intercepts) are similar across groups, then MIMIC analysis would be appropriate and would allow for a straightforward interpretation of results. If the model parameters are dissimilar, then the stepwise approach to SMM presented in Table 5.3 should be used to test factor means.

We should also consider Type I error rate and power for the MIMIC and SMM approaches. Simulations by Hancock and colleagues (2000) demonstrated that SMM controls adequately for Type I error across a range of conditions in which factorial invariance does or does not hold. The MIMIC approach controlled adequately for Type I error under the more limited conditions of equal sample sizes across groups and/or equal determinants of covariance matrices across groups. Type I error rates were inflated when the larger sample had weaker loadings, and were too conservative when the smaller sample had weaker loadings. The power of the Wald test to detect differences in latent means appeared to be similar across the MIMIC and SMM approaches. Kaplan and George (1995) found power to be somewhat greater for larger models and for equal sample sizes across groups. They found that power was not strongly affected by noninvariance when sample sizes were equal. Researchers would be well served to design studies to have approximately equal sample sizes across groups whenever possible.

For some studies, researchers may opt to use the MIMIC approach rather than apply the structured means model or take a data-driven strategy for choosing approaches. An advantage of the MIMIC model is its simplicity, and it is this simplicity that allows for the testing of hypotheses that are difficult to formulate with SMM. For example, although not illustrated in the current chapter, the testing of interactions is relatively straightforward with the MIMIC approach, whereas it is less clear how to conduct this test using an SMM approach.

We have attempted to provide a fairly comprehensive introduction to SEM analyses of between-group differences in latent means. Although these methods have been available for quite some time, applied studies using latent means analysis have appeared rather infrequently in the literature. Perhaps this is because few resource books on SEM have illustrated these techniques; exceptions include sections in books by Bollen (1989) and, more recently, Byrne (1998; see also similar books for EQS and AMOS users) and Kaplan (2000). Conceptual introductions to latent means methods with illustrative examples are presented in Hancock's introductory article (1997) and his somewhat more technical chapter (2004). Addressing more specific aspects of these methods, Byrne and colleagues (1989) and Vandenberg and Lance (2000) discuss methods for assessing partial factorial invariance, while Cole and colleagues (1993) offer a comparative treatment of MANOVA and SEM approaches for multivariate group comparisons. Finally, a well-explained application of multiple-group latent means analysis can be found in Aiken and colleagues (1994).

ACKNOWLEDGMENTS

We would like to thank Gregory Hancock and Ralph Mueller for their careful editing and thoughtful feedback on this chapter. The chapter benefited from discussions we had with Roger Millsap about evaluating partial invariance. We also thank Wen-juo Lo for his assistance with writing SEM program syntax and checking our examples.

APPENDIX A: GLOBAL FIT INDICES

A variety of global fit indices are available for assessing fit. We will limit our focus to three indices: chi-square statistic (χ^2), standardized root mean square residual (SRMR), and root mean square error of approximation (RMSEA). We demonstrate how to calculate these indices using the SMM approach for the first dataset. All model parameters are constrained between groups except the intercepts for factors. We discuss these indices in that they may be calculated differently with various SEM programs. In the chapter, we report fit indices calculated with EQS.

The approximate chi-square, T, is a function of the sample size N and of the sample maximum likelihood discrepancy function, $\hat{F}$, which is minimized in estimating model parameters. For a single-group model, $T = n\ \hat{F}$. Minor discrepancies in the value of T occur across SEM programs depending on whether n is defined as equal to $N - 1$ (e.g., EQS and LISREL) or N (e.g., Mplus). For a model with J multiple groups,

$T = \left(\sum_{j=1}^{J} n_j \right) \hat{F}_{MG}$. $\hat{F}_{MG}$ is the sample discrepancy function for multiple groups computed by averaging the fit functions across groups, weighted by their relative group sample sizes. (See Bollen, 1989, for a detailed presentation of the multisample fit function with mean structure.) Comparable to the one-group *T*, n_j (i.e., *n* for group *j*) may be defined as $N_j - 1$ (e.g., EQS and LISREL) or as N_j. *T* is distributed approximately in large samples as a chi-square with *df* degrees of freedom if the assumptions underlying the test are met (e.g., multivariate normality of the measured variables in the population and specification of the correct model). For our example based on the data in Table 5.1, the $\hat{F}_{MG} = .10498$ so $T = [(200\text{-}1) + (200\text{-}1)]\ (.10498) = 41.78$. Accordingly, the multiple-group model and, in particular, the within-group and between-group constraints imposed on the model cannot be rejected based on the approximate chi-square test, $\chi^2(33) = 41.78, p = .14$.

The computation of the standardized root mean square residual is straightforward

$$\text{SRMR} = \sqrt{\frac{\left[\sum_{i}^{p}\sum_{j}^{i}(s_{ij} - \hat{\sigma}_{ij})/s_i s_j\right]^2}{p(p+1)/2}} \qquad \text{(A.1)}$$

where s_{ij} is the sample covariance between measures *i* and *j*, $\hat{\sigma}_{ij}$ is the reproduced covariance between measures *i* and *j*, s_i and s_j are the sample standard deviations for measures *i* and *j*, and *p* is the number of measures. It should be noted that the SRMR, as defined, is sensitive to lack of fit to the sample covariance matrix, but not to the sample means.

The population RMSEA for a single group is $\sqrt{F/df}$, where *F* is a population discrepancy function that is minimized to obtain maximum likelihood estimates. According to Steiger and Lind (1980), *F* is the iteratively reweighted generalized least squares (IRGLS) discrepancy function, which may be minimized to obtain maximum likelihood estimates. However, the standard maximum likelihood discrepancy function is sometimes substituted for the IRGLS function. The population RMSEA can be estimated for a single sample as follows:

$$\text{RMSEA} = \sqrt{\frac{(T - df)/n}{df}} \qquad \text{(A.2)}$$

where T is the large sample chi-square statistic based on the sample discrepancy function and n is defined comparably as it was defined to compute T. The sample RMSEA is set to zero if the value under the square root is negative.

Steiger (1998) argued that the population RMSEA for a multiple-group model should have a divisor representing the degrees of freedom averaged across groups in that the fit function is averaged across groups in multiple-group analyses. Accordingly, the population RMSEA would be defined as $\sqrt{F/(df/J)}$, where J is the number of groups. At the time of the writing of this chapter, EQS had not yet incorporated Steiger's correction; that is, EQS did not average across groups by dividing df by J. However, most recently this adjustment was incorporated into EQS 6.1 (Build 88). The population RMSEA, thus, can be estimated two different ways:

$$\text{RMSEA}_{\text{averaged } df} = \sqrt{\frac{(T - df)/\sum_{j=1}^{J} n_j}{df/J}} \quad \text{or} \tag{A.3}$$

$$\text{RMSEA}_{\text{total } df} = \sqrt{\frac{(T - df)/\sum_{j=1}^{J} n_j}{df}}.$$

If the value under the square root is negative, RMSEA is reported as 0. In summary, the value for the RMSEA may differ across programs depending on whether (a) T is based on the standard maximum likelihood or the IRGLS discrepancy function, (b) the definition of n_j as $N_j - 1$ or N_j in the computation of T, and (c) for multiple-groups analyses, the decision to employ an averaged df or total df for the analysis. Based on the T for the IRGLS discrepancy function, $n_j = N_j - 1$, and the averaged df, the RMSEA is computed as

$$\text{RMSEA}_{\text{averaged } df} = \sqrt{\frac{(40.96 - 33)/(199 + 199)}{33/2}} = .035.$$

In comparison, based on the standard maximum likelihood T, $n_j = N_j - 1$, and the total df, the RMSEA is computed as

$$\text{RMSEA}_{\text{total } df} = \sqrt{\frac{(41.78 - 33)/(199 + 199)}{33}} = .026.$$

Note that RMSEA confidence intervals are similarly affected.

APPENDIX B: SYNTAX FOR SMM AND MIMIC ANALYSES FOR DATASET 1

Programs are presented here for SMM and MIMIC analyses for Dataset 1. The SMM syntax specifies the most complex and constrained form of the multiple-groups model; other models presented can generally be specified by removing intercepts and/or constraints. The between-group constraints on factor variances, factor covariances, and error variances (shown in bold print in EQS and Mplus programs) are typically not made in practice. Instructions are included for removing these constraints following each program.

To adapt syntax for the partial invariance example for Dataset 2, first remove the constraints noted above from the SMM program. Second, replace the data with the covariance matrix and means presented in Table 5.2. Third, follow the stepwise instructions under Model Specification in Table 5.3. For example, for the combined-groups model under Model Specification for Step 1, the intercepts are removed from the model, and the metric of the factors is fixed by setting factor variances to 1 in all groups and freely estimating the factor loadings for V1 and V6.

SYNTAX FOR SMM ANALYSES FOR DATASET 1

EQS Program

```
/TITLE
   SMM FOR DATA SET 1 — GROUP 1
/SPECIFICATIONS
   VARIABLES=6; CASES=200; MATRIX=COV; ANAL=MOMENT;
   GROUPS=2;
/EQUATIONS
   V1 = 1F1 + *V999 + E1;
   V2 = *F1 + *V999 + E2;
   V3 = *F1 + *V999 + E3;
   V4 = *F2 + *V999 + E4;
   V5 = *F2 + *V999 + E5;
   V6 = 1F2 + *V999 + E6;
   F1 = 0V999 + D1;
   F2 = 0V999 + D2;
```

```
/VARIANCES
   E1 TO E6 = *; D1 to D2 = *;
/COVARIANCES
   D1, D2=*;
/PRINT
   FIT=ALL; COV=YES;
/MATRIX
  138.00
   45.58   80.49
   35.19   23.56   56.34
   45.13   32.00   16.64  232.17
   35.33   10.49   10.56   79.74  149.16
   73.34   28.73   33.21  117.30   79.90  324.36
/MEANS
     50.40   79.60   98.88    74.06    49.12   120.10
/END
/TITLE
   SMM FOR DATA SET 1 — GROUP 2
/SPECIFICATIONS
   VARIABLES=6; CASES=200; MATRIX=COV; ANAL=MOMENT;
/EQUATIONS
   V1 = 1F1 + *V999 + E1;
   V2 = *F1 + *V999 + E2;
   V3 = *F1 + *V999 + E3;
   V4 = *F2 + *V999 + E4;
   V5 = *F2 + *V999 + E5;
   V6 = 1F2 + *V999 + E6;
   F1 = *V999 + D1;
   F2 = *V999 + D2;
/VARIANCES
   E1 TO E6 = *; D1 TO D2 = *;
/COVARIANCES
   D1,D2=*;
/MATRIX
  127.61
   58.49   76.81
   29.09   19.72   54.29
   45.84   28.94   31.82  223.09
   20.33   13.27    5.81   62.25  135.72
   50.64   55.45   30.15  126.74   62.16  337.37
/MEANS
   53.70   81.32  101.82    77.69    51.62   123.37
/CONSTRAINTS
```

```
    (1,V2,F1)=(2,V2,F1);
    (1,V3,F1)=(2,V3,F1);
    (1,V4,F2)=(2,V4,F2);
    (1,V5,F2)=(2,V5,F2);
    (1,V1,V999)=(2,V1,V999);
    (1,V2,V999)=(2,V2,V999);
    (1,V3,V999)=(2,V3,V999);
    (1,V4,V999)=(2,V4,V999);
    (1,V5,V999)=(2,V5,V999);
    (1,V6,V999)=(2,V6,V999);
    (1,E1,E1)=(2,E1,E1);
    (1,E2,E2)=(2,E2,E2);
    (1,E3,E3)=(2,E3,E3);
    (1,E4,E4)=(2,E4,E4);
    (1,E5,E5)=(2,E5,E5);
    (1,E6,E6)=(2,E6,E6);
    (1,D1,D1)=(2,D1,D1);
    (1,D2,D2)=(2,D2,D2);
    (1,D1,D2)=(2,D1,D2);
/END
```

Note. To remove the constraints in bold print, delete these statements.

SIMPLIS Program

```
! SMM FOR DATA SET 1
GROUP1:DAYCARE
OBSERVED VARIABLES:
V1 V2 V3 V4 V5 V6
COVARIANCE MATRIX:
138.00
 45.58  80.49
 35.19  23.56  56.34
 45.13  32.00  16.64  232.17
 35.33  10.49  10.56  179.74  149.16
 73.34  28.73  33.21  117.30   79.90  324.36
MEANS:
50.40  79.60  98.88  74.06  49.12  120.10
SAMPLE SIZE:
200
LATENT VARIABLES:
F1 F2
```

```
RELATIONSHIPS:
V1 = CONST + 1*F1
V2 = CONST + F1
V3 = CONST + F1
V4 = CONST + F2
V5 = CONST + F2
V6 = CONST + 1*F2
ITERATIONS = 500
GROUP2: HOMECARE
COVARIANCE MATRIX:
127.61
 58.49  76.81
 29.09  19.72  54.29
 45.84  28.94  31.82  223.09
 20.33  13.27   5.81   62.25  135.72
 50.64  55.45  30.15  126.74   62.16  337.37
MEANS:
53.70  81.32  101.82  77.69  51.62  123.37
SAMPLE SIZE:
200
RELATIONSHIPS:
F1 = CONST
F2 = CONST
END OF PROBLEM
```

Note. To remove the constraints on factor variances, factor covariances, and error variances, add the following statements before the END OF PROBLEM statement:

```
Set Error Variance of V1-V6 FREE
Set Covariance of F1 and F2 FREE
Set Variance of F1-F2 FREE
```

Mplus Program

```
TITLE: SMM FOR DATA SET 1
DATA:
FILE IS "A:\SMM.TXT";
TYPE = MEANS COVARIANCE;
NOBSERVATIONS = 200 200;
NGROUPS = 2;
VARIABLE:
```

```
NAMES ARE V1 V2 V3 V4 V5 V6;
GROUPING = GROUP (1=DAYCARE 2=HOMECARE);
ANALYSIS:
TYPE = MEANSTRUCTURE MGROUP;
ESTIMATOR = ML;
MODEL:
F1 BY V1@1 V2* V3*;
F2 BY V4* V5* V6@1;
F1 (1);
F2 (2);
F1 WITH F2 (3);
V1 (4);
V2 (5);
V3 (6);
V4 (7);
V5 (8);
V6 (9);
```

Note. To remove the constraints in bold print, delete these statements. The data for this program are stored in SMM.TXT (a text document) on the A drive. The data are stored as follows: means for the first group in the first row, lower left triangle of the covariance matrix for the first group in the next six rows; means for the second group in the next row, lower left triangle of the covariance matrix for the second group in the last six rows.

SYNTAX FOR MIMIC ANALYSES FOR DATASET 1

EQS Program

```
/TITLE
   MIMIC ANALYSIS FOR DATA SET 1
/SPECIFICATIONS
   VARIABLES=7; CASES=400; METHODS=ML; MATRIX=COV;
/EQUATIONS
   V1 = 1F1 + E1;
   V2 = *F1 + E2;
   V3 = *F1 + E3;
   V4 = *F2 + E4;
   V5 = *F2 + E5;
   V6 = 1F2 + E6;
   F1 = *V7 + D1;
```

```
   F2 = *V7 + D2;
/VARIANCES
   E1 TO E6 =*; V7=*; D1 TO D2=*;
/COVARIANCES
   D1,D2=*;
/PRINT
   FIT=ALL; COV=YES;
/MATRIX
   135.19
   53.33  79.21
   34.48  22.86  57.33
   48.37  31.96  26.83  230.35
   29.82  12.93   9.99   73.08  143.64
   64.53  43.40  34.00  124.69   72.90  332.72
    0.83   0.43   0.73    0.91    0.62    0.82  0.25
/END
```

SIMPLIS Program

```
!MIMIC ANALYSIS FOR DATA SET 1
OBSERVED VARIABLES:
   V1 V2 V3 V4 V5 V6 V7
COVARIANCE MATRIX:
   135.19
    53.33  79.21
    34.48  22.86  57.33
    48.37  31.96  26.83  230.35
    29.82  12.93   9.99   73.08  143.64
    64.53  43.40  34.00  124.69   72.90  332.72
     0.83   0.43  10.73    0.91    0.62    0.82  0.25
SAMPLE SIZE:
   400
LATENT VARIABLES:
   F1 F2
RELATIONSHIPS:
   V1 = 1*F1
   V2 = F1
   V3 = F1
   V4 = F2
   V5 = F2
   V6 = 1*F2
   F1 = V7
```

```
   F2 = V7
SET COVARIANCE OF F1 AND F2 FREE
END OF PROBLEM
```

Mplus Program

```
TITLE: MIMIC FOR DATA SET 1
DATA:
   FILE IS "A:\MIMIC.TXT";
   TYPE = COVARIANCE;
   NOBSERVATIONS = 400;
VARIABLE:
   NAMES ARE V1 V2 V3 V4 V5 V6 V7;
ANALYSIS:
   ESTIMATOR = ML;
MODEL:
   F1 BY V1@1 V2* V3*;
   F2 BY V4* V5* V6@1;
   F1 WITH F2;
   F1 F2 ON V7*;
```

Note. The covariance matrix for this program is stored in MIMIC.TXT (a text document) on the A drive.

APPENDIX C: COMPARISONS BASED ON EQUIVALENT MODELS FOR ASSESSING BETWEEN-GROUP DIFFERENCES IN FACTOR MEANS

Steps Involved in Comparison	*Specification of Equivalent Models*	
	Less Constrained Model	*More Constrained (Nested) Model*
Step 1 versus Step 2a	**Model for step 1** Factor variances and covariances • Freely estimate variances and covariances of all factors in all groups. Factor loadings • For each factor, choose one measure that is a function of that factor. Fix loading between measure and factor to 1 in all groups. ⇒ Freely estimate remaining loadings in all groups. Intercepts for factors and measures • Intercepts are not included in model.	**Model for step 2a** Factor variances and covariances • Same as model for step 1. Factor loadings • For each factor, choose one measure that is a function of that factor. Fix loading between measure and factor to 1 in all groups. ⇒ Constrain remaining loadings to be equal between groups. Intercepts for factors and measures • Same as model for step 1.
Step 2 versus Step 3a	**Model for step 2** Disturbance variances and covariances[a] • Freely estimate variances and covariances of disturbances in all groups. Factor loadings • For each factor, choose one measure that is a function of that factor. Fix loading between measure and factor to 1 in all groups. • If comparison is based on model from step 2a, constrain all remaining factor loadings to be equal between groups. If comparison is based on model from step 2b, constrain a subset of the remaining factor loadings to be equal between groups based on the results of the substeps of 2b.	**Model for step 3a** Disturbance variances and covariances[a] • Same as model for step 2. Factor loadings • Same as model for step 2.

(Table continues)

Steps Involved in Comparison	Specification of Equivalent Models	
	Less Constrained Model	*More Constrained (Nested) Model*
Step 2 versus Step 3a	Intercepts for measures • For each factor, a loading for a measure was fixed to 1 in one group. Constrain intercept for this measure to be equal between groups. ⇒ Freely estimate remaining intercepts for measures in all groups.	Intercepts for measures • For each factor, a loading for a measure was fixed to 1 in one group. Constrain intercept for this measure to be equal between groups. ⇒ Constrain remaining intercepts for measures to be equal between groups.
	Intercepts for factors[b] • Constrain intercepts for factors to 0 in one group. • Allow intercepts for factors in other groups to be freely estimated.	Intercepts for factors[b] • Same as model for step 2.

Note: The models described in this table are equivalent to the models described in Table 5.3. A paragraph preceded by the symbol ⇒ describes the model parameters evaluated by the model comparison.

[a]The variances and covariances among the disturbance terms are the variances and covariances among the factors. With some SEM programs, factor variances and covariances are directly specified rather than disturbance variances and covariances.

[b]The factor intercepts are the factor means. In some SEM programs, the factor means are directly specified.

REFERENCES

Aiken, L. S., Stein, J. A., & Bentler, P. M. (1994). Structural equation analyses of clinical subpopulation differences and comparative treatment outcomes: Characterizing the daily lives of drug addicts. *Journal of Consulting and Clinical Psychology, 62,* 488–499.

Bentler, P. (1995). *EQS structural equations program manual.* Encino, CA: Multivariate Software Inc.

Bollen, K. A. (1989). *Structural equations with latent variables.* New York: Wiley.

Bollen, K. A., & Lennox, R. (1991). Conventional wisdom on measurement: A structural equation perspective. *Psychological Bulletin, 110,* 305–314.

Byrne, B. M. (1998). *Structural equation modeling with LISREL, PRELIS, and SIMIPLIS: Basic concepts, applications, and programming.* Mahwah, NJ: Erlbaum.

Byrne, B. M., Shavelson, R. J., & Muthén, B. (1989). Testing for the equivalence of factor covariance and means structures: The issue of partial measurement invariance. *Psychological Bulletin, 105,* 456–466.

Chou, C.-P., & Bentler, P. M. (1990). Model modification in covariance structure modeling: A comparison among likelihood ratio, Lagrange multiplier, and Wald tests. *Multivariate Behavioral Research, 25,* 115–136.

Cohen, J. (1988). *Statistical power analysis for the behavioral sciences* (2nd ed.). Hillsdale, NJ: Erlbaum.

Cohen, P., Cohen, J., Teresi, M., Marchi, M., & Velez, C.N. (1990). Problems in the measurement of latent variables in structural equations causal models. *Applied Psychological Measurement, 14,* 183–196.

Cole, D. A., Maxwell, S. E., Arvey, R., & Salas, E. (1993). Multivariate group comparisons of variable systems: MANOVA and structural equation modeling. *Psychological Bulletin, 114,* 174–184.

Green, S. B., & Thompson, M. S. (2003a). Structural equation modeling in clinical research. In M. C. Roberts & S. S. Illardi (Eds.), *Methods of research in clinical psychology: A handbook* (pp. 138–175). London: Blackwell.

Green, S. B., & Thompson, M. S. (2003b, April). *Understanding discriminant analysis/MANOVA through structural equation modeling.* Paper presented at the annual meeting of the American Educational Research Association, Chicago.

Hancock, G. R. (1997). Structural equation modeling methods of hypothesis testing of latent variable means. *Measurement and Evaluation in Counseling and Development, 20,* 91–105.

Hancock, G. R. (2001). Effect size, power, and sample size determination for structured means modeling and mimic approaches to between-groups hypothesis testing of means on a single latent construct. *Psychometrika, 66,* 373–388.

Hancock, G. R. (2003). Fortune cookies, measurement error, and experimental design. *Journal of Modern and Applied Statistical Methods, 2,* 293–305.

Hancock, G. R. (2004). Experimental, quasi-experimental, and nonexperimental design and analysis with latent variables. In D. Kaplan (Ed.), *The Sage handbook of quantitative methodology for the social sciences* (pp. 317–334). Thousand Oaks, CA: Sage.

Hancock, G. R., Lawrence, F. R., & Nevitt, J. (2000). Type I error and power of latent mean methods and MANOVA in factorially invariant and noninvariant latent variable systems. *Structural Equation Modeling: A Multidisciplinary Journal, 7*, 534–556.

Kano, Y. (2001). Structural equation modeling for experimental data. In R. Cudeck, S. du Toit, & D. Sörbom (Eds.), *Structural equation modeling: Present and future. Festschrift in honor of Karl Jöreskog* (pp. 381–402). Lincolnwood, IL: Scientific Software International, Inc.

Kaplan, D. (2000). *Structural equation modeling: Foundations and extensions.* Thousand Oaks, CA: Sage.

Kaplan, D., & George, R. (1995). A study of the power associated with testing factor mean differences under violations of factorial invariance. *Structural Equation Modeling: A Multidisciplinary Journal, 2*, 101–118.

Meredith, W. (1993). Measurement invariance, factor analysis and factor invariance. *Psychometrika, 58*, 525–544.

Steiger, J. H. (1998). A note on multisample extensions of the RMSEA fit index. *Structural Equation Modeling: A Multidisciplinary Journal, 5*, 411–419.

Steiger, J. H., & Lind, J. C. (1980, May). *Statistically based tests for the number of common factors*. Paper presented at the annual Spring Meeting of the Psychometric Society, Iowa City, IA.

Vandenberg, R. J., & Lance, C. E. (2000). A review and synthesis of the measurement invariance literature: Suggestions, practices, and recommendations for organizational research. *Organizational Research Methods, 3*, 4–70.

CHAPTER 6

USING LATENT GROWTH MODELS TO EVALUATE LONGITUDINAL CHANGE

Gregory R. Hancock and Frank R. Lawrence

Researchers are frequently interested in understanding how some aspect of an individual changes over time. The focus of their investigation might be on general outcomes such as behavior, performance, or values; or it could be on more specific aspects such as substance abuse, depression, communication skills, attitudes toward disabled veterans, or advancement of math aptitude. Regardless of an investigation's focus, the real attraction in longitudinal studies is in understanding how change comes about, how much change occurs, and how the change process may differ across individuals.

Various methods can be used to analyze longitudinal data (see, e.g., Collins & Sayer, 2001; Gottman, 1995). Among the more traditional methods is analysis of variance (ANOVA), multivariate analysis of variance (MANOVA), analysis of covariance (ANCOVA), multivariate analysis of covariance (MANCOVA), and auto-regressive and cross-lagged multiple regression. No one method is necessarily considered superior, but each has strengths and shortcomings that researchers should be aware of in order to select the analytic method best suited for the particular research context.

Structural Equation Modeling: A Second Course, 171–196

Selection of an appropriate analytic method revolves around two central issues. The first is the nature of the hypotheses being tested, while the second concerns the underlying assumptions necessary to apply a particular method. The analytic method must be compatible with both the hypotheses being tested and the data, so as to be capable of providing evidence to tentatively support or refute the researcher's hypotheses. Unfortunately, traditional longitudinal data-analytic techniques can present challenges on both fronts. With regard to the data, technical assumptions such as sphericity underlying repeated-measures ANOVA (see, e.g., Yandell, 1997) are rarely met in practice in the social sciences. Also, traditional methods tend to operate at the group level, providing potentially interesting aggregated results but failing to address hypotheses regarding the nature and determinants of change at the level of the individual. Thus, when such incompatibilities of methods with data and hypotheses arise, conclusions from traditional methods can become somewhat circumspect (see, e.g., Rogosa & Willett, 1985).

To help overcome some of the limitations of more traditional analytic approaches to the assessment of change over time, a class of useful methods has emerged from the area of structural equation modeling (SEM). Such methods, falling under the general heading of *latent growth modeling* (LGM), approach the analysis of growth[1] from a somewhat different perspective than the aforementioned methods. Specifically, LGM techniques can describe individuals' behavior in terms of reference levels (e.g., initial amount) and their developmental trajectories to and from those levels (e.g., linear, quadratic). In addition, they can determine the variability across individuals in both reference levels and trajectories, as well as provide a means for testing the contribution of other variables or constructs to explaining those initial levels and growth trajectories (Rogosa & Willett, 1985; Short, Horn, & McArdle, 1984). In doing so, LGM methods simultaneously focus on changes in covariances, variances, and mean values over time (Dunn, Everitt, & Pickles, 1993; McArdle, 1988), thus utilizing more information available in the measured variables than do traditional methods.

Consider Figure 6.1, a theoretical representation of four individuals' growth patterns over four equally spaced time intervals. A number of elements of this figure are noteworthy. First, the growth trajectory of all individuals displayed is linear; that is, for equal time periods a given individual is growing the same amount. Thus, David grows as much between time 1 and time 2 as he does between time 3 and time 4. We begin by assuming that growth follows a linear trend. This linear form may or may not be a realistic representation of growth for a particular variable; growth may actually be of a different functional nature (e.g., quadratic or logarithmic). Given that growth processes are often mea-

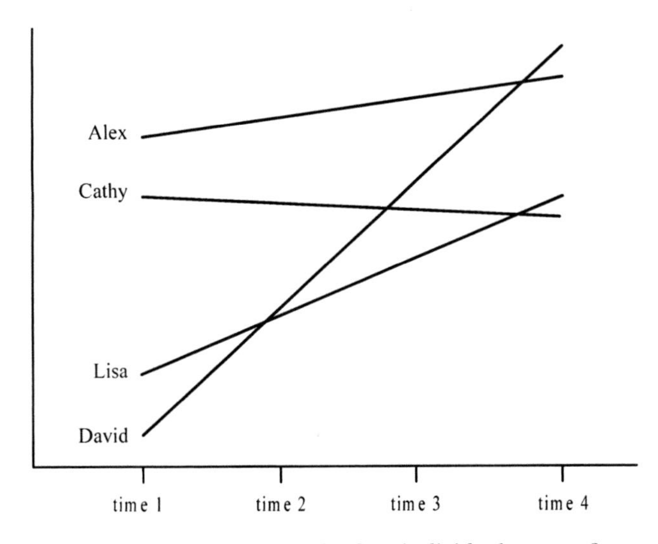

Figure 6.1. Linear growth for four individuals across four equally spaced time points.

sured over a fairly restricted portion of the span of development, even complex growth patterns can be well approximated by simpler models (Willett & Sayer, 1994). That said, after developing LGM concepts with this simple linear model, we discuss the relaxation of the assumption of growth linearity and methods for exploring other developmental patterns.

A second key point regarding Figure 6.1 is that, in contrast to what is often assumed in more traditional methods, the four individuals represented do not develop at the same rate. David, for example, starts below the others but finishes with the highest value because he progresses at such a rapid rate. Alex, on the other hand, starts at the highest level but does not grow much from that point, while Cathy actually declines over the time period represented. Thus, all individuals differ in their initial levels of behavior as well as in their rates of growth from that initial level; assuming the same degree of change over time for all individuals is clearly unrealistic for these data. Fortunately, LGM methods typically only assume that all individuals' growth follows the same functional form, reflecting a deeper assumption that growth is grounded in some social and/or biological mechanism. In this case that mechanism is linear; as such, growth over this span may be summarized by, for example, a unique *intercept* and *slope*. An individual's intercept describes the amount of the variable possessed at the initial measurement point, while the slope captures information about how much that individual changes for each time interval following the initial measurement point.

Third, as a result of individual differences in these intercepts and slopes, changes occur in the relations among individuals' data across measurement occasions. For example, at time 1, David has the least amount of the variable of interest and Alex has the most; at time 4, David has the most and Cathy has the least. As a result of individuals' relative positions shifting across time, correlations between scores at different times are not identical. In addition, the variance of scores at different times also fluctuates; scores are more spread out at the first and last times, and are less variable at the intermediate measurement points. These changes in correlation and variance combine to illustrate a violation of the sphericity assumption underlying repeated-measure analysis of variance, a method commonly used in the evaluation of longitudinal data.

In short, then, a method is required that does not assume all individuals change at the same rate, thereby transcending the limitations of methods bound to this and other related assumptions. The purpose of the current chapter is to illustrate the use of LGM techniques, which overcome many traditional methods' limitations and thus serve as an extremely useful research tool for the explication of longitudinal data. Such methods have been used to analyze change in alcohol, cigarette, and marijuana use (Curran, Stice, & Chassin, 1997; Duncan & Duncan, 1994), tolerance of deviant behavior (Willett & Sayer, 1994), antisocial behavior (Patterson, 1993), client resistance during parent training therapy (Stoolmiller, Duncan, Bank, & Patterson, 1993), and family functioning (Willett, Ayoub, & Robinson, 1991). Our intent is to describe this methodological tool in enough conceptual and practical detail for readers with a basic understanding of SEM to be able to apply it. We start with the most common linear case, and then expand to address nonlinearity, external predictors, and a host of variations and extensions to the basic model. Those readers wishing to explore LGM methodology further are encouraged to seek out the aforementioned references as well as some excellent recent texts (e.g., Duncan, Duncan, Strycker, Li, & Alpert, 1999; Moskowitz & Hershberger, 2002).

LGM METHODOLOGY FOR LINEAR DEVELOPMENTAL PATTERNS

To facilitate the introduction of LGM methods, consider the hypothetical situation in which high school students' mathematics proficiency has been assessed using the same instrument at the beginning of each school year. Thus, the researcher has data at an initial point (9th grade) and at three equally spaced subsequent intervals (as was depicted in Figure 6.1), in this case 10th, 11th, and 12th grade (note that data are gathered from individuals at the same measurement occasions). Having multiple data points such as these allows one to evaluate more accurately the functional form of the growth under examination; Stoolmiller (1995) recommends a careful consideration of the phenomenon under study to determine a suitable number and spacing of measurements.

Once data have been gathered, a preliminary assessment of several individuals' growth patterns, coupled with a sound theoretical understanding of the nature of the behavior being investigated, often yields a reasonable assumption as to the functional form of the growth under investigation (e.g., linear, quadratic, etc.). In the current hypothetical case, assume that past research and theory dictate that math proficiency scores should follow a generally linear pattern over the time period studied, and that the initial point of measurement (time 1) is a useful point of developmental reference. As such, each individual's score can be expressed by the following functional form:

$$\begin{aligned}\text{score at time } t = {} & \text{initial score} + (\text{change in score per unit time}) \\ & \times (\text{time elapsed}) + \text{error}.\end{aligned}$$

As the initial score is simply an intercept, and the change in score per unit time is simply a slope, this expression could be expanded into a system of equations for each individual's score from time 1 (V1) through time 4 (V4).

$$\begin{aligned}
&\text{time 1:} \quad V1 = \text{intercept} + (\text{slope}) \times (0) + E1 \\
&\text{time 2:} \quad V2 = \text{intercept} + (\text{slope}) \times (1) + E2 \\
&\text{time 3:} \quad V3 = \text{intercept} + (\text{slope}) \times (2) + E3 \\
&\text{time 4:} \quad V4 = \text{intercept} + (\text{slope}) \times (3) + E4.
\end{aligned}$$

Thus, a student's observed math proficiency score at any time is believed to be a function of his or her own intercept and slope, as well as error. Note that the terms *intercept* and *slope* have the same interpretation here as they do in a simple regression model. That is, the intercept represents the expected value of the outcome at the point where the predictor variable is zero. Hence, the intercept is the expected value for V1 when elapsed time is zero. This can easily be seen from the system of equations. The first equation exemplifies this concept by fixing elapsed time to zero, thereby placing the intercept at that initial location. We can build on that interpretation by noting that the intercept appears in every subsequent equation. Thus, we can say that the intercept allows the initial amount of math proficiency to be "frozen" in to the measured variable, while the slope describes change in proficiency beyond that initial score.

In linear LGM, these intercept and slope terms are treated as latent variables (factors) on which individuals may vary; as such these are not assumed to be measured directly. Certainly an individual's V1 measure provides an estimate of initial math proficiency, but, as the time 1 equation above indicates, this is not without error. Similarly, while one could estimate for each individual the slope of the best-fit line describing math proficiency over time, the rate of change possessed by an individual is treated as a latent trait. Observed scores, however, are not expected to be a perfect function of the latent growth constituents; the E terms in the equations represent error. There are many sources of error that appear in data and the modeling of data, including measurement error arising from instrument or rater unreliability and model misspecification arising from incorrect assumptions regarding the functional form of change. Thus, error may be regarded as the degree of deviation between the observed outcome and that which we would expect from the latent growth portion of the model. The system of equations represented above, which represents a theory about the data generation process for math proficiency, can be expressed in the model presented in Figure 6.2, shown with both covariance and mean structure parameters.

With regard to the covariance structure, note that in this model all paths to the variables (i.e., loadings) are fixed. Of particular theoretical

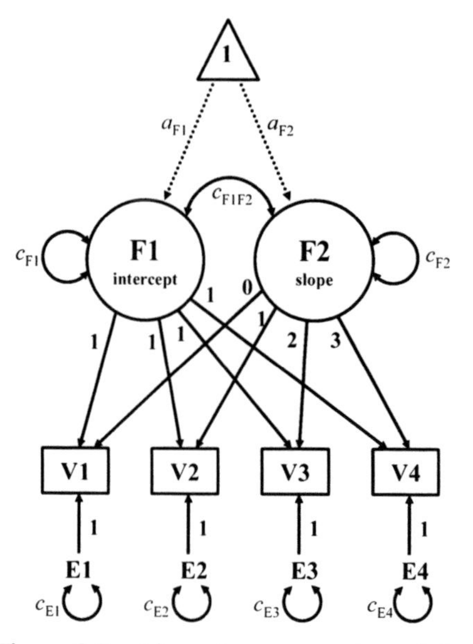

Figure 6.2. Linear latent growth model with covariance and mean structure.

importance is that all paths from the slope factor have been fixed.[2] This indicates that the researcher wishes to test if a model representing linear growth fits the data satisfactorily. This also leaves only variance and covariance parameters to be estimated within the covariance structure: all four error variances (c_{E1}, c_{E2}, c_{E3}, c_{E4}), which are usually only of peripheral interest[3]; both factor variances (c_{F1} and c_{F2}); and the factor covariance (c_{F1F2}). The intercept factor variance (c_{F1}) gives the researcher a sense of how much diversity exists in initial math proficiency scores, where a smaller variance would indicate that students start with rather similar math proficiency in ninth grade and a large intercept variance indicates students are quite different in terms of ninth-grade math proficiency. The slope factor variance (c_{F2}) conveys the diversity in rates of change in math proficiency throughout high school, where a smaller variance would indicate that students grow at a fairly similar rate throughout high school and a large slope variance indicates students change at greatly varied rates. Finally, the factor covariance (c_{F1F2}) captures the extent to which the rate of growth in math proficiency is related to the initial amount of math proficiency. A positive covariance implies that more proficient students tended to grow more while less proficient students tended to grow less; conversely, a negative covariance implies that less proficient students tended to grow more while more proficient students tended to grow less.

While the covariance structure of the latent growth model contains information about individual differences in growth characteristics, the mean structure portion contains information about growth at the aggregate level. As detailed in the chapter by Thompson and Green in this volume, factor means can be estimated through the introduction of a pseudovariable that assumes a constant value of 1 for all subjects. The pseudovariable has no variance, and as such it cannot covary with (or have a meaningful causal effect on) any measured variable or factor. Nonetheless, it is placed into diagrams and structural models as though it were a variable, and its inclusion in a latent growth model as depicted in Figure 6.2 allows for intercept and slope factor means to be estimated.

Paths from the pseudovariable (depicted as a 1 inside a triangle) to F1 and F2 are denoted as a_{F1} and a_{F2}, respectively, indicating the latent intercept and slope factor means. In general, a high value of a_{F1} implies that students start with high math proficiency in 9th grade *on average*, and a low value implies that students tend to start with low math proficiency; it does not, however, communicate anything about individual differences, which was conveyed by c_{F1}. Furthermore, a positive value of a_{F2} implies that students have a positive rate of growth in math proficiency across the high school years *on average*, while a negative value implies that students tend to have a negative rate of growth; it does not say anything about individual differences in growth rate, which was conveyed by c_{F2} and which could imply (if sufficiently relatively large) that some students increase in math proficiency while others decline over time.[4]

The linear latent growth model in Figure 6.2, with both covariance and mean structure, has implications for the pattern of variances, covariances, and means that should be observed if the hypothesis of linear growth in math proficiency is correct. The model-implied covariance matrix and mean vector, $\hat{\boldsymbol{\Sigma}}$ and $\hat{\boldsymbol{\mu}}$, would be

$$\hat{\boldsymbol{\Sigma}} = \begin{bmatrix} c_{F1} + c_{E1} & & & \\ c_{F1} + c_{F1F2} & c_{F1} + c_{F2} + 2c_{F1F2} + c_{E2} & & \\ c_{F1} + 2c_{F1F2} & c_{F1} + 2c_{F2} + 3c_{F1F2} & c_{F1} + 4c_{F2} + 4c_{F1F2} + c_{E3} & \\ c_{F1} + 3c_{F1F2} & c_{F1} + 3c_{F2} + 4c_{F1F2} & c_{F1} + 6c_{F2} + 5c_{F1F2} & c_{F1} + 9c_{F2} + 6c_{F1F2} + c_{E4} \end{bmatrix}$$

$$\hat{\boldsymbol{\mu}} = \begin{bmatrix} a_{F1} & a_{F1} + a_{F2} & a_{F1} + 2a_{F2} & a_{F1} + 3a_{F2} \end{bmatrix}$$

Thus, the model in Figure 6.2 would be fit to the sample covariance matrix **S** and the vector of sample means **m**, choosing estimates for the parameters in $\hat{\boldsymbol{\Sigma}}$ and $\hat{\boldsymbol{\mu}}$ so as to minimize the discrepancy of the observed moments **S** and **m** from the expected moments $\hat{\boldsymbol{\Sigma}}$ and $\hat{\boldsymbol{\mu}}$. As there are 14 pieces of data in **S** and **m** and 9 parameters to estimate in $\hat{\boldsymbol{\Sigma}}$ and $\hat{\boldsymbol{\mu}}$, the model will be overidentified with 5 degrees of freedom. The data–model fit is evaluated using the model-implied covariance matrix and mean vector, which represent the expected moments to be compared with the observed moments of our sample data. If the fit between the observed and expected moments is deemed poor, which would be signaled by unsatisfactory data–model fit indices (e.g., root mean square error of approximation [RMSEA]; standardized root mean residual [SRMR]; Comparative Fit Index, [CFI]), then the hypothesis that growth in math proficiency over high school follows a reasonably linear form is rejected and further theoretical and/or exploratory work is likely required. If, however, the data–model fit is deemed satisfactory, the researcher has gathered information supporting the hypothesis of linearity for growth in math proficiency (or, more accurately, the researcher has failed to disconfirm said hypothesis). Following the establishment of reasonable data–model fit, then, the interpretation of the parameters of interest becomes permissible, that is, of the intercept and slope factor variances, their covariance, and their latent means. An example of this linear latent growth model follows next.[5]

Example of a Linear Latent Growth Model

Following from the previous hypothetical context, imagine that a standardized test intending to measure math proficiency is administered to the same n = 1,000 girls in 9th, 10th, 11th, and 12th grades, yielding the following observed moments:

$$\mathbf{S} = \begin{bmatrix} 204.11 & & & \\ 139.20 & 190.50 & & \\ 136.64 & 135.19 & 166.53 & \\ 124.86 & 130.32 & 134.78 & 159.87 \end{bmatrix}$$

$$\mathbf{m} = \begin{bmatrix} 32.97 & 36.14 & 40.42 & 43.75 \end{bmatrix}.$$

As LGM is simply a constrained version of confirmatory factor analysis, with a mean structure, the model in Figure 6.2 may be imposed upon these data using any common structural equation modeling software. The Appendix contains appropriate syntax for this example for the EQS, SIMPLIS, and Mplus software packages, each of which yields output, including data–model fit statistics, parameter estimates, and parameter estimate test statistics.

For this example, the data–model fit is excellent overall: CFI=1.000, SRMR = .010, and RMSEA = .026 (with a 90% confidence interval of .000 - .056). This implies that the hypothesis of linear growth in math proficiency over the high school years is reasonable, and allows further interpretation of specific model parameter estimates. For the mean structure, the estimate of the latent intercept mean is $\hat{a}_{F1}$ = 32.863; this value is statistically significantly greater than zero (z = 76.354), but this merely reflects the fact that the math proficiency scale itself contains all positive numbers. The estimate of the latent slope mean is $\hat{a}_{F2}$ = 3.654; this value is also statistically significantly greater than zero (z = 34.480), implying that growth in math proficiency throughout high school tended to be in the positive direction (i.e., increased). As the mean structure merely conveys average growth information for this linear model, we may consult the covariance structure parameters to gain a sense of individual differences in intercept and slope. The estimate of the intercept variance is $\hat{c}_{F1}$ = 148.707, which is statistically significantly greater than zero (z = 17.483). From this we may infer that a model positing that all ninth graders start with the same initial amount of math proficiency would be rejected, and that diversity in initial amount of math proficiency exists. The estimate of the slope variance is $\hat{c}_{F2}$ =3.609, which is also statistically significantly greater than zero (z = 4.968). This implies that a model positing that all 9th graders have the same rate of change in math proficiency throughout high school would be rejected, and that diversity in growth rates does exist. Finally, the estimated covariance between the intercept and slope factors is $\hat{c}_{F1F2}$ =−7.214, which is statistically significantly less than zero (z = −3.952). Further interpretation of these parameter estimates follows.

Consider first the intercept parameter values of $\hat{a}_{F1}$ = 32.863 and $\hat{c}_{F1}$ = 148.707, the latter of which may be transformed to an estimated standard deviation for the intercept factor of $(148.707)^{1/2}$ = 12.195. If one may reasonably assume normality for latent math proficiency in ninth grade, then the estimated latent mean and standard deviation could be used to create the distributional representation for ninth-grade latent math proficiency shown in Figure 6.3. More interestingly, we could do the same for the slope parameter values of $\hat{a}_{F2}$ = 3.654 and $\hat{c}_{F2}$ = 3.609, the latter of which may be transformed to an estimated standard deviation for the slope factor of $(3.609)^{1/2}$ = 1.900. Again assuming normality, the estimated latent mean and standard deviation could be used to create the distributional representation for growth in latent math proficiency, also shown in Figure 6.3. We may infer from this second distribution that approximately 68% of students grow annually between 1.709 and 5.509 points along the latent math proficiency continuum, which has the same metric as the measured math proficiency variables. Assuming normality, we may also infer that approximately 3% of students either do not change over time, or in fact decline in latent math proficiency during high school.

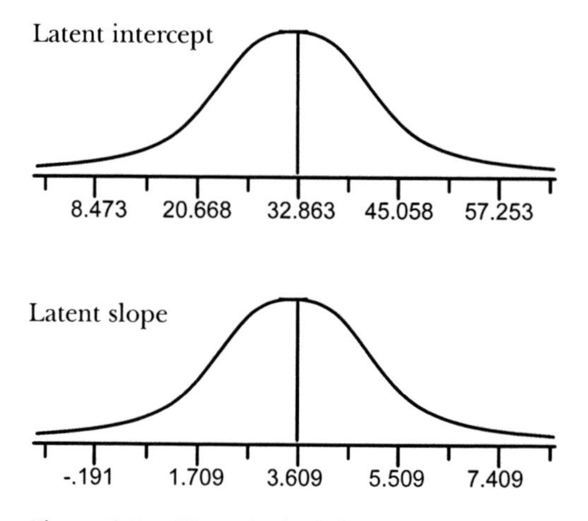

Figure 6.3. Hypothetical distribution for latent intercept and latent slope factors.

Considering the relation between the intercept and slope factors, the estimated covariance of $\hat{c}_{F1F2} = -7.214$ corresponds to an estimated correlation between the intercept and slope factors of –.311. This implies that students who start with high math proficiency tend to grow at a slower rate, while students with lower initial math proficiency tend to grow at a faster rate. This result could be indicative of a ceiling effect of the math proficiency instrument itself, and/or a curricular emphasis targeted toward improving those students with lower initial math proficiency rather than challenging those students with higher initial math proficiency.

Variations on the Linear Latent Growth Model

Changes in growth metric. If a researcher wishes to assess the rate of change per an alternate unit of time, this may be accomplished simply by rescaling the slope factor loadings. For example, had semi-annual growth been of interest in the previous math proficiency example, the growth factor loadings for V1 through V4 could have been fixed to 0, 2, 4, and 6, respectively. This change would have no effect on data–model fit, but would merely rescale the parameters associated with the growth factor F2. Specifically, the estimated factor mean $\hat{a}_{F2}$, which was 3.654 math proficiency units per year, would become 1.827 math proficiency units per half year. As F2 now represents semi-annual growth, the estimated variance $\hat{c}_{F2}$, which was originally 3.609 for annual growth, would become one-fourth of its value (.902). The covariance of F2 with the intercept, $\hat{c}_{F1F2}$, would change from –7.214 to one-half of its value (–3.607).

Changes in reference point. The previous development and example had the initial amount of the variable of interest serving as the intercept. However, other reference points could also be chosen for the intercept factor. For example, F2 loadings of –3, –2, –1, and 0 would make the intercept factor F1 represent the amount of math proficiency in 12th grade rather than 9th. One could also choose F2 loadings of –1.5, –.5, .5, and 1.5, thus making the intercept factor F1 represent the amount of math proficiency halfway between 10th and 11th grade. Even a point outside of the measured range could be selected as the reference point for the intercept factor F1; the F2 loadings of 2, 3, 4, and 5 would make seventh grade the temporal reference point for math proficiency. The researcher may choose any theoretically interesting developmental juncture as the temporal reference for F1, although caution certainly must be exercised when it lies outside of the range of observed measurements. Finally, note that changes in location of F1 will affect the estimated mean $\hat{a}_{F1}$, the estimated variance $\hat{c}_{F1}$, and the estimated covariance $\hat{c}_{F1F2}$.

Unequally spaced time points. Although equally spaced time points have been used in the previous explanations for the linear model, a linear growth system could still be tested with unequally spaced time points. The only difference would be the choice of path coefficients from the slope factor to the measured variables. For example, if data were gathered at 1 year, 3 years, and 6 years after an initial measurement point, the paths from the F2 slope factor to V1 through V4 could simply be fixed to 0, 1, 3, and 6, respectively, to yield a growth rate in an annual metric (assuming the intercept factor F1 to be the initial amount).

LGM METHODOLOGY FOR NONLINEAR DEVELOPMENTAL PATTERNS

Latent growth models certainly need not be restricted to linear forms. When theory dictates a specific nonlinear form, or if a researcher wishes to compare models with linear and nonlinear components (e.g., as in trend analysis in ANOVA), several options exist. First, if one has a theoretical reason to suspect a specific nonlinear functional form for individuals' longitudinal change, then loadings for F2 may be chosen accordingly. For example, across a set of four equally spaced time points, if a researcher theorizes that change is decreasing with each interval, F2 loadings for V1 through V4 could be set to 0, 1, 1.5, and 1.75, respectively, such that each interval's change is half that of the immediately preceding interval. Similarly, F2 loadings for V1 through V4 could be set to 0, 1, $2^{1/2} = 1.414$, and $3^{1/2} = 1.732$, respectively, where growth in the outcome follows the square root of time.

Another option, which could be considered somewhat more exploratory, would be to assume that a single functional form underlies individuals' growth, but that such form is not known to the researcher a priori. In this case the researcher may impose a latent growth model with an unspecified trajectory. For the four equally spaced measurements shown in Figure 6.2, the loadings from the growth factor F2 could be designed as 0, 1, b_{V3F2}, and b_{V4F2}, where the last two parameters are to be estimated rather than fixed to specific values. In such a model the amount of change for an individual in the initial interval from time 1 to time 2 becomes a yardstick against which other change is gauged. Thus, the values of b_{V3F2} and b_{V4F2} represent the amount of change from time 1 to time 3, and from time 1 to time 4, respectively, relative to the amount from time 1 to time 2. For example, consider estimated values of $\hat{b}_{V3F2} = 1.8$ and $\hat{b}_{V4F2} = 2.2$, which, although increasing numerically, indicate a general decline in the rate of change over time. The value of $\hat{b}_{V3F2} = 1.8$ implies that that change from time 1 to time 3 is 1.8 times that from time 1 to time 2, or that the growth from time 2 to time 3 is only .8 times the initial interval. Similarly, the value of 2.2 implies that that change from time 1 to time 4 is 2.2 times that from time 1 to time 2. This means that the growth from time 2 to time 4 is only 1.2 times the initial interval; furthermore, because the growth from time 2 to time 3 was .8 times the initial interval, we may infer that growth from time 3 to time 4 is .4 times the initial interval.

For both the specified and unspecified alternatives for modeling nonlinearity presented above, one may compare multiple models to assess competing hypotheses regarding growth's functional form. In each case, for example, one may fit a linear growth model as well as the nonlinear alternative. For the specified approach, with F2 loadings fixed to specific numerical values reflecting a hypothesized nonlinear trajectory, the linear and nonlinear models do not have a hierarchical relation; thus, comparison may proceed using data–model fit indices such as the Akaike Information Criterion (AIC). For the unspecified strategy, the linear model is nested within that model; as such, the two models share a hierarchical relation and thus may be compared statistically using a χ^2 difference (likelihood ratio) test.

If a researcher suspects that growth is governed by multiple specific functional forms, such as having both linear and quadratic components, a model like that depicted in Figure 6.4 may be fit to the data. In this model F1 is the intercept factor representing the initial amount of the outcome at time 1, F2 is the linear growth factor representing the expected rate of change per unit time, and F3 is the quadratic growth factor capturing the degree of quadratic curvature for each individual. The additional parameters as a result of F3's inclusion include the latent mean a_{F3}, where a positive value indicates upward concavity (i.e., a general tendency for individuals' change to taper off over time), and a negative value indicates downward concavity (i.e., a general tendency for individuals' change to accelerate over time). The latent variance c_{F3} captures the diversity in magnitude of quadratic curvature, where a zero value would indicate a common degree of curvature (that reflected by a_{F3}) for all individuals. Lastly, the latent covariances c_{F1F3} and c_{F2F3} may be estimated, representing the degree to which quadratic curvature relates to individuals' initial amount of the outcome and their linear rate of longitudinal growth.[6] Nonlinear models with multiple functional forms are discussed by Stoolmiller (1995), and an excellent example of this approach is illustrated by Stoolmiller and colleagues (1993).

The concept of additional factors to capture nonlinearity may also be adapted to situations where different growth patterns are believed to occur during different time periods. Imagine, for example, five equally spaced time points in which a linear pattern is hypothesized to govern

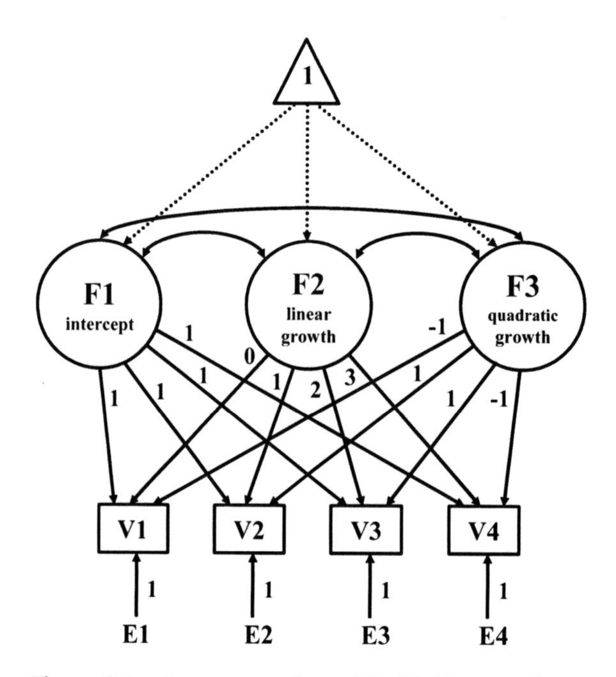

Figure 6.4. Latent growth model with linear and quadratic growth components.

change from the first to third time points, and a potentially different linear pattern from the third through fifth. Reasons for such a model typically involve an event occurring, such as a transition from middle school to high school, or perhaps some kind of treatment intervention. The effect is to yield a discontinuity in the overall growth pattern, making it resemble two lines spliced together—a *spline*. For such *discontinuity* designs there are different options, depending upon the researcher's interest.

Consider the top of Figure 6.5, in which a single individual's linear trajectory is depicted as a solid line from time 1 to time 3 (phase 1), followed by a second linear trajectory from time 3 to time 5 as another solid line (phase 2). If the researcher wishes to view the second phase as entirely separate from the first, that is, as if time 3 were itself an intercept, then the dotted line shows the frame of reference for growth during the second phase. The corresponding model appears in the bottom of Figure 6.5, where F2 is intended to represent linear growth during the first phase and F3 is linear growth during the second phase. Notice that the paths from F2 stop changing in value at time 3, holding at a value of 2 for the remaining measurement points. This effectively makes the combination of F1 and F2 serve as the intercept for the second growth phase, freezing

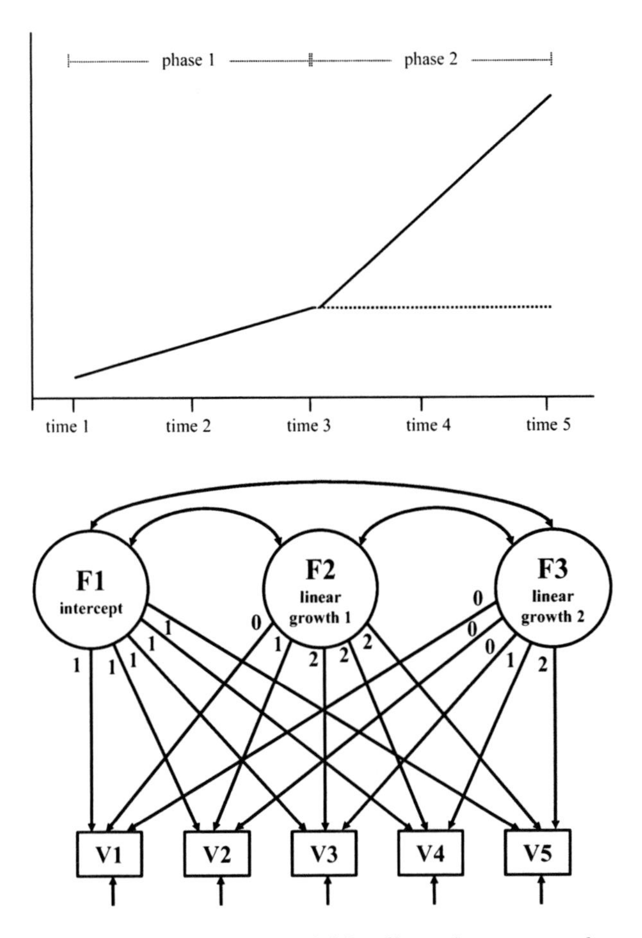

Figure 6.5. Piecewise model for discontinuous growth.

in the level of growth attained by time 3 as the point of departure for the second growth factor F3. During the first three time points F3 lay dormant, with paths of 0; at times 4 and 5, however, the linear growth from time 3 is modeled with the expected linear weights of 1 and 2. Because growth is treated in separate, although potentially related pieces, this approach is sometimes referred to as a *piecewise* growth model.

All of the expected latent parameters exist for this model for both the covariance structure (c_{F1}, c_{F2}, c_{F3}, and c_{F1F2}, c_{F1F3}, c_{F2F3}) and mean structure (a_{F1}, a_{F2}, a_{F3}; not depicted in Figure 6.5). Notice that a special case of this model is that in which a single linear process governs the entire time span; such a model would have variances $c_{F2} = c_{F3}$, covariances $c_{F1F2} = c_{F1F3}$ and $c_{F2F3} = (c_{F2}c_{F3})^{1/2}$ (i.e., F2 and F3 are perfectly correlated), and latent means $a_{F2}=a_{F3}$. Thus, a model with a single continuous process may be compared to the discontinuous model using a χ^2 difference (likelihood ratio) test.[7]

As a variation on the piecewise model, the researcher may wish to frame growth in the second phase somewhat differently. Specifically, one may consider the continuation of the first growth phase as one's baseline, and then model growth in the second phase as potentially *additive* to that process that is already underway. Consider the top of Figure 6.6, where the solid lines depict the same individual's two phases of growth as in Figure 6.5, but where the dotted line now indicates the linear mechanism set into motion in the first phase. The corresponding model in the bottom of Figure 6.6 shows F2 to be the same as in a continuous growth model

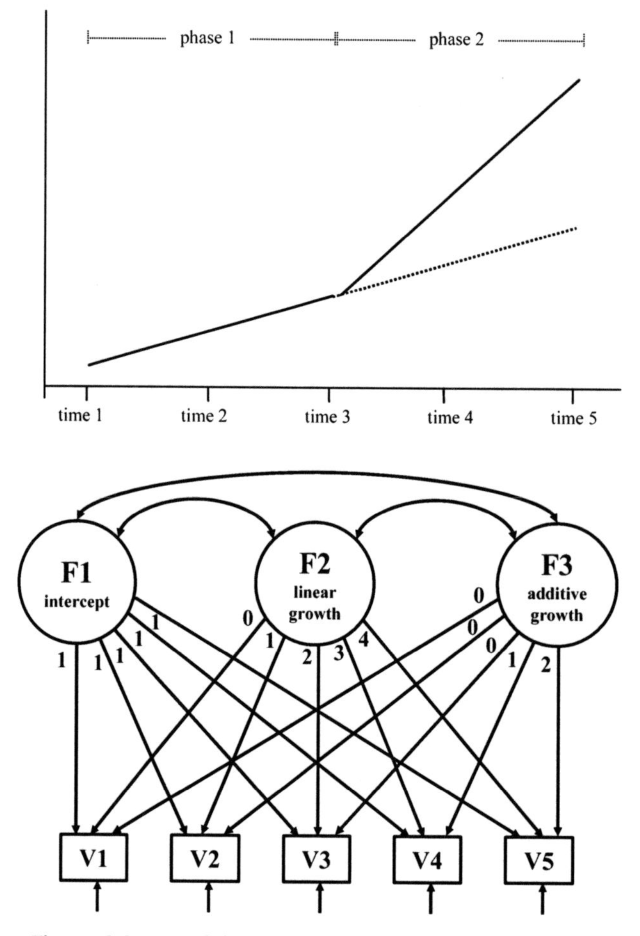

Figure 6.6. Model for additive growth.

spanning all time points, and F3 to be a growth mechanism that does not engage until the second phase, just as in Figure 6.5. A key difference between F3 in Figures 6.5 and 6.6, however, is their frame of reference. Whereas in Figure 6.5 the F3 factor was essentially an entirely new linear growth process, in Figure 6.6 it is an additive process relative to the continuation of the first phase's growth mechanism. Thus, in Figure 6.5 if F3 were irrelevant then it implies that growth has stopped (flattened), while in Figure 6.6 an irrelevant F3 factor means growth continues just as it did in the first phase. This model again has all of the expected covariance and mean structure parameters, but with their interpretation adjusted according to the additive role for F3. Notice again that the single continuous process model is a special case of this model, and perhaps more obviously so than in the piecewise case of Figure 6.5. Quite simply, if variance $c_{F3} = 0$, covariances $c_{F1F3} = c_{F2F3} = 0$, and latent mean $a_{F3} = 0$, one has the traditional linear latent growth model.

ACCOMMODATING EXTERNAL VARIABLES IN LATENT GROWTH MODELS

One of the unique and powerful advantages of utilizing LGM methodology over traditional methods is its ability to incorporate predictors of the latent growth factors, thereby attempting to explain individual differences in latent trajectories. For example, imagine a researcher was interested in the potential effect of mothers' years of education on high school girls' initial math proficiency as well as on the degree to which the girls' math proficiency changes over time. Such questions can be addressed directly by introducing the appropriate predictor variable into the model. Consider the model in Figure 6.7, from which the mean structure has been omitted for simplicity.

Assuming the model fits satisfactorily to the covariance matrix relating V1 through V5 (i.e., including mother's education), one may interpret the new structural parameter estimates. In particular, the $\hat{b}_{F1V5}$ estimate describes the sign and magnitude of the relation between mothers' education and initial math proficiency in the sample examined; the test statistic for this path facilitates inference regarding this relation in the population. As for the proportion of variability in F1 explained by mothers' education, the researcher may compute this as one minus the squared value of the standardized path coefficient from D1 to F1 (which is part of standard SEM computer output). Similarly, the $\hat{b}_{F2V5}$ estimate describes the sign and magnitude of the relation between mothers' education and growth in math proficiency in the sample examined, while the test statistic for this path facilitates inference at the population level. To determine the pro-

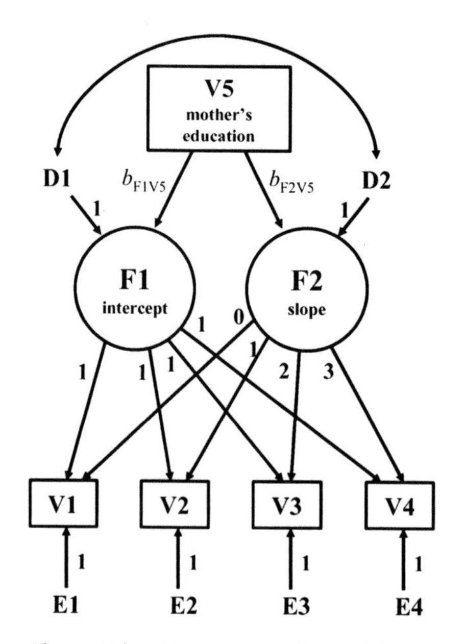

Figure 6.7. Latent growth model with predictor of intercept and slope factors.

portion of variability in F2 explained by mothers' education, the researcher may compute one minus the squared value of the standardized path from D2 to F2.

In general, there are many interesting and highly useful variations to the inclusion of external predictors of latent growth factors. In the above example, the lone predictor mother's education was included. Certainly researchers may wish to include multiple measured predictors in their models in order to assess their impact on growth factors, as is illustrated by numerous papers cited herein (e.g., Patterson, 1993; Stoolmiller et al., 1993). Particularly interesting are predictors within the piecewise and additive growth model framework, where the latter can be used to assess determinants of individual differences in responsiveness to an intervention. Researchers also have the option of using predictors that are latent in nature; that is, an exogenous factor (with its own measured indicators) may be included as a predictor of the growth factors. For example, imagine that a researcher wished to investigate the effect of ninth-grade academic self-concept on initial math proficiency and growth in that proficiency. Rather than using a single measure of academic self-concept as a predictor, the researcher may define a self-concept factor from multiple indicators (e.g., questionnaire items), and build this factor as a causal element into the growth model.[8]

Further examples of measured (as opposed to latent) external predictors in a growth model include group code (e.g., dummy) variables indicating group membership. For example, differences between males and females in initial math proficiency and proficiency growth could be assessed by creating a dichotomous sex variable as a predictor in a dataset containing high school students of both sexes. Statistically significant paths from that dummy variable to the intercept or slope factor would indicate sex differences in initial amount and/or growth in math proficiency in high school. As an extension, if differences among more than two populations are desired, a set of $k - 1$ group code variables may be used as predictors to facilitate inference about differences among the k populations of interest.

VARIATIONS, EXTENSIONS, AND CONCLUSIONS

While the models discussed in the current chapter were chosen to illustrate the basic principles of LGM methods, many additional variations and extensions of these models may be imposed upon one's data as well. Following along the lines of group comparisons, as accomplished above with group code variables, researchers may actually build separate models for different groups of interest if sample size permits. For example, one may wish to compare the effect of two treatments over time. Constraining growth parameters to be equal across groups, and then examining the results of statistical tests of those equality constraints, would facilitate inference about longitudinal differences in the treatments under investigation (for a variation on this with additive growth models, see Muthén & Curran, 1997).

Multiple groups can also be used in a *cohort sequential design* in order to "accelerate" assessment of the longitudinal process. For example, imagine wishing to assess linear change in alcohol consumption among adolescents from ages 12 through 16. Rather than following one cohort of students for 4 years, one could follow, say, three cohorts for 2 years (see Duncan, Duncan, & Hops, 1996). Specifically, the first cohort could be measured at ages 12, 13, and 14; the second at ages 13, 14, and 15; and the third at ages 14, 15, and 16. By fixing growth factor loadings to create a commonly defined intercept (e.g., 0, 1, 2; 1, 2, 3; and 2, 3, 4), and by constraining all common parameters across groups (i.e., latent means, variances, covariances, and corresponding error variances), only 2 years would be required to approximate parameters for the 4-year process.

Latent growth modeling may even be used to compare groups whose membership is not known. *Latent growth mixture models* combine principles of LGM with those of finite mixture models (see Gagné, Chapter 7, this volume) to test hypotheses positing multiple latent growth classes. For example, a model with three classes might be found to fit the data better than models with one or two, suggesting, say, a class of individuals who start low and change very little over time, another class that starts slightly higher and grows considerably, and a third class that starts high and stays high over time. These classes might differ in terms of one or more latent parameters, although differences in mean structure parameters might allow the greatest ability to detect distinct latent growth classes. More detail on such innovations may be found in seminal work by Muthén (e.g., 2001a, 2004).

Note that latent growth models may also have multiple measures at each point in time. As discussed by Willett and Sayer (1996), for example, measures from multiple domains (e.g., reading and math) may have their growth modeled simultaneously within a single model, allowing for the assessment of relations among growth factors across the domains of interest (e.g., how rate of reading growth relates to rate of growth in math). Alternatively, if one has multiple measures at each time point that are believed to be indicators of the same construct (e.g., multiple self-concept questionnaire items), then the researcher is likely more interested in growth in the latent construct underlying those measures rather than in the individual measures themselves. In such cases one may analyze a *second-order latent growth model* (Hancock, Kuo, & Lawrence, 2001; Sayer & Cumsille, 2001), also known as a *curve-of-factors model* (McArdle, 1988), where the growth factors are modeled as second-order factors influencing the first-order constructs whose longitudinal change is of interest. Details of such models, including measurement invariance across time points and requirements of the mean structure, are presented elsewhere (see Hancock et al., 2001; Sayer & Cumsille, 2001), as is the interesting case where subjects' developmental levels necessitate that different indicators be used at different time points in the growth model (Hancock, Buehl, & Ployhart, 2002).

Other LGM innovations are both exciting and too numerous to address, including models with multilevel (e.g., Duncan, Duncan, Li, & Strycker, 2002) or even categorical (e.g., Muthén, 2001b) data. New developments by leading LGM methodologists are constantly underway, further increasing researchers' options for addressing longitudinal change. It is our hope that the introduction offered here has clearly illustrated the benefits of LGM, and in turn will help researchers to utilize these methods in working toward a fuller explication of longitudinal growth in variables from their own areas of study.

APPENDIX: SOFTWARE SYNTAX FOR LINEAR LATENT GROWTH MODEL EXAMPLE

EQS (6.1)

```
/TITLE
 LINEAR LATENT GROWTH MODEL FOR MATH PROFICIENCY
/SPECIFICATIONS
 CASES=1000; VAR=4; MA=COV; AN=MOMENTS;
/LABELS
 V1=MATHSC9; V2=MATHSC10; V3=MATHSC11; V4=MATHSC12;
 F1=INTERCEPT; F2=SLOPE;
/EQUATIONS
 V1 = 1F1 + 0F2 + E1;
 V2 = 1F1 + 1F2 + E2;
 V3 = 1F1 + 2F2 + E3;
 V4 = 1F1 + 3F2 + E4;
 F1 = *V999 + D1;
 F2 = *V999 + D2;
/VARIANCES
 E1 TO E4 = *;
 D1 TO D2 = *;
/COVARIANCE
 D1,D2 = *;
/MATRIX
 204.11
 139.20 190.05
 136.64 135.19 166.53
 124.86 130.32 134.78 159.87
/MEANS
 32.97 36.14 40.42 43.75
/PRINT
 FIT=ALL;
/END
```

SIMPLIS (8.7)

```
LINEAR LATENT GROWTH MODEL FOR MATH PROFICIENCY
OBSERVED VARIABLES
MATHSC9 MATHSC10 MATHSC11 MATHSC12
COVARIANCE MATRIX
```

```
204.11
139.20 190.05
136.64 135.19 166.53
124.86 130.32 134.78 159.87
MEANS
32.97 36.14 40.42 43.75
SAMPLE SIZE = 1000
LATENT VARIABLES
INTERCEPT SLOPE
RELATIONSHIPS
MATHSC9 = 1*INTERCEPT + 0*SLOPE
MATHSC10 = 1*INTERCEPT + 1*SLOPE
MATHSC11 = 1*INTERCEPT + 2*SLOPE
MATHSC12 = 1*INTERCEPT + 3*SLOPE
INTERCEPT = CONST
SLOPE = CONST
PATH DIAGRAM
END OF PROBLEM
```

Mplus (3.11)

Data file: MATHPROF.DAT

```
32.97   36.14  40.42  43.75
204.11
139.20 190.05
136.64 135.19 166.53
124.86 130.32 134.78 159.87
```

Input file, option 1 (general format)

```
TITLE:       LINEAR LATENT GROWTH MODEL
             FOR MATH PROFICIENCY

DATA:        FILE IS C:\MATHPROF.DAT;
             TYPE IS COVARIANCE MEANS;
             NOBS IS 1000;

VARIABLE:    NAMES ARE y1-y4;

ANALYSIS:    TYPE IS MEAN;

MODEL:       INTERCEP BY y1-y4@1;
             SLOPE BY y1@0 y2@1 y3@2 y4@3;
             INTERCEP* SLOPE*;
```

```
          [y1-y4@0 INTERCEP* SLOPE*];
          INTERCEP WITH SLOPE*;

OUTPUT:  SAMP;
```

Input file, option 2 (LGM-specific format)

```
TITLE:     LINEAR LATENT GROWTH MODEL
           FOR MATH PROFICIENCY

DATA:      FILE IS C:\MATHPROF.DAT;
           TYPE IS COVARIANCE MEANS;
           NOBS IS 1000;

VARIABLE:  NAMES ARE y1-y4;

MODEL:     INTERCEP SLOPE | y1@0 y2@1 y3@2 y4@3;

OUTPUT:    SAMP;
```

AUTHORS' NOTE

Parts of this chapter appeared as Lawrence and Hancock (1998), and are reproduced here with the permission of the American Counseling Association.

NOTES

1. In the current chapter the term "growth" will be used in generic reference to change over time, regardless of whether that change is positive or negative.
2. Note that fixing the path from F2 to V1 to a value of 0 is equivalent to omitting the path altogether. Including the path in Figure 6.2 with a coefficient of 0 was done simply to illustrate the slope factor's impact on V1 relative to that on all other measured variables.
3. Note that part of the researcher's hypothesis regarding growth could involve specifying the equality of the variances of the variables' errors (i.e., $c_{E1}=c_{E2}=\ldots$). Such a hypothesis could be made if the researcher believed that measurement error in all variables, as well as the extent to which individuals deviated from the estimated growth function, were consistent across time points. The imposition of such an equality constraint creates a more restrictive, though conceptually simpler explanation of the data (see, e.g., Dunn et al., 1993, p. 128; Patterson, 1993).
4. Two elements regarding this mean structure may seem inconsistent with latent means modeling in common practice (e.g., Thompson & Green, Chapter 5, this volume). First, one notices the absence of paths from the pseudovariable to the measured variables, creating variable-specific intercept terms. Second, no factor mean needs to be fixed for identification

purposes. In LGM, two properties make these requirements unnecessary. First, rather than each variable having its own intercept term, as in latent means models, one common intercept factor is built directly into the model (making variable-specific intercept terms superfluous). Second, because so many loadings are constrained, the model is identified and the factors are given the scale associated with all of the indicators simultaneously (which, in LGM, have the same units).

5. It is worth pointing out that a linear latent growth model with only three time points would have six parameters estimated in its covariance structure, and two in its mean structure. While the model as a whole would have 1 degree of freedom, the covariance structure itself would be saturated. As such, the covariance structure alone does not pose a rejectable test of the hypothesis of linearity in individual growth; only the mean structure, from whence the single degree of freedom emanates, facilitates a test of linearity at the group (mean) level. Thus, three time points represents somewhat of a lower bound on fitting a latent growth model, while four time points represents a lower bound on rejectability of hypotheses regarding a single specified functional form.
6. Note that the nonlinear model depicted in Figure 6.4, with four time points, would have 10 parameters estimated in its covariance structure and three in its mean structure. Thus, while the model as a whole would have 1 degree of freedom, the covariance structure itself would be saturated and thus not pose a rejectable test of the hypothesis of nonlinearity. However, by comparing the linear and nonlinear models one may infer the reasonableness of including the hypothesized nonlinear (i.e., quadratic) component. More generally, as expected, the number of measured time points will limit the number of nonlinear components to the latent growth model.
7. An interesting extrapolation of piecewise growth models is to consider latent change between each pair of adjacent points as interesting and worthy of modeling directly. Variations of these *latent difference score models* are addressed in work by McArdle (2001) and Steyer and colleagues (e.g., Steyer, Eid, & Schwenkmezger, 1997; Steyer, Partchev, & Shanahan, 2000).
8. The predictors discussed herein, whether measured or latent, are examples of *time-independent covariates*, whose values are assumed to be stable over time. In the case of mother's education, the assumption was that the mother received no additional education after her daughter's first math proficiency measurement in ninth grade. However, one can certainly imagine predictors whose values change over time; such variables may be termed *time-dependent covariates* and are modeled as covarying predictors of the corresponding measured variables directly. In this manner the growth that is modeled is that above and beyond the variation accounted for by the covariate at each time point.

REFERENCES

Bollen, K. A., & Curran, P. J. (2005). *Latent curve models: A structural equation approach.* New York: Wiley.

Collins, L. M., & Sayer, A. G. (Eds.). (2001). *New methods for the analysis of change.* Washington, DC: American Psychological Association.

Curran, P. J., Stice, E., & Chassin, L. (1997). The relation between adolescent alcohol use and peer alcohol use: A longitudinal random coefficients model. *Journal of Consulting and Clinical Psychology, 65*, 130–140.

Duncan, S. C., & Duncan, T. E. (1994). Modeling incomplete longitudinal substance use data using latent variable growth curve methodology. *Multivariate Behavioral Research, 29*, 313–338.

Duncan, S. C., Duncan, T. E., & Hops, H. (1996). Analysis of longitudinal data within accelerated longitudinal designs. *Psychological Methods, 1*, 236–248.

Duncan, T. E., Duncan, S. C., Li, F., & Strycker, L. A. (2002). Multilevel modeling of longitudinal and functional data. In D. S. Moskowitz & S. L. Hershberger (Eds.), *Modeling intraindividual variability with repeated measures data: Methods and applications* (pp. 171–202). Mahwah, NJ: Erlbaum.

Duncan, T. E., Duncan, S. C., Strycker, L. A., Li, F., & Alpert, A. (1999). *An introduction to latent variable growth curve modeling: Concepts, issues, and applications.* Mahwah, NJ: Erlbaum.

Dunn, G., Everitt, B., & Pickles, A. (1993). *Modelling covariances and latent variables using EQS*. London: Chapman & Hall.

Gottman, J. M. (Ed.). (1995). *The analysis of change*. Mahwah, NJ: Erlbaum.

Hancock, G. R., Buehl, M., & Ployhart, R. E. (2002, April). *Second-order latent growth models with shifting indicators*. Paper presented at the annual meeting of the American Educational Research Association, New Orleans, LA.

Hancock, G. R., Kuo, W., & Lawrence, F. R. (2001). An illustration of second-order latent growth models. *Structural Equation Modeling: A Multidisciplinary Journal, 8*, 470–489.

Lawrence, F. R., & Hancock, G. R. (1998). Assessing change over time using latent growth modeling. *Measurement and Evaluation in Counseling and Development, 30*, 211–224.

McArdle, J. J. (1988). Dynamic but structural equation modeling with repeated measures data. In J. R. Nesselroade & R. B. Cattell (Eds.), *Handbook of multivariate experimental psychology* (*Vol. 2*, pp. 561–614). New York: Plenum Press.

McArdle, J. J. (2001). A latent difference score approach to longitudinal dynamic structural analyses. In R. Cudeck, S. du Toit, & D. Sorbom (Eds.), *Structural equation modeling: Present and future—A Festschrift in honor of Karl Jöreskog* (pp. 342–380). Lincolnwood, IL: Scientific Software International, Inc.

Moskowitz, D. S., & Hershberger, S. L. (Eds.). (2002). *Modeling intraindividual variability with repeated measures data: Methods and applications*. Mahwah, NJ: Erlbaum.

Muthén, B. (2001a). Latent variable mixture modeling. In G. A. Marcoulides & R. E. Schumacker (Eds.), *New developments and techniques in structural equation modeling* (pp. 1–33). Mahwah, NJ: Erlbaum.

Muthén, B. (2001b). Second-generation structural equation modeling with a combination of categorical and continuous latent variables: New opportunities for latent class/latent growth modeling. In L. M. Collins & A. Sayer (Eds.), *New methods for the analysis of change* (pp. 291–322). Washington, DC: American Psychological Association.

Muthén, B. (2004). Latent variable analysis: Growth mixture modeling and related techniques for longitudinal data. In D. Kaplan (Ed.), *The Sage hand-*

book of quantitative methodology for the social sciences (pp. 345–368), Thousand Oaks, CA: Sage.

Muthén, B., & Curran, P. (1997). General longitudinal modeling of individual differences in experimental designs: A latent variable framework for analysis and power estimation. *Psychological Methods, 2*, 371–402.

Patterson, G. R. (1993). Orderly change in a stable world: The antisocial trait as a chimera. *Journal of Consulting and Clinical Psychology, 61*, 911–919.

Rogosa, D., & Willett, J. B. (1985). Understanding correlates of change by modeling individual differences in growth. *Psychometrika, 50*, 203–228.

Sayer, A.G., & Cumsille, P.E. (2001). Second-order latent growth models. In L. Collins & A. G. Sayer (Eds.), *New methods for the analysis of change* (pp. 177–200). Washington, DC: American Psychological Association.

Short, R., Horn, J. L., & McArdle, J. J. (1984). Mathematical-statistical model building in analysis of developmental data. In R. N. Emde & R. J. Harmon (Eds.), *Continuities and discontinuities in development* (pp. 371–401). New York: Plenum Press.

Steyer, R., Eid, M., & Schwenkmezger, P. (1997). Modeling true intraindividual change: True change as a latent variable. *Methods of Psychological Research Online, 2*, 21–33.

Steyer, R., Partchev, I., & Shanahan, M. J. (2000). Modeling true intraindividual change in structural equation models: The case of poverty and children's psychosocial adjustment. In T. D. Little, K. U. Schnabel, & J. Baumert (Eds.), *Modeling longitudinal and multilevel data: Practical issues, applied approaches, and specific examples* (pp. 109–126). Mahwah, NJ: Erlbaum.

Stoolmiller, M. (1995). Using latent growth curve models to study developmental processes. In J. M. Gottman (Ed.), *The analysis of change* (pp. 103–138). Mahwah, NJ: Erlbaum.

Stoolmiller, M., Duncan, T., Bank, L., & Patterson, G. R. (1993). Some problems and solutions in the study of change: Significant patterns in client resistance. *Journal of Consulting and Clinical Psychology, 61*, 920–928.

Willett, J. B., Ayoub, C. C., & Robinson, D. (1991). Using growth modeling to examine systematic differences in growth: An example of change in the functioning of families at risk of maladaptive parenting, child abuse, or neglect. *Journal of Consulting and Clinical Psychology, 59*, 38–47.

Willett, J. B., & Sayer, A. G. (1994). Using covariance structure analysis to detect correlates and predictors of individual change over time. *Psychological Bulletin, 116*, 363–381.

Willett, J. B., & Sayer, A. G. (1996). Cross-domain analysis of change overtime: Combining growth modeling and covariance structure analysis. In G. A. Marcoulides & R. E. Schumacker (Eds.), *Advanced structural equation modeling. Issues and techniques* (pp. 125–157). Mahwah, NJ: Erlbaum.

Yandell, B. S. (1997). *Practical data analysis for designed experiments*. London: Chapman & Hall.

CHAPTER 7

MEAN AND COVARIANCE STRUCTURE MIXTURE MODELS

Phill Gagné

Mixture modeling is becoming an increasingly useful tool in applied research settings. At the most basic end of the continuum, such methods might be used to determine whether a single univariate dataset arose from one population or from a mixture of multiple populations differing in their univariate distributions (e.g., means and/or variances). More advanced applications of mixture modeling are used to assess potential mixtures of populations that have different multivariate distributions (e.g., mean vectors and/or covariance matrices). Tests of mixtures can also be conducted using samples in which a possible mixture exists of multiple populations differing in latent variable distributions, which in turn yield differences in measured variable distributions.

We will continue with our introduction to mixture modeling by briefly revisiting several nonmixture analyses that should be quite familiar. Such a review will help to set up our discussion of mixture analyses and to contrast the mixture procedures with their nonmixture analogues. From there, we will provide a general definition of a mixture model and then move into a conceptual and mathematical discussion of select mixture

Structural Equation Modeling: A Second Course, 197–224

models, with an eventual focus on the application of mixture models to structural equation modeling (SEM). The chapter concludes with examples of using the statistical software program Mplus (v3.01; Muthén & Muthén, 2004) to conduct a mixture measured-variable path analysis and a mixture confirmatory factor analysis.

POPULATIONS SPECIFIED A PRIORI

Among the simplest of statistical analyses are those for making comparisons of population means. Analysis of variance (ANOVA) is commonly used to test the null hypothesis of equal means among $J \geq 2$ populations measured on a single variable, with the population membership of each observation in a sample known *a priori*. The extension of ANOVA to two or more dependent variables is multivariate analysis of variance (MANOVA), which is used to determine whether there are differences in the centroids (i.e., vectors of means) of two or more populations defined *a priori*.

Somewhat more complex are the analyses of relations among observed variables. Pearson zero-order correlations are used to test relations between two observed variables with no causal inferences made. Multiple regression features the construction of an equation for predicting a manifest criterion variable from one or more manifest predictor variables in a sample of data. The observations in the sample are typically assumed to be from a single population, but that is not a necessary assumption. In samples comprised of two or more populations, with population membership defined *a priori*, multiple regression can be used to test the equality of parameters (e.g., slopes) across populations. A straightforward application is to investigate whether the nature/extent of the relation between the one predictor and the criterion variable differs across populations. Another application of multiple regression is to evaluate a possible difference in the population means on the criterion variable, while taking the effects of the predictor variable(s) into account, a process that is equivalent to analysis of covariance (ANCOVA).

In confirmatory factor analysis (CFA), relations among observed variables are explained by positing that they are caused by one or more latent variables. A CFA model, generally speaking, represents relations in one population from which all of the observations are assumed to have originated, but CFA can be readily extended to the comparison of multiple populations when population membership is known *a priori*. Such a test might be conducted, for example, to determine whether the pattern of factor loadings is invariant across populations (see, e.g., Byrne, Shavelson, & Muthén, 1989). In a special case of CFA called latent growth modeling (LGM; see, e.g., Duncan, Duncan, Strycker, Li, & Alpert, 1999), measurements are taken of the same variables in the same sample at multiple time points, with the intercept and slope of change over time modeled as latent variables that influence the observed variables. LGM can be extended to compare the means and variances of the intercept factor and the slope factor across multiple populations when population membership is known *a priori*.

In measured-variable path analysis (MVPA), the models depict the structural and nonstructural relations among observed variables, but each observed variable is viewed as perfectly representing a latent variable. When the latent factors are indicated by multiple imperfectly measured manifest variables, latent-variable path analysis (LVPA) is used to analyze the hypothesized structural and nonstructural relations in the model. MVPA and LVPA typically assume that the data come from a single population, but they can be extended to situations in which the data are from multiple populations when each observation's population membership is known *a priori*. For MVPA, tests involving multiple populations would typically focus on invariance in the structural relations, while hypotheses involving multiple populations in LVPA could be about invariance in the structural relations as well as the measurement portion of the model (see, e.g., Kline, 2004).

POPULATIONS NOT SPECIFIED

We have touched on several analyses that test the equality of various parameters in multiple populations with the benefit of information about population membership for each observation in the sample. For such analyses, obtaining and statistically comparing subsample parameters are relatively straightforward processes. When population membership is not known, or when it is not even known whether a mixture of populations exists within a sample, similar statistical questions can be addressed, but the analyses are a bit more complicated.

Before detailing such analyses, we should establish a definition of the term *mixture analysis*. In a manner of speaking, any sample that is made up of observations from two or more populations can be thought of as a mixed sample. In ANOVA, for example, the full sample is known to be drawn from multiple populations and population membership is known for each observation, so parameters for each population can be estimated and compared directly. Because population membership is known *a priori*, ANOVA is not formally considered a mixture analysis. A mixture analysis is called upon when there is a research question to be addressed that involves the parameter estimates of potentially multiple populations but the dataset lacks information about population membership. A mixture analysis is therefore an analysis that estimates parameters for a given number of populations in a single dataset without the availability of a classification variable or other such *a priori* information about population membership with which to sort the data (see, e.g., Everitt & Hand, 1981; Lindsay, 1995; McLachlan & Peel, 2000; Titterington, Smith, & Makov, 1985).

As a simple example of the application of a mixture analysis, consider Figure 7.1, which illustrates two different univariate datasets. The observations contributing to the distribution in Figure 7.1a all came from the same normally distributed population; describing the data with a single mean and variance (i.e., a single distribution) is therefore appropriate. In Figure 7.1b, we see that the distribution of the sample as a whole seems to be bimodal. A single normal curve would not fit the data as well as fitting two normal distributions with estimated means that correspond to the two peaks seen in the bimodal distribution and with estimated variances that may (but do not necessarily) differ between the two distributions.

If the data yielding Figure 7.1b had included population membership, then a statistical comparison of population means and/or variances would be fairly straightforward, because parameters could be estimated for each population using the observations known to be from each population. Without *a priori* knowledge of population membership, however, a mixture analysis would be needed to determine the nature of the mixture and the parameter estimates of interest. Applied to the data in Figure 7.1b, a mixture analysis would, in a manner of speaking, "reverse engineer" the estimated population distributions by finding the two sets of parameter estimates (mean and variance of Population 1 and mean and variance of Population 2) that best fit the data. The two-population model could then be compared to the one-population model via a measure of data–model fit to determine which is the more desirable model.

Two very important assumptions are made in the above example and must be made in all mixture analyses. First, conducting a mixture analysis typically involves fitting the data to multiple models that differ in the hypothesized number of populations, but any one mixture model needs to have the hypothesized number of populations specified. Second, in addition to a specific number of populations, a mixture model needs to be provided a distributional form for the potential population distributions. In the above example, the assumption is made that all potential population distributions are normal. Other distributions, however, can be applied in mixture analyses, and it is not necessary to assume that each hypothesized distribution has the same form. In this chapter, we limit our discussion to normal distributions as many statistical analyses assume normality in the populations.

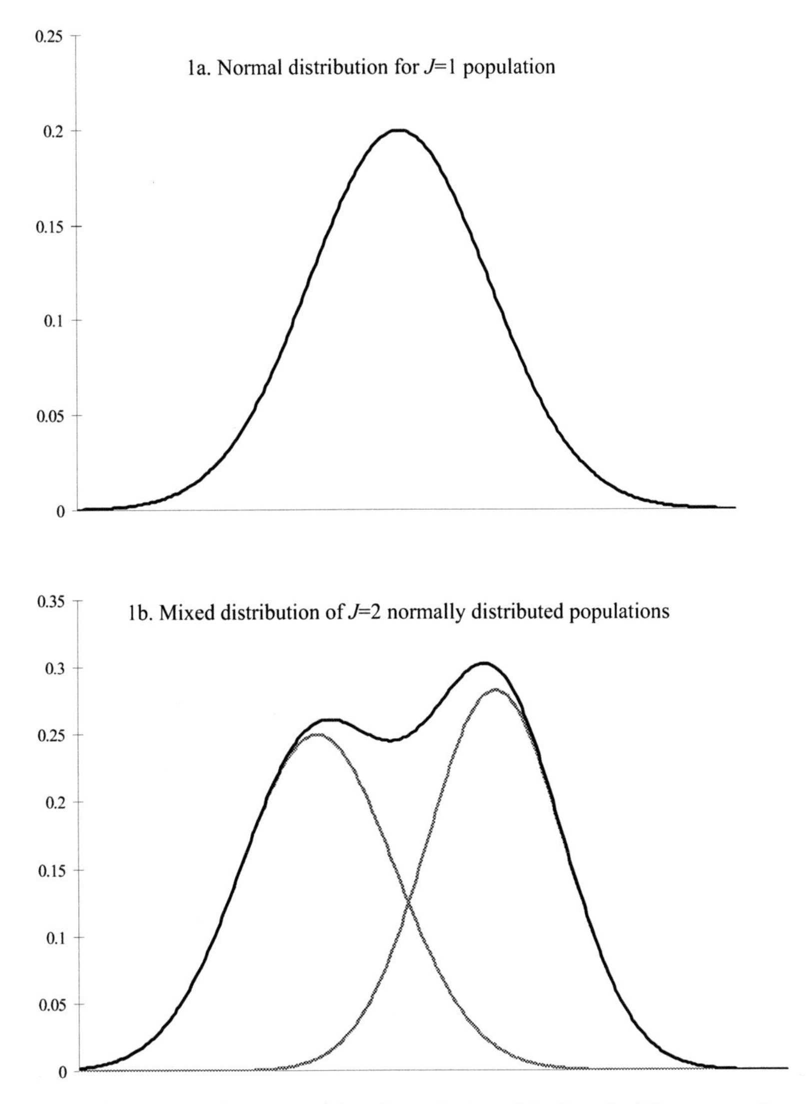

Figure 7.1. Distributions arising from $J=1$ and $J=2$ underlying normal populations.

The remainder of this chapter is spent describing specific mixture analyses. We discuss scenarios as relatively simple as a univariate mixture with unknown population membership all the way up to testing for possible mixtures of latent variable distributions when population membership is not available *a priori*. The methods described in the previous section (i.e., ANOVA, multiple regression, CFA, etc.) are assumed to be familiar or at least easily refreshed. The analyses that follow, however, are assumed to be new, so much more detail is provided for each.

MEASURED VARIABLE MIXTURE MODELS

Univariate Mixtures

As a simple case, suppose you gather a sample of observations and obtain data on a single continuous variable (X). Suppose also that you expect the population distribution to be normal, but instead, the distribution in your sample ends up being bimodal. At this point, you have a couple of options. You can insist that sampling error and/or an insufficient sample size caused the apparent deviation from normality and maintain that the data are actually from a single normally distributed population. You could instead be more statistically rigorous and compare a single-population model to a mixture model using the procedure described below.

The fit of a single normal distribution in a sample can be determined by using the sample mean and variance as estimates of the parameters in the normal probability density function

$$f(x_i) = (1/\sqrt{2\pi\sigma^2})e^{(-.5)(x_i-\mu)^2/(\sigma^2)}, \tag{1}$$

where x_i is the value of an observation on variable X, μ is the mean of variable X in the population, σ^2 is the variance of X in the population, and $f(x_i)$ is proportional to the probability of an observation with value x_i in the normally distributed population. The likelihood (L_i) for the ith observation within that one population is computed by substituting its value of X into the above equation and evaluating $f(x_i)$. For a sample of N observations, assuming independence of observations, the joint likelihood is computed as

$$\prod_{i=1}^{N} L_i, \tag{2}$$

and their log-likelihood is computed as

$$\ln\left(\prod_{i=1}^{N} L_i\right) = \sum_{i=1}^{N} [\ln(L_i)] = \Psi. \tag{3}$$

Using Ψ, three measures of data–model fit can be conveniently calculated, only one of which will be introduced at the moment (the other two are used for model comparison, and we have only one model so far). The one measure of fit we can make use of at this point is χ^2, where

$$\chi^2 = -2\Psi, \tag{4}$$

which can be used to test the null hypothesis of perfect fit of a distribution at a desired level. The degrees of freedom for this χ^2 are computed as N–q, where q is the number of parameters to be estimated in order to fit the model; in the current example, $q = 2$ (the sample mean and the sample variance used to estimate the population parameters in the density function).

Evaluating the model of a single underlying distribution is relatively straightforward, because parameter estimates are readily available: The full sample is hypothesized to follow the same normal distribution, so the sample can provide estimates for that distribution. If a mixture of samples is postulated to be from, say, two normally distributed populations, without knowledge of population membership, then parameters for two separate distributions need to be estimated. A mixing proportion that quantifies the proportion of the dataset that is from each of the distributions must also be estimated. Equation 1 is applied for the parameters of each of the two hypothesized populations to give the respective likelihoods of the ith observation, given the different sets of population parameters:

$$L_{i1} = (1/\sqrt{2\pi\sigma_1^2})e^{(-.5)(x_i - \mu_1)^2/(\sigma_1^2)} \tag{5}$$

and

$$L_{i2} = (1/\sqrt{2\pi\sigma_2^2})e^{(-.5)(x_i - \mu_2)^2/(\sigma_2^2)}, \tag{6}$$

with values of μ and σ^2 to be estimated for each distribution.

Because the full sample is postulated to be a mixture from these two distributions, with population membership not known, the likelihood for the ith observation is a weighted sum of its likelihood in each distribution,

$$L_i = \varphi L_{i1} + (1 - \varphi)L_{i2}, \tag{7}$$

where φ is the mixing proportion, directly interpreted as the proportion of the full sample that was drawn from Population 1. The values of the means, variances, and the mixing proportion are estimated via an iterative process seeking to maximize Ψ, where (combining Equations 3 and 7)

$$\Psi = \sum_{i=1}^{N}[\ln(L_i)] = \sum_{i=1}^{N}[\ln(\varphi \cdot L_{i1} + (1-\varphi)\cdot L_{i2})] \,. \qquad (8)$$

It is important to note that the value of the mixing proportion, φ, is estimated while the other parameters are being estimated. Population membership is not known, so the proportion of the sample that came from each population is also not known and could take on any value between 0 and 1 (but not including 0 or 1, because either of those would eliminate one of the two populations and thereby defeat the purpose of trying to fit a model that has two populations).

With Ψ maximized, a χ^2 statistic can be computed for the model to evaluate its absolute fit, with $q = 5$ in the two-population situation (two means, two variances, and one mixing proportion). Our intention, however, is to determine whether postulating two normal distributions operating in the sample yields a more desirable model than a single normal distribution, so a comparison between these two models must be made. Note that the likelihood associated with the two-population model cannot be less than the likelihood from the single-population model, so the question of fit is not, "Are the data more likely, given two distributions than one?" but instead is, "Is the increase in the likelihood worth the cost of having to estimate more parameters and thereby having a less parsimonious model?"

Now that we have two models to compare, it is more meaningful to describe the two measures of fit alluded to above that are used for model comparison, AIC and BIC. Before we discuss the role of these information criteria in mixture modeling, note that the use of a χ^2 difference test (likelihood ratio test) was not introduced as a way to compare mixture models, nor will it be described later as a tool for making such a comparison. The χ^2 difference test cannot be used, because the transition from, say, a three-population mixture model to a two-population mixture model can be accomplished via either of two ways that differ in terms of change in degrees of freedom. For a model with three hypothesized populations, setting the third population's proportion φ_3 equal to 0 eliminates the third population via only one restriction. The third population can also be removed from the model by leaving φ_3 unrestricted but setting all of the other parameters in the third population equal to the corresponding parameters in one of the other two populations. Thus, the number of degrees of freedom associated with such a χ^2 difference test is indeterminate.

The information criteria measures AIC (Akaike, 1987) and BIC (Schwarz, 1978) can be computed rather easily for each model separately, using quantities we have already obtained:

$$\text{AIC} = -2\Psi + 2q \tag{9}$$

$$\text{BIC} = -2\Psi + \ln(N)q. \tag{10}$$

As measures of comparative data–model fit, AIC and BIC provide a convenient way to examine the relative fit of the mixture model against the model that has just one population. The AIC values for the models are compared, and the model with the smaller value is considered the preferred model. The application of BIC for testing relative fit is done in the same manner, with the same decision rule. When AIC and BIC disagree about which is the preferred model, the difference will generally be that AIC favors a more complex model than BIC favors. A thorough study of the index which is preferred in such situations has not been conducted in the context of mixture models, so a disagreement between the information criteria in a mixture modeling context should be resolved either by relying on the theoretical viability of the involved models or by treating both of the models as having comparable utility.

A researcher can hypothesize more than two distributions in a mixture analysis by incorporating additional terms into the equation for L_i. For each population, a new φ is needed, so in the current univariate example, each additional population requires three additional parameters to be estimated: a mean; a variance; and a mixing proportion. Furthermore, restrictions of statistical or theoretical interest can be made on the parameter estimates. For example, restricting the variances of all of the populations to be equal while not constraining the means to be equal is similar to a one-way ANOVA but without predetermined populations.

Regression Mixture Models

Suppose you have a sample in which X and Y are measured for the purposes of simple linear regression ($y = \alpha + \beta x + e$). Concerned that these data might be a mixture of populations that have different intercept parameters and/or slope parameters for the regression of Y on X, you decide to conduct a mixture regression analysis. As with nonmixture regression, it is assumed in mixture regression that the residuals are normally distributed. The likelihood for an individual observation, given the hypothesis of a single set of parameters for the full sample, is thus

$$L_i = (1/\sqrt{2\pi\sigma_e^2})e^{(-.5)(y_i-\alpha-\beta x_i)^2/(\sigma_e^2)}. \quad (11)$$

The error variance, σ_e^2, replaces the observed score variance of the univariate likelihood (Equation 1), and the error of prediction, $e_i = y_i - \alpha - \beta x_i$, replaces the deviation score of the univariate likelihood, with α, β, and σ_e^2 estimated from the sample. Equation 3 is employed to use the individual likelihoods to determine the likelihood of the full sample as before, $\Psi = \sum_{i=1}^{N} [\ln(L_i)]$. The evaluation of fit proceeds in the same manner as in the univariate case, using Equation 4, $\chi^2 = -2\Psi$, to test a single model, while for later model comparison purposes, we can compute information criteria, AIC $= -2\Psi + 2q$ and BIC $= -2\Psi + \ln(N)q$.

If a mixture of samples from two populations is postulated in a regression context, then two regression equations are estimated simultaneously, so two likelihoods need to be combined for each observation. Based on the general formula (Equation 11), we obtain

$$L_{i1} = (1/\sqrt{2\pi\sigma_{e_1}^2})e^{(-.5)(y_i-\alpha_1-\beta_1 x_i)^2/(\sigma_{e_1}^2)}, \quad (12)$$

and

$$L_{i2} = (1/\sqrt{2\pi\sigma_{e_2}^2})e^{(-.5)(y_i-\alpha_2-\beta_2 x_i)^2/(\sigma_{e_2}^2)}, \quad (13)$$

which correspond, respectively, to the likelihood based on the estimated parameters for Population 1 and the likelihood of an observation based on the estimated parameters for Population 2. As with a mixture for a single measured variable, the likelihood for observation i in the mixture model of two populations is computed per Equation 7, as a weighted sum of the two likelihoods, $L_i = \varphi L_{i1} + (1 - \varphi)L_{i2}$, where the mixing proportion, φ, is again the proportion of the sample drawn from Population 1. Assessing the fit of the mixture relative to the single population is again a matter of computing Ψ for each model and using it to compute AIC and BIC for each model.

A straightforward test of the difference between intercepts and/or slopes of the populations can be conducted by placing statistically or theoretically meaningful restrictions on the parameters in a regression mixture model. For example, restricting the analysis to have equal estimates of β and σ_e^2 across populations leaves the intercept as the only potential source of difference across the populations. If the mixture model is favored, then the sample is deemed a mixture of populations that differ in their adjusted average value of Y. In this way, the equivalent of ANCOVA has been conducted, with population membership unknown.

MEAN AND COVARIANCE STRUCTURE MIXTURE MODELS

After handling a univariate mixture model and a regression mixture model, we are now ready to use a variable that is not observed to distinguish among populations that are not observed. I am referring to mixture modeling when the data could be a mixture of multiple populations that might differ in their distributions of latent variables, in the relations between latent variables and manifest variables, and/or in the relations among the latent variables. In this section, we review the foundations of a single-population CFA as the introduction to the presentation of mixture CFA. A general latent mixture modeling framework will subsequently be rendered and then built upon for the discussion of mixture MVPA and mixture LVPA.

CFA Mixture Models

For the purpose of mixture model comparison, the single-population CFA model is often estimated, because it represents the hypothesis that a single population gave rise to the sample. In such a model, the ith observation's vector of values, $\mathbf{x}_i$, on the p manifest variables indicating the m factors, is the function

$$\mathbf{x}_i = \boldsymbol{\tau} + \boldsymbol{\Lambda}\boldsymbol{\xi}_i + \boldsymbol{\delta}_i, \tag{14}$$

where $\boldsymbol{\tau}$ is a $p \times 1$ vector of variable intercept terms, the $m \times 1$ vector $\boldsymbol{\xi}_i$ contains the values for the ith observation on the theoretical latent variables in $\boldsymbol{\xi}$, the factor loadings (i.e., the unstandardized slopes for the theoretical regression of X on each ξ) are contained in the $p \times m$ matrix $\boldsymbol{\Lambda}$, and $\boldsymbol{\delta}_i$ is a $p \times 1$ vector of residuals for the ith observation. For this general CFA model, the first moment implied by the model is

$$\hat{\boldsymbol{\mu}}_x = E[\mathbf{x}_i] = \boldsymbol{\tau} + \boldsymbol{\Lambda}\boldsymbol{\kappa}, \tag{15}$$

where $\boldsymbol{\kappa}$ is the $m \times 1$ vector of factor means ($\boldsymbol{\kappa}$ is a scalar if there is only one factor). The second moment implied by the model is

$$E[(\mathbf{x}_i - \hat{\boldsymbol{\mu}}_X)(\mathbf{x}_i - \hat{\boldsymbol{\mu}}_X)'] = \hat{\boldsymbol{\Sigma}} = \boldsymbol{\Lambda\Phi\Lambda}' + \boldsymbol{\Theta}, \tag{16}$$

where $\boldsymbol{\Phi}$ is the $m \times m$ factor variance–covariance matrix and $\boldsymbol{\Theta}$ is the $p \times p$ variance–covariance matrix of residuals (δ).

Assuming multivariate normality (specifically, p-variate normality), the parameters in $\boldsymbol{\tau}$, $\boldsymbol{\kappa}$, $\boldsymbol{\Lambda}$, $\boldsymbol{\Phi}$, and $\boldsymbol{\Theta}$ in the single-population model are estimated in the full sample by maximizing the likelihood function

$$\prod_{i=1}^{N} (2\pi)^{-p/2} \,|\hat{\boldsymbol{\Sigma}}|^{-1/2}\, e^{(-.5)(\mathbf{x}_i - \hat{\boldsymbol{\mu}}_X)'\hat{\boldsymbol{\Sigma}}^{-1}(\mathbf{x}_i - \hat{\boldsymbol{\mu}}_X)}, \tag{17}$$

which is the product across observations of each observation's manifest variable values ($\mathbf{x}_i$) entered into the p-variate normal distribution with model-implied mean vector $\hat{\boldsymbol{\mu}}_x$ and variance–covariance matrix $\hat{\boldsymbol{\Sigma}}$. This maximization is equivalently accomplished using the maximum likelihood fit function F, where

$$\hat{F} = [\ln|\hat{\boldsymbol{\Sigma}}| + tr(\mathbf{S}\hat{\boldsymbol{\Sigma}}^{-1}) - \ln|\mathbf{S}| - p] + (\mathbf{m} - \hat{\boldsymbol{\mu}}_x)'\hat{\boldsymbol{\Sigma}}^{-1}(\mathbf{m} - \hat{\boldsymbol{\mu}}_x), \tag{18}$$

expressed using summary statistics in the vector $\mathbf{m}$ of observed means and matrix $\mathbf{S}$ of observed variances and covariances (e.g., Bollen, 1989). For samples comprised of J populations for which population membership is known *a priori*, parameters in all J subsamples' respective model-implied moments ($\hat{\boldsymbol{\Sigma}}_j$ and $\hat{\boldsymbol{\mu}}_{xj}$) are estimated by maximizing the likelihood function,

$$\prod_{j=1}^{J}\prod_{i=1}^{n_j} (2\pi)^{-p/2} \,|\hat{\boldsymbol{\Sigma}}_j|^{-1/2}\, e^{(-.5)(\mathbf{x}_i - \hat{\boldsymbol{\mu}}_{xj})'\hat{\boldsymbol{\Sigma}}_j^{-1}(\mathbf{x}_i - \hat{\boldsymbol{\mu}}_{xj})}, \tag{19}$$

or equivalently via the multisample maximum likelihood fit function, G, where

$$\hat{G} = \sum_{j=1}^{J}\left(\frac{n_j}{N}\right)\left\{\left[\ln|\hat{\boldsymbol{\Sigma}}_j| + tr(\mathbf{S}_j\hat{\boldsymbol{\Sigma}}_j^{-1}) - \ln|\mathbf{S}_j| - p\right] + \left[(\mathbf{m}_j - \hat{\boldsymbol{\mu}}_{xj})'\hat{\boldsymbol{\Sigma}}_j^{-1}(\mathbf{m}_j - \hat{\boldsymbol{\mu}}_{xj})\right]\right\}. \tag{20}$$

Note that in Equations 19 and 20, the model-implied mean vector and the model-implied variance–covariance matrix have been replaced by their population-specific analogs, computed from equations that are the population-specific versions of Equations 15 and 16, respectively, where

$$\hat{\boldsymbol{\mu}}_{xj} = \boldsymbol{\tau}_j + \boldsymbol{\Lambda}_j \boldsymbol{\kappa}_j, \tag{21}$$

and

$$\hat{\boldsymbol{\Sigma}}_j = \boldsymbol{\Lambda}_j \boldsymbol{\Phi}_j \boldsymbol{\Lambda}_j' + \boldsymbol{\Theta}_j. \tag{22}$$

Without an *a priori* way of defining population membership, a mixture CFA is needed for situations in which the presence of two or more populations is hypothesized. Suppose we want to compare the single-population model to a model that posits a mixture of two populations that have different sets of parameters. Because there are two populations, one mixing proportion needs to be estimated, and for each of the two populations, there is a set of model parameters (i.e., in $\boldsymbol{\tau}$, $\boldsymbol{\kappa}$, $\boldsymbol{\Lambda}$, $\boldsymbol{\Phi}$, and $\boldsymbol{\Theta}$) to be estimated. All of these quantities are estimated simultaneously by maximizing the product across all observations of Equation 7, $L_i = \varphi L_{i1} + (1-\varphi)L_{i2}$, where the likelihoods are now

$$L_{i1} = f(\mathbf{x}_i \mid \boldsymbol{\tau}_1, \boldsymbol{\kappa}_1, \boldsymbol{\Lambda}_1, \boldsymbol{\Phi}_1, \text{and } \boldsymbol{\Theta}_1) \tag{23}$$

and

$$L_{i2} = f(\mathbf{x}_i \mid \boldsymbol{\tau}_2, \boldsymbol{\kappa}_2, \boldsymbol{\Lambda}_2, \boldsymbol{\Phi}_2, \text{and } \boldsymbol{\Theta}_2). \tag{24}$$

The joint likelihood of all observations in this two-population case, assuming independence, becomes

$$\prod_{i=1}^{N} \left[\sum_{j=1}^{2} \varphi_j f(\mathbf{x}_i \mid \boldsymbol{\tau}_j, \boldsymbol{\kappa}_j, \boldsymbol{\Lambda}_j, \boldsymbol{\Phi}_j, \boldsymbol{\Theta}_j) \right] \tag{25}$$

or, somewhat less compactly (via substitution),

$$\prod_{i=1}^{N} \left\{ \sum_{j=1}^{2} \varphi_j (2\pi)^{-p/2} \left| \left(\boldsymbol{\Lambda}_j \boldsymbol{\Phi}_j \boldsymbol{\Lambda}_j' + \boldsymbol{\Theta}_j \right) \right|^{-1/2} e^{(-.5)(\mathbf{x}_i - \boldsymbol{\tau}_j - \boldsymbol{\Lambda}_j \boldsymbol{\kappa}_j)' \left(\boldsymbol{\Lambda}_j \boldsymbol{\Phi}_j \boldsymbol{\Lambda}_j' + \boldsymbol{\Theta}_j \right)^{-1} (\mathbf{x}_i - \boldsymbol{\tau}_j - \boldsymbol{\Lambda}_j \boldsymbol{\kappa}_j)} \right\}. \tag{26}$$

Equation 26 is a CFA-specific, two-population version of the following general formula for a J-population latent variable mixture model:

$$\prod_{i=1}^{N}\left\{\sum_{j=1}^{J}\varphi_j(2\pi)^{-p/2}\left|\hat{\boldsymbol{\Sigma}}_j\right|^{-1/2}\mathrm{e}^{(-.5)(\mathbf{x}_i-\hat{\boldsymbol{\mu}}_j)'\hat{\boldsymbol{\Sigma}}_j^{-1}(\mathbf{x}_i-\hat{\boldsymbol{\mu}}_j)}\right\}, \tag{27}$$

with $\hat{\boldsymbol{\Sigma}}_j$ substituted per Equation 22 and $\hat{\boldsymbol{\mu}}_j$ substituted per Equation 21.

Note that in addition to the standard CFA, the mixture model depicted in Equation 26 incorporates an analysis of the factor mean structure. This is a necessary feature of a latent variable mixture model, because a latent variable mixture of populations can be quite difficult to detect if, in each population, all factors in the model have the same means as their corresponding factors in the other populations. For mixture modeling, as is the case when population membership is known *a priori*, the estimation of the mean structure requires that the mean(s) of one population's factor(s) be fixed to 0, making that the reference point for the other populations' factor mean(s). Also like the single-population situation, the intercepts and factor loadings are traditionally assumed (and fixed) to be invariant across populations in a mixture model when the mean structure is analyzed.

Although the likelihood equation for the mixture CFA model is quite a bit more elaborate than for the univariate and regression mixture models, the assessment of an individual model's fit and the comparison of models are conducted in the same manner. The χ^2-value can be used to test the hypothesis of perfect fit of each model. It is important to keep in mind that the degrees of freedom for the χ^2 of a model with q parameters are still $N - q$, not the $p(p + 1)/2 - q$ degrees of freedom associated with the fit of an isolated CFA model (without mean structure). The factor model involved in the mixture does need to be at least just-identified in the standard sense of a CFA model. The mixture model, however, is a separate model that draws information from the sum of N log-likelihoods with q parameters estimated in the likelihood equations; thus, the degrees of freedom for the mixture model χ^2 are $N - q$. Comparison of the AIC values and BIC values for the two (or more) models yields a decision about the relative fit of the models, which would enable a latent mixture model to be assessed in a relative manner instead of just by itself.

Various restrictions on the CFA mixture model yield different types of mixture tests. To test for a mixture at the latent variable level, the factor loadings, error variances, and intercepts are typically fixed to be equal across populations, while some combination of factor means, factor variances, and factor covariances are freely estimated. Note that with variances constrained across populations while the factor means are freely estimated across populations, the mixture analysis is basically a structured means model but with unknown population membership for the observations (see, e.g., Hancock, 2004; Thompson & Green, Chapter 5, this volume).

Measured-Variable Path Analysis Mixture Models

For MVPA mixture models, the means of the t exogenous variables are modeled to be the intercepts. The w endogenous variables are modeled as a function of the exogenous variables and potentially as a function of the other endogenous variables,

$$\mathbf{y}_i = \boldsymbol{\tau}_j + \boldsymbol{\Gamma}_j \mathbf{x}_i + \mathbf{B}_j \mathbf{y}_i + \boldsymbol{\varepsilon}_{ij} \,. \tag{28}$$

The estimated mean vector for the endogenous variables may therefore be shown to be

$$E[\mathbf{y}_i] = \hat{\boldsymbol{\mu}}_{yj} = (\mathbf{I} - \mathbf{B}_j)^{-1} \boldsymbol{\tau}_j + (\mathbf{I} - \mathbf{B}_j)^{-1} \boldsymbol{\Gamma}_j \hat{\boldsymbol{\mu}}_{xj} \,, \tag{29}$$

where $\mathbf{I}$ is the identity matrix, $\mathbf{B}_j$ is a $w \times w$ matrix of the effects of the endogenous manifest variables on each other, and $\boldsymbol{\Gamma}_j$ is a $t \times w$ matrix of the effects of the exogenous manifest variables on the endogenous manifest variables in the model. The data vector in Equation 25, there labeled as $\mathbf{x}_i$, can be expanded to contain values for the Y-variables and for the X-variables. With the Y-values appearing first in the column vector and the X-values below them, the jth population's model-implied mean vector would first have the values computed per Equation 29 followed by the implied means of the X-variables as per Equation 21.

For MVPA mixture models, the jth population's $p \times p$ model-implied variance–covariance matrix, $\hat{\boldsymbol{\Sigma}}_j$, where $p = t + w$, and where

$$\hat{\boldsymbol{\Sigma}}_j = E[(\mathbf{x}_i - \hat{\boldsymbol{\mu}}_j)(\mathbf{x}_i - \hat{\boldsymbol{\mu}}_j)'] \,, \tag{30}$$

can be divided into four submatrices. The upper left submatrix is the $w \times w$ model-implied variance–covariance matrix for just the endogenous variables, $\hat{\boldsymbol{\Sigma}}_{y_j y_j}$, and is determined as

$$\hat{\boldsymbol{\Sigma}}_{y_j y_j} = (\mathbf{I} - \mathbf{B}_j)^{-1} (\boldsymbol{\Gamma}_j \boldsymbol{\Phi}_j \boldsymbol{\Gamma}_j' + \boldsymbol{\Theta}_j)(\mathbf{I} - \mathbf{B}_j)^{-1'} \,, \tag{31}$$

where $\boldsymbol{\Theta}_j$ is the $w \times w$ variance–covariance matrix of the residuals, ε_{ij} (Jöreskog & Sörbom, 1988). The upper right submatrix is a $w \times t$ matrix of covariances between the endogenous variables and the exogenous variables, $\hat{\boldsymbol{\Sigma}}_{y_j x_j}$, the equation for which is

$$\hat{\boldsymbol{\Sigma}}_{y_j x_j} = (\mathbf{I} - \mathbf{B}_j)^{-1} \boldsymbol{\Gamma}_j \boldsymbol{\Phi}_j \,. \tag{32}$$

The lower left submatrix is simply the transpose of $\hat{\boldsymbol{\Sigma}}_{y_j x_j}$,

$$\hat{\Sigma}_{x_j y_j} = \mathbf{\Phi}_j \mathbf{\Gamma}_j'(\mathbf{I} - \mathbf{B}_j)^{-1\prime}. \tag{33}$$

The matrix $\mathbf{\Phi}_j$ is the $t \times t$ variance–covariance matrix of the exogenous variables, making it equal to its transpose and also making it the only quantity in the lower right submatrix of the overall variance-covariance matrix

$$\hat{\Sigma}_{x_j x_j} = \mathbf{\Phi}_j. \tag{34}$$

Each of these submatrices is arranged as described in one $p \times p$ matrix, $\hat{\Sigma}_j$, to form the jth population's model-implied variance-covariance matrix to be used in the general equation for MVPA mixture models. Recall that as the parameters in the model-implied mean vectors and the model-implied variance–covariance matrix are being estimated for each of the J populations, the mixing proportions are also being estimated in the iterative process of maximizing the likelihood function of the data.

Latent-Variable Path Analysis Mixture Models

In LVPA mixture models, the exogenous factors have multiple manifest indicators while being modeled to have a causal influence on one or more endogenous factors, which themselves have multiple manifest indicators. Exogenous factors may covary amongst themselves, endogenous factors may influence other endogenous factors, and the disturbances of the endogenous factors may covary. If a mixture model is to be estimated, then Equation 25 can again be called upon, with the appropriate substitutions for $\hat{\boldsymbol{\mu}}_j$ and $\hat{\Sigma}_j$ in order to estimate the parameters for each of the J populations and to estimate the mixing proportions.

The measurement model governing the manifest indicators of the exogenous factors is as shown previously in Equation 14, while the model-implied means of the exogenous factors' indicators are determined as per Equation 15. The measurement model governing the manifest indicators of the endogenous factors is

$$\mathbf{y}_i = \boldsymbol{\tau}_{yj} + \boldsymbol{\Lambda}_{yj}\boldsymbol{\eta}_i + \boldsymbol{\varepsilon}_i, \tag{35}$$

where $\boldsymbol{\eta}_i$ is the ith person's $w \times 1$ model-implied vector of values on the endogenous latent variables, established by the structural relations

$$\boldsymbol{\eta}_i = \boldsymbol{\alpha}_j + \mathbf{\Gamma}_j \xi_i + \mathbf{B}_j \boldsymbol{\eta}_i + \zeta_i, \tag{36}$$

where $\boldsymbol{\alpha}_j$ is the jth population's $w \times 1$ vector of intercepts for the endogenous factors and $\boldsymbol{\zeta}_i$ is the ith person's $w \times 1$ model-implied vector of disturbances (errors) for the endogenous latent variables. The model-implied mean vector for the Y-variables is

$$\hat{\boldsymbol{\mu}}_{yj} = \boldsymbol{\tau}_{yj} + \boldsymbol{\Lambda}_j \boldsymbol{\kappa}_{\eta j}, \tag{37}$$

where $\boldsymbol{\kappa}_{\eta j}$ is the model-implied mean vector of the endogenous latent variables for the jth population, which may be shown to be computed as

$$\boldsymbol{\kappa}_{\eta j} = (\mathbf{I} - \mathbf{B}_j)^{-1}\boldsymbol{\alpha}_j + (\mathbf{I} - \mathbf{B}_j)^{-1}\boldsymbol{\Gamma}_j \boldsymbol{\kappa}_{\xi j}. \tag{38}$$

The model-implied variance–covariance matrix is similar in form to that of mixture MVPA in that the $p \times p$ matrix can be considered in four distinct submatrices: the variance–covariance matrix for the endogenous variables ($\hat{\boldsymbol{\Sigma}}_{y_j y_j}$); the covariances between the endogenous variables and the exogenous variables ($\hat{\boldsymbol{\Sigma}}_{y_j x_j}$); the transpose of that matrix ($\hat{\boldsymbol{\Sigma}}_{x_j y_j}$); and the variance–covariance matrix of the exogenous variables ($\hat{\boldsymbol{\Sigma}}_{x_j x_j}$). For three of the four submatrices, the shift from measured- to latent-variable path analysis mixture modeling is simply a matter of including matrices of factor loadings. The model-implied variance–covariance matrix for the endogenous variables,

$$\hat{\boldsymbol{\Sigma}}_{y_j y_j} = \boldsymbol{\Lambda}_{yj}(\mathbf{I} - \mathbf{B}_j)^{-1}(\boldsymbol{\Gamma}_j \boldsymbol{\Phi}_j \boldsymbol{\Gamma}_j' + \boldsymbol{\Psi}_j)(\mathbf{I} - \mathbf{B}_j)^{-1\prime}\boldsymbol{\Lambda}_{yj}' + \boldsymbol{\Theta}_j, \tag{39}$$

is similar in form to $\hat{\boldsymbol{\Sigma}}_{y_j y_j}$ for MVPA (Equation 31), but for LVPA, multiple nonunity factor loadings of manifest variables on the endogenous factor(s) are incorporated by premultiplying Equation 31 by $\boldsymbol{\Lambda}_{yj}$ and postmultiplying by its transpose (Jöreskog & Sörbom, 1988). Also included is $\boldsymbol{\Psi}_j$, the $w \times w$ variance–covariance matrix of the factor disturbances, ζ_{ij}, while $\boldsymbol{\Theta}_j$ is the variance–covariance matrix of the y residuals, ε_{ij} (Jöreskog & Sörbom, 1988). For $\hat{\boldsymbol{\Sigma}}_{y_j x_j}$, endogenous and exogenous variables are crossed, so instead of using $\boldsymbol{\Lambda}_{yj}$ and its transpose, we use $\boldsymbol{\Lambda}_{yj}$ with the transpose of the loadings of the exogenous factor indicators,

$$\hat{\boldsymbol{\Sigma}}_{y_j x_j} = \boldsymbol{\Lambda}_{yj}(\mathbf{I} - \mathbf{B}_j)^{-1}\boldsymbol{\Gamma}_j \boldsymbol{\Phi}_j \boldsymbol{\Lambda}_{xj}', \tag{40}$$

the transpose of which gives the submatrix,

$$\hat{\boldsymbol{\Sigma}}_{x_j y_j} = \boldsymbol{\Lambda}_{xj}\boldsymbol{\Phi}_j\boldsymbol{\Gamma}_j{}'(\mathbf{I}-\mathbf{B}_j)^{-1}{}'\boldsymbol{\Lambda}_{yj}{}' , \tag{41}$$

where $\boldsymbol{\Gamma}_j$ is now a $t \times w$ matrix of the effects of the latent factors on the endogenous latent variables in the jth population, and $\mathbf{B}_j$ is now a $w \times w$ matrix of the effects of the endogenous latent variables on each other. The model-implied variance–covariance matrix for the exogenous variables ($\hat{\boldsymbol{\Sigma}}_{x_j x_j}$) is identical to $\hat{\boldsymbol{\Sigma}}_j$ in CFA,

$$\hat{\boldsymbol{\Sigma}}_{x_j x_j} = \boldsymbol{\Lambda}_{xj}\boldsymbol{\Phi}_j\boldsymbol{\Lambda}_{xj}{}' + \boldsymbol{\Theta}_j . \tag{42}$$

To analyze data–model fit for mixture path analysis (measured- or latent-variable), we use the same quantities as before. Individual data–model fit is tested with a χ^2 that has $N - q$ degrees of freedom, and models postulating a different number of underlying populations can be compared to each other with AIC and BIC. The model with the smallest AIC is the best-fitting model according to AIC, and the model with the smallest BIC is the best-fitting model according to BIC.

PROGRAMMING MIXTURE ANALYSES

As one might gather from the previous sections, mixture modeling is not something for which one would want to do the calculations by hand. These analyses are unfortunately not available in such commonly used statistical software packages as SPSS and SAS, and most specialized software packages for SEM (e.g., EQS, LISREL, AMOS) do not currently have mixture modeling as an option. Mplus, however, can be used to conduct the mixture analyses described in this chapter. For the Mplus code to conduct univariate or regression mixture analyses, the reader is simply referred to the Mplus user's manual (Muthén & Muthén, 2004). Mplus code for mixture MVPA and mixture CFA, although also available from the Mplus user's manual, are presented and detailed herein. Code for mixture LVPA, as well as for mixtures involving latent growth models (e.g., Muthén, 2001), are left for the reader to compose based on the information contained in the Mplus manual and on the mixture MVPA and mixture CFA examples presented here. We will begin the section on using software to conduct mixture analyses by elaborating upon the need for a mean structure within those models.

A Practical Consideration in Mixture Analyses: The Role of Mean Structure

With mixtures of unknown populations, we have discussed a variety of parameters that can be estimated, ranging from the mean of a single manifest variable to means, variances, and relations among latent variables. When conducting any of these analyses, it is of critical practical importance that all of the unknown populations differ in their means. If the means do not differ, then a search for multiple populations could encounter an indeterminacy problem. An algorithm attempting to estimate such a (non)mixture would likely step toward one solution on a given iteration but step toward a different one on the next iteration, and thereby rarely converge on a single satisfactory solution.

As a simple example of this, consider a univariate mixture of two unknown populations, both of which have normal distributions with the same mean but different variances. Drawing these two distributions on the same axes would result in one of the distributions being hidden, so to speak, by the distribution with the greater variance. The presence of skewness or kurtosis in the distribution with the smaller variance might yield a unique contribution to the overall pattern, but the degree of skewness or kurtosis would have to be substantial in order for the overall pattern to give the appearance of multiple distributions atop each other rather than a single distribution with a large variance. Two populations that have different means, on the other hand, will typically present two different peaks, rendering such a mixture easier to detect. The variances are important, because large variances can obfuscate relatively small mean differences (recall the effect of variance on statistical power in t tests and ANOVA), as can a substantial amount of skew. But with no mean separation at all, it is difficult, if not impossible, to get a satisfactory solution for a mixture model.

When a mixture is hypothesized in the latent variable(s) of a model, the presence of mean separation in the manifest variables is still imperative. The algorithm has only manifest variable data with which to work. Without a mean separation at the manifest level, the algorithm cannot easily "see" different populations in the data, so it numerically has no clear and unique way to estimate multiple latent distributions. Mean differences in manifest variables across multiple populations can result from differences in latent means, differences in the factor loadings, or differences in the intercepts. In analyses with population membership defined *a priori*, factor loadings and intercepts are typically held equal across populations in analyses. In mixture modeling, however, estimating the latent mean difference while constraining all loadings and intercepts to be equal can result in low convergence rates and appreciable bias in the parameter estimates of converged solutions if the standardized difference between latent means is not large (Gagné, 2004; Lubke, Muthén, & Larsen, 2002). Freeing one or more intercepts to vary across populations improves both convergence and the accuracy of parameter estimates, with notable improvement resulting from freeing even one of the intercepts.

Mixture MVPA

Figure 7.2 provides a relatively simple mixture measured-variable path model. The exogenous variable V1 is modeled to have a causal influence on two endogenous variables V2 and V3, each of which also has a path coming into it from its error term. V3 is modeled to influence V4, which has a separate error term affiliated with it. Suppose a researcher has in mind the hypothesis that the form of this model is correct for two populations, but the path coefficients and the intercepts/means are thought to differ across the two populations. Suppose also that although two populations are thought to contribute to the single sample of data, no information is available with which to categorize the observations as belonging to one or the other population. This scenario calls for a mixture MVPA model with a mean structure.

Figure 7.2 also provides code for conducting the analysis of interest in Mplus (v3.01). The first two lines of the Mplus code are just the title and data file location, respectively. The "variable" command contains two lines: The first names the input variables from the data file, and the sec-

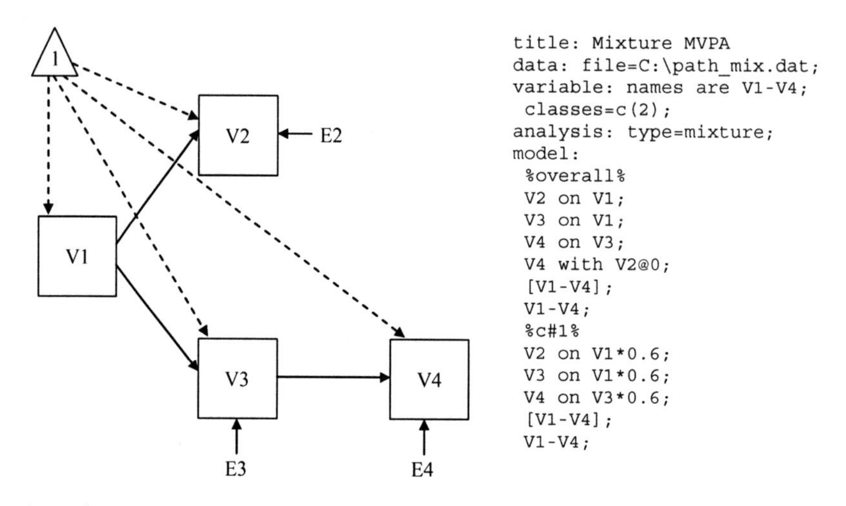

1
V1
V2
E2
V3
V4
E3
E4
title: Mixture MVPA
data: file=C:\path_mix.dat;
variable: names are V1-V4;
classes=c(2);
analysis: type=mixture;
model:
%overall%
V2 on V1;
V3 on V1;
V4 on V3;
V4 with V2@0;
[V1-V4];
V1-V4;
%c#1%
V2 on V1*0.6;
V3 on V1*0.6;
V4 on V3*0.6;
[V1-V4];
V1-V4;

Figure 7.2. Path model and Mplus syntax for mixture MVPA.

ond creates and names the categorical mixture variable (in this case "c") and provides the number of classes to be modeled. In the "analysis" line, Mplus is told that a mixture model is to be conducted; without this line, Mplus would be expecting a classification variable from the original dataset to designate population membership. The combination of "type=mixture" and "classes=c(2)" tells Mplus that there is an hypothesized mixture model with two populations (which means one mixing proportion will be estimated in addition to two sets of parameters for MVPA).

Although Mplus does need to be told explicitly that a mixture analysis should be conducted, it does not need to be told the specific type of statistical analysis to conduct (i.e., regression, CFA, etc.). The analysis is determined by the code provided in the "model" command. The subheading "%overall%" provides the model to be applied to all observations in the sample. The first three lines of this command establish the path model. The default in Mplus is to estimate the covariances among the errors for variables that are exclusively endogenous, that is, endogenous variables with no causal outputs. In this example, that means Mplus will estimate the residual covariance between E2 and E4 (V3 is an endogenous variable, but it is modeled to cause V4, so it can be thought of as not exclusively endogenous). This covariance, however, is not part of the hypothesized model, so "V4 with V2@0" is included in order to fix that path to 0. The next two lines tell Mplus to estimate the means and variances, respectively, associated with the variables in the model.

If the "model" command included only an "%overall%" subheading, then the path coefficients, intercepts, and variances would be constrained equal across the populations. The researcher, however, does not want these parameter estimates to be constrained equal in this example. To remove the constraints, a class subheading is used to tell Mplus to do something different with the class(es) receiving a subheading.

In this example, Class 1 has been designated specifically in order to free the cross-population constraints on the parameter estimates. The paths, intercepts, and variances are meant to be free to differ from those in Class 2, so the lines of code corresponding to the path model, the intercepts, and the variances are repeated under the class subheading "%c#1%" (note that Class 2 could have been designated instead, as in the next example below). There is no need to repeat "V4 with V2@0" in the class-specific portion of the program, because the residual covariance has been fixed to 0 in the "%overall%" heading, and the researcher wants that to be the case in both classes. The estimated solution is unaffected by the choice of latent class(es) modeled in addition to the "%overall%" subheading, except that label switching may sometimes occur (i.e., the designation of "Class 1" or "Class 2" may switch as a function of the class number specified in the code, but a given group will have the same values for its parameter estimates regardless of whether it is labeled as "1" or "2" in the solution). For illustration purposes, the algorithm will use different start values (as accomplished by the asterisk followed by a number) in each class for the parameter estimates that are freed across classes, thereby giving the algorithm a little bit of a push, so to speak, in its effort to find a potential mixture.

Mplus does not require a specific output command or "End" statement. There are several output options for additional statistical or diagnostic information, but the default output setting includes basic characteristics of the iteration process, the likelihood and information criteria for data–model fit considerations, and all of the obvious parameter estimates for the path model. The order of the commands as they are typed into the Mplus syntax, although presented in a logical sequence herein, is not important: Mplus will run the commands in the appropriate order, regardless of their sequence in the code.

Mixture CFA

Suppose you are asked to apply mixture modeling to a CFA model that has a single Math Ability factor with four subscales from a math placement exam as its indicators (algebra, geometry, trigonometry, and calculus). Of specific interest is a comparison of a single-population model to a mixture model positing two populations, with no grouping variable for categorizing the examinees *a priori*. Estimating and fitting the single-population model calls for a straightforward CFA, but a mixture CFA is needed to obtain parameter estimates and fit information for the two-population model.

Figure 7.3 provides the model under investigation and the Mplus code necessary to conduct the mixture analysis. The first two lines of code provide the title and data file, respectively, as they did in the measured-variable path analysis example. Also similar to the path analysis example, the "variable" command in the CFA program provides names for the variables in the dataset and then establishes the categorical variable for the mixture, while the "analysis" command is used to tell Mplus that we have a mixture situation.

The code in the "model" command looks quite a bit different than the mixture MVPA code. Here the "%overall%" subheading is used only to define the factor model. The code provides a metric by fixing the first factor loading in each class to 1.0; Mplus would do this by default, but for illustration purposes, the code for specifying this constraint is included. Nothing specific is required to obtain variance and intercept estimates, because when conducting CFA, Mplus automatically estimates the vari-

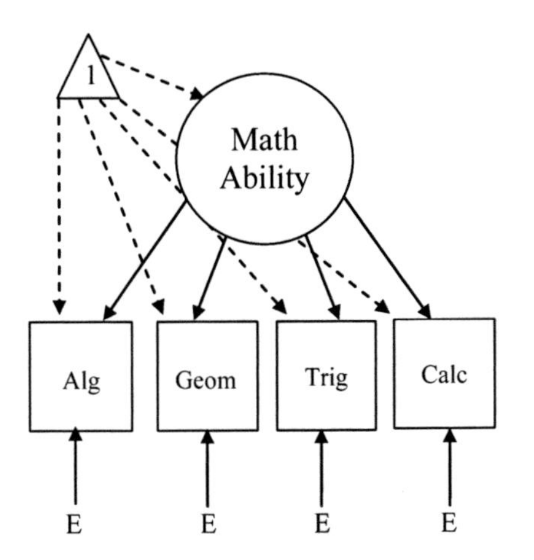


```
title: Mixture CFA
data: file=C:\cfa_mix.dat;
variable: names are V1-V4;
 classes=c(2);
analysis: type=mixture;
model:
 %overall%
 F1 by V1@1.0 V2-V4;
 %c#2%
 F1;
 V1-V4;
```

Figure 7.3. Confirmatory factor model and Mplus syntax for example of CFA mixture.

ances and intercepts, with exogenous variables (latent or manifest) having their variances estimated while endogenous variables have residual variances estimated.

The class-specific portion of the “model” command (this time choosing Class 2 just for illustration purposes) is used to free specific default cross-population equality constraints. Mplus has been asked to free the cross-class constraint on the factor variance and the error variances of the manifest variables. We are also interested in having the factor means differ across populations, which Mplus handles automatically by fixing to 0 the factor mean of one population and letting the factor mean(s) of the other population(s) be freely estimated. Note, however, that the default cross-population loading and intercept constraints are remain in place.

Using the mixture CFA model and code provided in Figure 7.3, Figure 7.4 contains excerpts of the output using a simulated dataset (available from the author). The output includes the program entered into Mplus, messages for ascertaining the success of the analysis, and the parameter estimates. Among the information in the Mplus output that is not provided in Figure 7.4 are summary information about the model entered for analysis (i.e., sample size, number of dependent variables, etc.), performance and diagnostic information about the estimation algorithm, and additional information regarding the estimation of the mixing proportions.

Under “INPUT INSTRUCTIONS,” Mplus repeats back the program that was submitted for the analysis. This is a good way to check to make sure Mplus read what was intended for reading (but remember to switch

```
INPUT INSTRUCTIONS
  title: Mixture CFA
  data: file=C:\cfa_mix.dat;
  variable:   names are V1-V4;
              classes=c(2);
  analysis: type=mixture;
  model:
    %overall%
      F1 by V1@1.0 V2-V4;
    %c#2%
      F1;
      V1-V4;

INPUT READING TERMINATED NORMALLY

THE MODEL ESTIMATION TERMINATED NORMALLY

TESTS OF MODEL FIT

Loglikelihood
            H0 Value                                        -3075.460
Information Criteria
            Number of Free Parameters                              19
            Akaike (AIC)                                     6188.921
            Bayesian (BIC)                                   6268.998
            Sample-Size Adjusted BIC                         6208.691
              (n* = (n + 2) / 24)
            Entropy                                             0.817
FINAL CLASS COUNTS AND PROPORTIONS FOR THE LATENT CLASSES BASED
ON THE ESTIMATED MODEL
        Latent Classes
            1                   270.05219                 0.54010
            2                   229.94781                 0.45990
```

Figure 7.4. Excerpt of Mplus output for example of mixture CFA.

back to the program code itself before making changes!). The message "INPUT READING TERMINATED NORMALLY" means that Mplus understood the code that was submitted. Note that this is no guarantee that the code does what it was meant to do; the message means only that Mplus recognized the code as a complete and properly formatted set of instructions for an analysis.

The message most pertinent to the success of the model estimation procedure follows a great deal of model and algorithm diagnostic information. In this present example, we have "THE MODEL ESTIMATION TERMINATED NORMALLY," with no associated warnings. Certain difficulties in model estimation, such as a specific variable causing a nonpositive definite covariance matrix, are not fatal to the estimation procedure.

```
MODEL RESULTS
                        Estimates        S.E.      Est./S.E.

Latent Class 1
   F1      BY
      V1                  1.000         0.000        0.000
      V2                  0.346         0.021       16.516
      V3                  0.393         0.021       18.432
      V4                  0.573         0.026       22.162

Means
      F1                  3.624         0.104       34.783

Intercepts
      V1                  4.792         0.132       36.199
      V2                  3.920         0.077       50.733
      V3                  2.835         0.081       35.139
      V4                  1.949         0.108       18.107

Variances
      F1                  1.272         0.259        4.914

Residual Variances
      V1                  0.209         0.116        1.795
      V2                  0.810         0.084        9.613
      V3                  0.795         0.074       10.759
      V4                  0.878         0.083       10.558

Latent Class 2
   F1      BY
      V1                  1.000         0.000        0.000
      V2                  0.346         0.021       16.516
      V3                  0.393         0.021       18.432
      V4                  0.573         0.026       22.162

Means
      F1                  0.000         0.000        0.000

Intercepts
      V1                  4.792         0.132       36.199
      V2                  3.920         0.077       50.733
      V3                  2.835         0.081       35.139
      V4                  1.949         0.108       18.107

Variances
      F1                  0.864         0.210        4.114

Residual Variances
      V1                  0.094         0.099        0.953
      V2                  0.885         0.079       11.195
      V3                  0.769         0.080        9.628
      V4                  1.059         0.118        8.956
```

Figure 7.4. Continued.

In such situations, "...ESTIMATION TERMINATED NORMALLY" will still appear, but it will be immediately followed by a warning message detailing a problem.

If Mplus understood the code and encountered no problems during model estimation, then the remaining information can be interpreted without reservation. Under "TESTS OF MODEL FIT," we find the relevant fit information for a CFA mixture model. Note that we do not find quantities such as the comparative fit index (CFI) or the root mean square error of approximation (RMSEA) that we would find for a single-population CFA. Instead, we find only values such as the AIC and BIC for this two-population mixture CFA model that can be compared to those of the single-population CFA model (results not depicted herein) to determine which of the models fits better.

The next section is "FINAL CLASS COUNTS AND PROPORTIONS FOR THE LATENT CLASSES BASED ON THE ESTIMATED MODEL." In this section, we have the values for the estimated mixing proportions, which for this example says that 54.010% of the sampled mixture is composed of observations from Population 1, while observations from Population 2 are the remaining 45.990% of the mixture. Note that these are estimates of the extent to which the sampled population is actually comprised of Population 1 and Population 2; these values are not interpreted to mean that 54.010% of the observations in the *sample* come from Population 1.

The last section of interest for us is "MODEL RESULTS." First for Class 1 and then for Class 2, Mplus provides the parameter estimates and standard errors for the factor loadings, factor mean difference, intercepts, factor variance, and error variances (for the manifest variables in the factor model). Inspection of the estimates for Class 2 tells us that the factor mean of Class 2 was fixed to 0, so the value of 3.624 listed as the factor mean for Class 1 is actually the factor mean difference $\kappa = \kappa_{current} - \kappa_{ref}$, where κ_{ref} is the reference factor mean that Mplus fixed to 0 (κ_2 in the present example) and $\kappa_{current}$ is the factor mean for the class listed (κ_1 in the present example).

SUMMARY AND CONCLUSION

Several familiar and commonly used statistical analyses are available for comparing population parameters when population membership is known for all of the observations in a sample. Differences in means, regression slopes, and structural relations can be investigated quite readily when observations can be categorized into their respective populations with perfect accuracy. With an added degree of difficulty, mixture analyses can be used to conduct statistical comparisons of populations when the population membership of the observations is not available for the analysis.

A sample with a univariate distribution that does not conform to distributional expectations can be tested for the possible presence of a mixture of two or more populations, the distributions of which could be combining to give the unexpected pattern seen in the data. The presence of heteroscedasticity in a regression analysis conducted under the assumption of a single underlying population might instead be the blending of multiple populations that have different relations between the predictor and criterion under investigation; a mixture regression analysis can be used to test such a hypothesis. Mixture CFA can be used to determine whether test items are biased as a function of differences in populations other than the more obvious population distinctions based on race, sex, or socioeconomic status. Mixture MVPA can be used to determine whether populations differ in their relations among measured variables, while mixture LVPA assesses potential population differences in latent relations.

Although growing somewhat, mixture modeling is currently not abundant in the applied literature, due in part to difficulties with model estimation and model convergence. There is, however, burgeoning methodological research into mixture modeling, addressing the issues of model estimation (e.g., Arminger, Stein, & Wittenberg, 1999; Jedidi, Ramaswamy, DeSarbo, & Wedel, 1996), model convergence (e.g., Gagné, 2004; Lubke et al., 2002), and the development of significance tests for comparing models with differing numbers of hypothesized populations (e.g., Lo, Mendell, & Rubin, 2001). Continued research efforts to resolve the methodological challenges that presently face mixture analyses will lead to the continued expansion of mixture modeling techniques into the applied research realm.

REFERENCES

Akaike, H. (1987). Factor analysis and AIC. *Psychometrika, 52,* 317–332.

Arminger, G., Stein, P., & Wittenberg, J. (1999). Mixtures of conditional mean and covariance structure models. *Psychometrika, 64,* 475–494.

Bollen, K. A. (1989). *Structural equations with latent variables*. New York: Wiley.

Byrne, B. M., Shavelson, R. J., & Muthén B. (1989). Testing for the equivalence of factor covariance and mean structures: The issue of partial measurement invariance. *Psychological Bulletin, 105,* 456–466.

Duncan, T. E., Duncan, S. C., Strycker, L. A., Li, F., & Alpert, A. (1999). *An introduction to latent variable growth curve modeling: Concepts, issues, and applications.* Mahwah, NJ: Erlbaum.

Everitt, B. S., & Hand. D. J. (1981). *Finite mixture distributions*. London: Chapman & Hall.

Gagné, P. (2004). *Generalized confirmatory factor mixture models: A tool for assessing factorial invariance across unspecified populations*. Unpublished doctoral dissertation, University of Maryland, College Park.

Hancock, G. R. (2004). Experimental, quasi-experimental, and nonexperimental design and analysis with latent variables. In D. Kaplan (Ed.), *The Sage handbook of quantitative methodology for the social sciences* (pp. 317-334). Thousand Oaks, CA: Sage.

Jedidi, K., Ramaswamy, V., DeSarbo, W. S., & Wedel, M. (1996). On estimating finite mixtures of multivariate regression and simultaneous equation models. *Structural Equation Modeling: A Multidisciplinary Journal, 3*, 266–289.

Jöreskog, K. G., & Sörbom, D. (1988). *LISREL 7: A guide to the program and applications* (2nd ed.). Chicago: SPSS, Inc.

Kline, R. B. (2004). *Principles and practice of structural equation modeling* (2nd ed.). New York: Guilford Press.

Lindsay, B. G. (1995). *Mixture models: Theory, geometry, and applications*. Hayward, CA: Institute of Mathematical Statistics.

Lo, Y., Mendell, N. R., & Rubin, D. B. (2001). Testing the number of components in a normal mixture. *Biometrika, 88*, 767–778.

Lubke, G., Muthén, B. O., & Larsen, K. (2002). *Empirical identifiability of factor mixture models*. Unpublished manuscript.

McLachlan, G., & Peel, D. (2000). *Finite mixture models*. New York: Wiley.

Muthén, B. O. (2001). Second-generation structural equation modeling with a combination of categorical and continuous latent variables: New opportunities for latent class-latent growth modeling. In L. M. Collins & A. G. Sayer (Eds.), *New methods for the analysis of change* (pp. 345-368). Washington, DC: American Psychological Association.

Muthén, L. K., & Muthén, B. O. (2004). *Mplus user's guide* (3rd ed.). Los Angeles: Authors.

Schwarz, G. (1978). Estimating the dimension of a model. *Annals of Statistics, 6*, 461–464.

Titterington, D. M., Smith, A. F. M., & Makov, U. E. (1985). *Statistical analysis of finite mixture distributions*. Chichester, UK: Wiley.

CHAPTER 8

STRUCTURAL EQUATION MODELS OF LATENT INTERACTION AND QUADRATIC EFFECTS

Herbert W. Marsh, Zhonglin Wen, and Kit-Tai Hau

Estimating the interaction between variables is an important concern in the social sciences. In educational psychology, for example, it is often hypothesized that the effect of an instructional technique will interact with characteristics of individual students (e.g., a special remediation program developed for slow learners may not be an effective instructional strategy for bright students). More generally, such aptitude–treatment interactions are posited in many areas of psychology. Developmental psychologists are frequently interested in how the effects of a given variable interact with age in longitudinal or cross-sectional studies. Many psychological theories explicitly hypothesize interaction effects. Thus, for example, some forms of expectancy–value theory hypothesize that resultant motivation is based on the interaction between expectancy of success and the value placed on success by the individual (e.g., motivation is high only if both probability of success and the value placed on the success are high). In self-concept research, the relation between an individual component of self-concept (e.g., academic, social, physical) and global self-

Structural Equation Modeling: A Second Course, 225–265

esteem is hypothesized to interact with the importance placed on a specific component of self-concept (e.g., if a person places no importance on physical accomplishments, then these physical accomplishments—substantial or minimal—are not expected to be substantially correlated with self-esteem). More generally, a variety of weighted-average models posit—at least implicitly—that the effects of a given set of variables will depend on the weight assigned to each variable in the set (i.e., the weight assigned to a given variable interacts with the variable to determine the contribution of that variable to the total effect).

For present purposes we treat moderation and interaction as synonymous, and use the term *interaction*. However, Baron and Kenny's (1986) classic distinction between moderation and mediation is important in understanding interaction effects. *Mediation* occurs when some of the effects of an independent variable (X) on the dependent variable (Y) can be explained in terms of another mediating variable (MED) that falls between X and Y in terms of the causal ordering. Thus, for example, the effects of mathematical ability prior to the start of high school (X) on mathematics achievement test scores at the end of high school (Y) are likely to be mediated in part by mathematics coursework during high school (MED). In contrast, *moderation* is said to have taken place when the size or direction of the effect of X on Y varies with the level of a moderating variable (MOD). Thus, for example, the effect of a remedial course in mathematics instead of regular mathematics coursework (X) on subsequent mathematics achievement (Y) may vary systematically depending on the student's initial level of mathematics ability (MOD); the effect of the remedial course may be very positive for initially less able students, negligible for average-ability students, and even detrimental for high-ability students who would probably gain more from regular or advanced mathematics coursework.

Analyses of quadratic effects are also important. A strong quadratic effect may give the appearance of a significant interaction effect. Thus, without the proper analysis of the potential quadratic effects, the investigator might easily misinterpret, overlook, or mistake a quadratic effect as an interaction effect (Lubinski & Humphreys, 1990). Whereas quadratic trends and other nonlinearity can complicate interpretations of interaction effects, in other situations the nonlinearity of relations may be the focus of the analyses. Thus, on the basis of theory or prior research, researchers may posit that the effect of the independent variable varies with the level of the independent variable itself. Examples are common:

- Nonlinear effects may be hypothesized between strength of interventions (or dosage level) and outcome variables such that benefits increase up to an optimal level, and then level off or even decrease beyond this optimal point.

- The effects of an intervention might be expected to be the highest immediately following the intervention and to decrease in a nonlinear manner over time following the intervention.
- Growth curves are typically nonlinear with the rate of change varying with time.
- At low levels of anxiety, increases in anxiety may facilitate performance but at higher levels of anxiety, further increases in anxiety may undermine performance.
- Self-concept may decrease with age for young children, level out in middle adolescence, and then increase with age into early adulthood.
- Levels of workload demanded by teachers may be positively related to student evaluations of teaching effectiveness for low to moderate levels of workload but may have diminishing positive effects or even negative effects for possibly excessive levels of workload.

Although the issue of nonlinear relations is very complicated, we limit discussion to the relatively simple case of quadratic effects.

In this chapter, we explore methods for the analysis of products between latent variables—latent interactions and latent quadratic effects. Although our focus is primarily on latent interactions between two different variables, the rationale generalizes to quadratic effects where the product is based on the same variable. We begin by briefly summarizing traditional methods for analyzing interactions between observed (manifest) variables, so that readers can better understand the alternative approaches to the latent interactions. We then move to alternative methods that attempt to estimate the nonlinear effects using traditional methods with observed variables. These methods include the replacement of the latent interaction with observed variables, and the formation of multiple groups. For example, factor scores replace the latent variables or the estimation is separated into stages as in the two-stage least squares (2SLS) method (Bollen, 1995).

Our main focus is on three recently developed approaches: the *centered constrained approach* (Algina & Moulder, 2001), based on the traditional constrained approach (Kenny & Judd, 1984; Jöreskog & Yang, 1996); the generalized appended product indicator approach (Wall & Amemiya, 2001), which we refer to as the *partially constrained approach*; and the *unconstrained approach* (Marsh, Wen, & Hau, 2004). Following from classic research by Kenny and Judd (1984), all three approaches treat interaction effects as latent variables inferred from multiple indicators that can be readily estimated with existing SEM software packages. For each of these approaches, we provide mathematical representations of interaction models, and illustrate these approaches with simulated examples using LISREL (Version 8.3). In the final section of the chapter we introduce some new and evolving approaches to the analysis of latent interaction and quadratic effects. These new approaches are unlikely to be used in applied research because they have not yet been integrated into existing SEM software packages, but are likely to be the basis of future developments in this rapidly evolving area of research.

TRADITIONAL (NONLATENT) APPROACHES TO INTERACTIONS BETWEEN OBSERVED VARIABLES

Consider the effect of observed variables X_1 and X_2 on outcome Y. Traditional approaches to the analysis of interactions between X_1 and X_2, as summarized below, depend on the nature of X_1 and X_2.

When the independent variables X_1 and X_2 are categorical variables that can take on a relatively small number of values, interaction effects can easily be estimated with traditional analysis of variance (ANOVA) procedures.

When the independent variables X_1 and X_2 are reasonably continuous variables, the following multiple regression with a product term is widely recommended to estimate both main and interaction effects (e.g., Aiken & West, 1991; Cohen & Cohen, 1983):

$$Y = \beta_0 + \beta_1 X_1 + \beta_2 X_2 + \beta_3 X_1 X_2 + e, \tag{1}$$

where β_1 and β_2 represent the main effects, β_3 represents the interaction effect, and e is a random disturbance term with zero mean that is uncorrelated with X_1 and X_2. In order to test whether the interaction effect is statistically significant, the test statistic is

$$t = \frac{\hat{\beta}_3}{\mathrm{SE}(\hat{\beta}_3)},$$

where $\mathrm{SE}(\hat{\beta}_3)$ is the standard error of estimated β_3.

When one independent variable is a categorical variable and the other one is a continuous variable, a suitable approach is to divide the total sample into separate groups representing the different levels of the categorical variable and then conduct separate multiple regressions on each group. The interaction effect is represented by the differences between unstandardized regression coefficients obtained with the separate groups (Aiken & West, 1991; Cohen & Cohen, 1983). Alternatively, researchers can represent the categorical variable with dummy coding and apply multiple regression (see Equation 1).

The critical feature common to each of these nonlatent approaches is that all of the dependent and independent variables are nonlatent variables inferred on the basis of a single indicator rather than latent variables inferred on the basis of multiple indicators. Particularly when there are multiple indicators of these variables, latent variable approaches described below provide a much stronger basis for evaluating the underlying factor structure relating multiple indicators to their factors, controlling for measurement error, increasing power, and, ultimately, providing more defensible interpretations of the interaction effects.

INTERACTION OF LATENT VARIABLES

When at least one latent independent variable is inferred from multiple indicators, there are many methods to analyze the interaction effect (see Schumacker & Marcoulides, 1998). Generally, structural equation modeling (SEM) provides many advantages over the use of analyses based on observed variables, but is more complicated to conduct. We begin our discussion with several approaches (factor score, 2SLS, multiple group approaches) that we consider to be partially—but not fully—latent variable approaches to the problem of latent interactions and quadratic effects. We then move on to fully latent variable approaches in which all of the dependent and independent variables are latent constructs inferred on the basis of multiple indicators.

Analyses of Factor Scores

Using factor scores of each latent variable to analyze a multiple regression equation (see Equation 1) is an intuitive approach to the analysis of interaction effects between latent variables (Yang, 1998). The underlying rationale is to replace the multiple indicators of each latent factor with a single factor score and then to use the factor scores as observed variables in subsequent analyses. The procedure has two steps. In the first step, the measurement model is estimated for the two independent variables and the one dependent variable, assuming these factors to be freely correlated. From this initial analysis, factor scores representing each latent variable are generated for each case (e.g., participants). In the second step, the structural equation model is estimated as if the latent variables were observed. The generation of factor scores is a standard option in many statistical packages, making this approach an attractive alternative. This method is very easy to understand and to conduct. However, like other methods in which latent variables are replaced by observed variables, measurement errors are ignored. In particular, because multiple indicators of the latent constructs are not incorporated into the analyses, there is an implicit assumption that the variables are measured without error so that parameter estimates are not corrected for measurement errors. Thus, important advantages of the latent variable approach to structural equations models are lost when each of the independent variables is based on a single score (be it a factor score, a scale score, or a response to a single item) rather than on multiple indicators of the latent construct. Hence, as is widely recognized (e.g., Bollen, 1989; Kaplan, 2000), the incorporation of multiple indicators of the latent variables helps to mitigate problems associated with measurement error, thus improving the parameter estimates in terms of reducing bias and sampling variability. The factor score approach has some of the advantages of fully latent variable approaches to assessing interaction effects (e.g., evaluating the factor structure relating multiple indicators and latent constructs). However, because the final analysis that includes the interaction effect is based on observed scores rather than latent constructs, the approach should really be considered as a nonlatent variable approach that suffers from most of the problems associated with other nonlatent approaches. On this basis, we do not consider the factor score approach to be a true latent variable approach and do not recommend its routine usage except, perhaps, for purposes of preliminary analyses.

Two-Stage Least Squares (2SLS)

When both of the interacting variables are latent variables, similar to Equation 1, the structural equation with a product term is

$$y = \alpha + \gamma_1\xi_1 + \gamma_2\xi_2 + \gamma_3\xi_1\xi_2 + \zeta. \tag{2}$$

Bollen (1995) proposed a 2SLS approach to address the problem of how to estimate the interaction effect. Bollen and Paxton (1998) described the rationale for 2SLS with the following simple model. Suppose that the dependent variable y is an observed variable and that the latent variables (ξ_1 and ξ_2) are each measured with two indicators (mean-centered) such that

$$x_1 = \xi_1 + \delta_1 \text{ and } x_2 = \lambda_2\xi_1 + \delta_2;$$
$$x_3 = \xi_2 + \delta_3 \text{ and } x_4 = \lambda_4\xi_2 + \delta_4; \quad (3)$$

where δ_i terms have means of zero, are uncorrelated with ξ_1 and ξ_2, and are uncorrelated with each other. From this we have $\xi_1 = x_1 - \delta_1$ and $\xi_2 = x_3 - \delta_3$, which can be substituted into Equation 2 such that

$$y = \alpha + \gamma_1(x_1 - \delta_1) + \gamma_2(x_3 - \delta_3) + \gamma_3(x_1 - \delta_1)(x_3 - \delta_3) + \zeta,$$

and can be rewritten as

$$y = \alpha + \gamma_1 x_1 + \gamma_2 x_3 + \gamma_3 x_1 x_3 + e, \quad (4)$$

where

$$e = -\gamma_1\delta_1 - \gamma_2\delta_3 - \gamma_3(x_1\delta_3 + x_3\delta_1 - \delta_1\delta_3) + \zeta$$

is a composite disturbance term. Equation 4 has the same form as Equation 1, involving only observable variables and a disturbance term. The disturbance e, however, is related to the independent variables x_1 and x_2 in Equation 4. This correlation of disturbance and explanatory variables violates a key assumption of ordinary least squares (OLS) regression. To solve the problem, Bollen (1995; see also Bollen & Paxton, 1998) introduced instrumental variables and a 2SLS estimator. Instrumental variables are observed variables that are correlated with the righthand-side variables in the multiple regression equation, but are uncorrelated with the disturbance term of that equation. In the present model, x_2, x_4, and x_2x_4 are instrumental variables. In the first step, x_1 is regressed on x_2, x_4, and x_2x_4 and the OLS predicted value $\hat{x}_1$ is formed. Similarly, x_3 and x_1x_3 are respectively regressed on x_2, x_4, and x_2x_4, and OLS predicted values $\hat{x}_3$ and $\widehat{x_1x_3}$ are formed. In the second step, x_1, x_3, and x_1x_3 in Equation 4 are replaced by $\hat{x}_1$, $\hat{x}_3$, and $\widehat{x_1x_3}$, respectively, but the composite disturbance term is unchanged. Because the predicted variables $\hat{x}_1$, $\hat{x}_3$, and $\widehat{x_1x_3}$ are all linear functions of x_2, x_4, and x_2x_4, they are uncorrelated with the disturbance term. The modified version of Equation 4 satisfies the assumption of OLS, and the coefficients can be estimated and tested by OLS methods.

Bollen's 2SLS approach is a well-known, simpler method for evaluating latent interactions and quadratic effects without involving structural equation models. Because it involves only OLS, 2SLS is easy to understand. Recent versions of SPSS and LISREL have 2SLS procedures that can be used to analyze interaction effects, thus facilitating the use of 2SLS to become a standard method. Importantly, the 2SLS approach does not depend on assumptions of normality of the latent variables that are problematic in many other approaches to the analysis of latent interaction effects (see subsequent discussion of normality assumptions).

Despite potentially attractive features of the 2SLS approach, its efficiency in comparison to alternative techniques has been challenged (e.g., Jöreskog, 1998; Klein & Moosbrugger, 2000; Schermelleh-Engel, Klein, & Moosbrugger, 1998). A particularly important limitation is that the dependent variable, *y*, is an observed variable based on a single indicator whereas in latent variable research the dependent variable will typically be a latent variable based on multiple indicators (see Jöreskog, 1998, for further discussion). If there are multiple indicators of the dependent variable, the researcher must choose one of the indicators as the basis of analysis or, perhaps, compute a total score based on the set of indicators. In this respect, the 2SLS approach is not a fully latent-variable approach to the analyses of latent interactions. Furthermore, Equation 4 involves only the indicators that are used to scale the latent variables ξ_1 and ξ_2. Different choices of *y*-indicators and scale indicators of latent variables ξ_1 and ξ_2 may lead to conflicting results.

Moulder and Algina (2002) also expressed serious reservations about the 2SLS-estimation method because they found that the method exhibited greater bias and lower power than the methods they considered that were based on latent variable interactions. Their greatest concern, however, was that when indicator reliability was low to moderate (less than .7), the size of the bias tended to increase with increasing sample size (*N*). On this basis, we do not recommend the 2SLS procedure, particularly when indicator reliability is low and when there are multiple indicators of the dependent variable.

Multiple Groups Analysis

When one of the independent variables is a latent variable based on multiple indicators and one is an observed categorical variable that can be used to form a relatively small number of groups, the invariance of the effect of the latent variable over the multiple groups provides an effective test of interaction effects (Bagozzi & Yi, 1989; Rigdon, Schumacker, & Wothke, 1998). This method is an extension of the multiple regression approach based on separate groups with observed variables.

However, when both of the interacting variables are reasonably continuous latent variables based on multiple indicators, the multisample approach is no longer appropriate. In particular, it is typically inappropriate to use values based on one of the continuous variables to form discrete, multiple groups that could be incorporated in the multisample approach. Potential problems include the artificiality or arbitrariness of the groups, the possibly small sample sizes of the different groups, the loss of information in forming discrete groups, the inability to control for measurement error in the grouping variable, the resultant increases in Type II error rates, and the problematic interpretations of the size and statistical significance of interaction effects (see, e.g., Cohen & Cohen, 1983; MacCallum, Zhang, Preacher, & Rucker, 2002; Ping, 1998). On this basis, we do not recommend the multiple group procedure unless one of the interacting variables is an observed categorical variable that can appropriately be represented by a relatively small number of discrete groups.

FULLY LATENT VARIABLE APPROACHES USING STRUCTURAL EQUATION ANALYSES

Structural equation analysis plays a very important role in analyses of interaction effects in which all independent variables are latent constructs inferred from multiple indicators. A variety of such approaches were demonstrated in the monograph edited by Schumacker and Marcoulides (1998). A new generation of approaches was subsequently developed, including the centered constrained approach (Algina & Moulder, 2001), the partially constrained approach (Wall & Amemiya, 2001), the quasi-maximum likelihood (QML) approach (Klein & Muthén, 2002), the two-step method of moments (2SMM) approach (Wall & Amemiya, 2000), and the unconstrained approach (Marsh et al., 2004). All of these recently developed approaches are introduced in this chapter, whereas centered constrained, partially constrained, and unconstrained approaches are illustrated in detail. Other recent proposed methods include Arminger and Muthén's (1998) Bayesian method and Blom and Christoffersson's (2001) method based on the empirical characteristic function of the distribution of the indicator variables. However, whereas both of these approaches have promising potential, it appears that they have seldom been used in practice because of their computational demands and a lack of readily accessible software.

The Evolution of the Constrained Approach

Although there are many approaches to estimating interaction effects between two latent variables, the most typical starting point is to constrain the loadings and variances of the product term. Kenny and Judd (1984) initially proposed this method for a simple model in which the dependent variable y is based on a single observed variable (although we subsequently drop this assumption to allow a latent dependent variable based on multiple indicators). In order to estimate the interaction effects of latent variables ξ_1 and ξ_2 on y, they formulated a latent regression equation with a product term

$$y = \gamma_1\xi_1 + \gamma_2\xi_2 + \gamma_3\xi_1\xi_2 + \zeta, \tag{5}$$

where y is an observed variable with a zero mean, ξ_1 and ξ_2 are latent variables with zero means, $\xi_1\xi_2$ is the latent interaction term between ξ_1 and ξ_2, and ζ is the disturbance. Equation 5 is similar to Equation 2, but does not include an intercept term. In Kenny and Judd's demonstration, just as in Equation 3, there were two observable indicators for each latent factor (x_1 and x_2 of ξ_1, and x_3 and x_4 of ξ_2) such that

$$x_1 = \xi_1 + \delta_1 \text{ and } x_2 = \lambda_2\xi_1 + \delta_2;$$

$$x_3 = \xi_2 + \delta_3 \text{ and } x_4 = \lambda_4\xi_2 + \delta_4;$$

where λ_1 and λ_3 were fixed to 1 for purposes of identification. Kenny and Judd suggested using product variables x_1x_3, x_1x_4, x_2x_3, and x_2x_4 as indicators of $\xi_1\xi_2$, together with y, x_1, x_2, x_3, and x_4, to estimate the model, and imposed many nonlinear constraints. For example,

$$x_2x_4 = \lambda_2\lambda_4\xi_1\xi_2 + \lambda_2\xi_1\delta_4 + \lambda_4\xi_2\delta_2 + \delta_2\delta_4, \tag{6}$$

where the loading of x_2x_4 on $\xi_1\xi_2$ is constrained to be $\lambda_2\lambda_4$. Altogether, Kenny and Judd posited 15 latent variables (e.g., $\xi_1\xi_2$, $\xi_1\xi_4$) and imposed many constraints on factor loadings and variances. They demonstrated their proposed technique using COSAN (Fraser, 1980) because it allowed the researcher to specify many nonlinear constraints at a time when this option was not generally available with other SEM packages. Nevertheless, the specification of these constraints was such a tedious task, and so prone to error, that this approach was rarely used in applied research.

Hayduk (1987) implemented the Kenny and Judd technique using LISREL, but he required even more additional latent variables to account for loadings and variances of the product indicators, and produced a large, unwieldy structural equation model. However, starting with Version 8, LISREL introduced nonlinear equality constraints for parameters that made the application of this approach to latent interactions less cumbersome. Using LISREL 8.0 to implement the Kenny and Judd technique, Jaccard and Wan (1995) introduced a latent variable η instead of observed variable y as the dependent variable in their model, thus making the constrained approach a fully latent-variable approach. That is, Equation 5 became

$$\eta = \gamma_1\xi_1 + \gamma_2\xi_2 + \gamma_3\xi_1\xi_2 + \zeta. \quad (7)$$

They constrained the residual variance for each observed product term such that, for example, Equation 6 was written as

$$x_2x_4 = \lambda_2\lambda_4\xi_1\xi_2 + \delta_{24}, \quad (8)$$

where

$$\text{var}(\delta_{24}) = \lambda_2^2 \text{ var}(\xi_1) \text{ var}(\delta_4) + \lambda_4^2 \text{ var}(\xi_2) \text{ var}(\delta_2) + \text{var}(\delta_2) \text{ var}(\delta_4). \quad (9)$$

Figure 8.1 illustrates the interaction model using indicator-products as the indicators of the product latent term where η has three indicators, ξ_1 and ξ_2 each have two indicators, and two paired indicator-products x_1x_3 and x_2x_4 are used as the indicators of $\xi_1\xi_2$. Note, however, that the necessary constraints on the parameters involving the product terms are not presented in Figure 8.1, but are described in subsequent discussion (see also the syntax for the constrained approach in Appendix 2).

Jöreskog and Yang (1996) provided a general model and thorough treatment of this approach. When observed variables are not mean-centered, they consider the Kenny-Judd model with constant intercept terms from Equation 2:

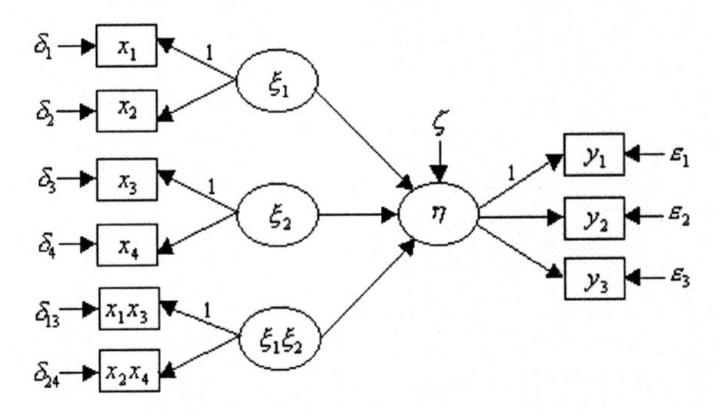

Figure 8.1. The interaction model using indicator products as the indicators of the product latent term. ξ_1 and ξ_2 are allowed to be correlated. The correlation between $\xi_1\xi_2$ and its component is fixed to be zero in the constrained approach, but freely estimated in partially constrained and unconstrained approaches.

$$y = \alpha + \gamma_1\xi_1 + \gamma_2\xi_2 + \gamma_3\xi_1\xi_2 + \zeta \tag{10}$$

and

$$\begin{aligned} x_1 &= \tau_1 + \xi_1 + \delta_1 \text{ and } x_2 = \tau_2 + \lambda_2\xi_1 + \delta_2; \\ x_3 &= \tau_3 + \xi_2 + \delta_3 \text{ and } x_4 = \tau_4 + \lambda_4\xi_2 + \delta_4. \end{aligned} \tag{11}$$

Under the supposition that ξ_1, ξ_2, ζ and all δ terms are multivariate normal with mean of zero, and each is uncorrelated with the other (except that ξ_1 and ξ_2 are allowed to be correlated), they proposed a model with a mean structure. The mean vector and covariance matrix of ξ_1, ξ_2, and $\xi_1\xi_2$ are, respectively:

$$\boldsymbol{\kappa} = \begin{bmatrix} 0 \\ 0 \\ \phi_{21} \end{bmatrix}, \quad \boldsymbol{\Phi} = \begin{bmatrix} \phi_{11} & & \\ \phi_{21} & \phi_{22} & \\ 0 & 0 & \phi_{11}\phi_{22} + \phi_{21}^2 \end{bmatrix}. \tag{12}$$

They further noted that even if ξ_1, ξ_2, and ζ are centered so as to have means of zero, $\kappa_3 = \phi_{21} = \text{cov}(\xi_1, \xi_2)$ will typically not be zero. Hence, a mean structure is always necessary, whether or not the independent observed variables include constant intercept terms (see Jöreskog, 1998; Yang, 1998). This implies that the Kenny-Judd (1984) model (or subsequent Jaccard-Wan [1995] model) without a mean structure is inappropriate unless $\text{cov}(\xi_1, \xi_2)$ is very small. For this reason, we note that intercepts of latent variables (i.e., κ terms) should always be included in latent interaction analyses. However, the additional constant intercept terms of indicators (i.e., τ terms) require many more nonlinear constraints than the Jaccard-Wan model, further complicating model specification.

Algina and Moulder (2001) revised the Jöreskog-Yang model so that independently observed variables are mean-centered as in Jaccard and Wan's model and have a mean structure as in the Jöreskog-Yang model (see Equation 12). Although their model is the same as the Jöreskog-Yang model except for mean centering, they found that the revised Jöreskog-Yang model (denoted here as the Algina-Moulder model) was more likely to converge. Even in cases when both models converged, their simulation results still favored their revised model. More recently, Moulder and Algina (2002) compared six methods for estimating and testing an interaction under a wide range of conditions: (1) ML (maximum likelihood) estimation of the Jaccard-Wan model, (2) ML estimation of the Jöreskog-Yang model, (3) ML estimation of the Algina-Moulder model, (4) Ping's two-step ML method, (5) Ping's revised two-step ML method, and (6) Bollen's 2SLS. Methods (4)–(6) involved a two-step procedure. In Ping's (1996) two-step ML method, a measurement model for defining the latent variables ξ_1 and ξ_2 is estimated in the first step. Parameters estimated from this step are treated as known parameters in the second step in which the model in Equation 5 is estimated. Ping's revised two-step method is a two-step procedure for estimating the Jöreskog-Yang model based on a similar logic (see Moulder & Algina, 2002). Moulder and Algina concluded that the ML estimation of the Algina-Moulder model was the most effective method for evaluating interactions, in terms of less bias, better control of Type I error rate, and higher power. For this reason, we recommend the Algina-Moulder model among the variety of constrained approaches, and refer to it simply as the *constrained approach*.

INTERACTION EFFECTS: CONSTRAINED, PARTIALLY CONSTRAINED, AND UNCONSTRAINED APPROACHES

For the sake of simplicity, suppose that each of the latent variables η, ξ_1 and ξ_2 is associated with three indicators—namely, y_1, y_2, y_3; x_1, x_2, x_3; and x_4, x_5, x_6, respectively. Further suppose that all of the independent indicators are centered so that they have zero means. The measurement equations are as below:

$$y_1 = \tau_{y1} + \eta + \varepsilon_1, y_2 = \tau_{y2} + \lambda_{y2}\eta + \varepsilon_2, y_3 = \tau_{y3} + \lambda_{y3}\eta + \varepsilon_3;$$
$$x_1 = \xi_1 + \delta_1, x_2 = \lambda_2\xi_1 + \delta_2, x_3 = \lambda_3\xi_1 + \delta_3;$$
$$x_4 = \xi_2 + \delta_4, x_5 = \lambda_5\xi_2 + \delta_5, x_6 = \lambda_6\xi_2 + \delta_6.$$

Indicators of the dependent variable (y_1, y_2, y_3) can either be mean-centered or not. The intercept terms τ_{y1}, τ_{y2}, and τ_{y3} of dependent indicators, however, must be included even if y_1, y_2, and y_3 are mean-centered, because η typically does not have a zero mean. The structural equation with the interaction term is (reiterating Equation 7):

$$\eta = \gamma_1\xi_1 + \gamma_2\xi_2 + \gamma_3\xi_1\xi_2 + \zeta, \tag{13}$$

where ξ_1, ξ_2, ζ, all δ terms, and all ε terms are multivariate normal (with mean zero), and are uncorrelated with each other (except that ξ_1 and ξ_2 are allowed to be correlated, and $\xi_1\xi_2$ is non-normal even if ξ_1 and ξ_2 are normal). Note that when y variables include the intercept terms (i.e., τ_y terms), the structural Equation 13 does not need an intercept term; otherwise, if both α and τ_y terms are added in the model, the model is not identified (for the details, see Jöreskog & Yang, 1996).

For all three (constrained, partially constrained, and unconstrained) approaches, matched pairs are used to form the product indicators. Thus, x_1, x_2, and x_3 may, for example, be matched with x_4, x_5, and x_6, respectively, to form the three matched-pairs x_1x_4, x_2x_5, and x_3x_6 (see the "Types of Product Indicators" section in this chapter).

Interaction Effects: Constrained Approach (see Appendix 2a)

The LISREL specification using 3 matched-pairs for the constrained approach (Algina & Moulder, 2001) is:

$$\begin{bmatrix} y_1 \\ y_2 \\ y_3 \end{bmatrix} = \begin{bmatrix} \tau_{y1} \\ \tau_{y2} \\ y_{y3} \end{bmatrix} + \begin{bmatrix} 1 \\ \lambda_{y2} \\ \lambda_{y3} \end{bmatrix} \eta + \begin{bmatrix} \varepsilon_1 \\ \varepsilon_2 \\ \varepsilon_3 \end{bmatrix},$$

$$\begin{bmatrix} x_1 \\ x_2 \\ x_3 \\ x_4 \\ x_5 \\ x_6 \\ x_1x_4 \\ x_2x_5 \\ x_3x_6 \end{bmatrix} = \begin{bmatrix} 1 & 0 & 0 \\ \lambda_2 & 0 & 0 \\ \lambda_3 & 0 & 0 \\ 0 & 1 & 0 \\ 0 & \lambda_5 & 0 \\ 0 & \lambda_6 & 0 \\ 0 & 0 & 1 \\ 0 & 0 & \lambda_8 \\ 0 & 0 & \lambda_9 \end{bmatrix} \begin{bmatrix} \xi_1 \\ \xi_2 \\ \xi_3 \end{bmatrix} + \begin{bmatrix} \delta_1 \\ \delta_2 \\ \delta_3 \\ \delta_4 \\ \delta_5 \\ \delta_6 \\ \delta_7 \\ \delta_8 \\ \delta_9 \end{bmatrix},$$

$$\eta = \gamma_1\xi_1 + \gamma_2\xi_2 + \gamma_3\xi_1\xi_2 + \zeta,$$

$$\boldsymbol{\kappa} = \begin{bmatrix} 0 \\ 0 \\ \kappa_3 \end{bmatrix}, \quad \boldsymbol{\Phi} = \begin{bmatrix} \phi_{11} & & \\ \phi_{21} & \phi_{22} & \\ 0 & 0 & \phi_{33} \end{bmatrix},$$

$$\boldsymbol{\Theta}_{\boldsymbol{\varepsilon}} = \text{diag}(\theta_{\varepsilon 1}, \theta_{\varepsilon 2}, \theta_{\varepsilon 3}),$$

$$\boldsymbol{\Theta}_{\boldsymbol{\delta}} = \text{diag}(\theta_{\delta 1}, \theta_{\delta 2}, \theta_{\delta 3}, \theta_{\delta 4}, \theta_{\delta 5}, \theta_{\delta 6}, \theta_{\delta 7}, \theta_{\delta 8}, \theta_{\delta 9}).$$

The necessary constraints are:

(i) $\lambda_8 = \lambda_2\lambda_5, \lambda_9 = \lambda_3\lambda_6$;

(ii) $\kappa_3 = \phi_{21}$;

(iii) $\phi_{33} = \phi_{11}\phi_{22} + \phi_{21}^2$;

(iv) $\theta_{\delta 7} = \phi_{11}\theta_{\delta 4} + \phi_{22}\theta_{\delta 1} + \theta_{\delta 1}\theta_{\delta 4}$, $\theta_{\delta 8} = \lambda_2^2\phi_{11}\theta_{\delta 5} + \lambda_5^2\phi_{22}\theta_{\delta 2} + \theta_{\delta 2}\theta_{\delta 5}$,

$\theta_{\delta 9} = \lambda_3^2\phi_{11}\theta_{\delta 6} + \lambda_6^2\phi_{22}\theta_{\delta 3} + \theta_{\delta 3}\theta_{\delta 6}$.

The proof of the above constraints is available in Appendix 1. It can be seen that these constraints are complicated and difficult to specify in the SEM syntax (e.g., see LISREL syntax in Appendix 2a).

Interaction Effects: Partially Constrained Approach (see Appendix 2b)

A fundamental and potentially untenable assumption underlying the constrained approach is the premise that ξ_1 and ξ_2 are normally distributed. If this is not the case, the assumption that $\phi_{31} = \text{cov}(\xi_1\xi_2, \xi_1) = 0$ is typically false, as are the assumptions that $\phi_{32} = 0$ and $\phi_{33} = \phi_{11}\phi_{22} + \phi_{21}^2$. Wall and Amemiya (2001) emphasized that the last row of $\mathbf{\Phi}$ in Equation 12 only holds under the assumption that ξ_1 and ξ_2 are normally distributed. They demonstrated that because constrained approaches are based on normality assumptions, estimated interaction effects applying the constrained approach to non-normal data provided systematically biased estimates of the interaction effects. Because of this critical problem with traditional constrained approaches, Wall and Amemiya proposed a generalized appended product indicator (partially constrained) procedure that did not impose constraints on $\mathbf{\Phi}$, but still constrained other parameters, as in other constrained procedures. Based on Wall and Amemiya's derivations, we can conclude that the partially constrained approach for the present simple case is to let all elements of $\mathbf{\Phi}$ be freely estimated, that is,

$$\mathbf{\Phi} = \begin{bmatrix} \phi_{11} & & \\ \phi_{21} & \phi_{22} & \\ \phi_{31} & \phi_{32} & \phi_{33} \end{bmatrix}.$$

All other parts of the partially constrained model and its constraints are the same as those in the constrained approach. The partially constrained approach, however, was designed also to deal with general polynomial effects (Wall & Amemiya, 2001). Their partially constrained procedure is more appropriate than the traditional constrained procedures when ξ_1 and ξ_2 are not normally distributed. Hence, the important advantage of the partially constrained approach is that it relaxes the assumption that ξ_1 and ξ_2 are normally distributed. However, because this model still requires all of the remaining nonlinear constraints that are in the constrained approach, the specification of this model is still complicated (see Appendix 2b) and this complication is likely to be an impediment to its routine use in applied research.

Interaction Effects: Unconstrained Approach (see Appendix 2c)

Marsh and colleagues (2004) evaluated an unconstrained approach in which the product of observed variables is used to form indicators of the latent interaction term, as in the constrained approach. Their approach, however, was fundamentally different in that they did not impose any complicated nonlinear constraints to define relations between product indicators and the latent interaction factor. They demonstrated that the unconstrained model is identified when there are at least two product indicators.

In the unconstrained approach, the interaction model is the same as that in the constrained approach (when the first-order effect terms are normally distributed) and in the partially constrained approach (when the first-order effect terms are not normally distributed) except that constraints are not imposed. Since the unconstrained approach and the partially constrained approach do not constrain parameters based on assumptions of normality, both of these approaches provide less biased estimates of the latent interaction effects than the constrained approach under widely varying conditions of non-normality. Marsh and colleagues' (2004) simulation studies showed that the unconstrained approach is comparable to the partially constrained approach in terms of goodness of fit, proportion of fully proper solutions, bias in estimation of first-order and interaction effects, and precision. Importantly, however, the unconstrained approach is much easier for applied researchers to implement, eliminating the need for the imposition of complicated, nonlinear constraints required by the partially constrained approach as well as the traditional constrained approaches. However, when the sample size is small and normality assumptions are met, the precision of the unconstrained approach is somewhat lower than the constrained approach.

INTERACTION EFFECTS: EXAMPLES OF CONSTRAINED, PARTIALLY CONSTRAINED, AND UNCONSTRAINED APPROACHES

We use simulated data to illustrate the estimation of the effect of the interaction between ξ_1 and ξ_2 on η with the constrained, partially constrained, and unconstrained approaches. In our simulated data, the structural equation with the interaction term is

$$\eta = \gamma_1\xi_1 + \gamma_2\xi_2 + \gamma_3\xi_1\xi_2 + \zeta,$$

where each of the latent variables η, ξ_1, and ξ_2 has three indicators. For present purposes, we considered only one sample generated from this population (see Marsh et al., 2004, for multisample simulation results). In order to provide a context for this example, assume that η is mathematics achievement, ξ_1 is prior mathematics ability, ξ_2 is mathematics motivation, and $\xi_1\xi_2$ is the interaction of prior mathematics ability and mathematics motivation. The sets of independent indicators x_1, x_2, x_3 and x_4, x_5, x_6, are mean-centered. The same data simulated from this underlying model (see Appendix 2) are used to demonstrate the application of the constrained, partially constrained, and unconstrained approaches.

Interaction Effects: Constrained Approach (see Appendix 2a)

The covariance matrix and mean vector of nine indicators and three (matched-pair) product indicators are presented in the LISREL syntax for the constrained approach (see Appendix 2a) and the path model and parameter estimates are shown in Figure 8.2. The estimated main effects are $\gamma_1 = 0.429$ (SE = 0.041, t = 10.372) and $\gamma_2 = 0.419$ (SE = 0.042, t = 9.929). The estimated interaction effect is $\gamma_3 = 0.238$ (SE = 0.043, t = 5.570). Both main effects and the interaction effect are statistically significant ($p < .05$). Hence, mathematics achievement is higher when prior mathematics ability is higher ($\gamma_1 = 0.429$), when mathematics motivation is higher ($\gamma_2 = 0.419$), and when the $\xi_1\xi_2$ term is higher ($\gamma_3 = 0.238$). Whereas both prior ability and motivation contribute substantially to subsequent achievement, the positive effect of mathematics ability is more substantial for highly motivated students. Equivalently, the positive effect of mathematics motivation is more substantial for students with higher levels of prior ability. Because simulated data were used, it is not surprising that the model fit the data very well: $\chi^2(65) = 72.6$ ($p = 0.242$), RMSEA=0.014, SRMR=0.049, NNFI=0.998, and CFI=0.998.

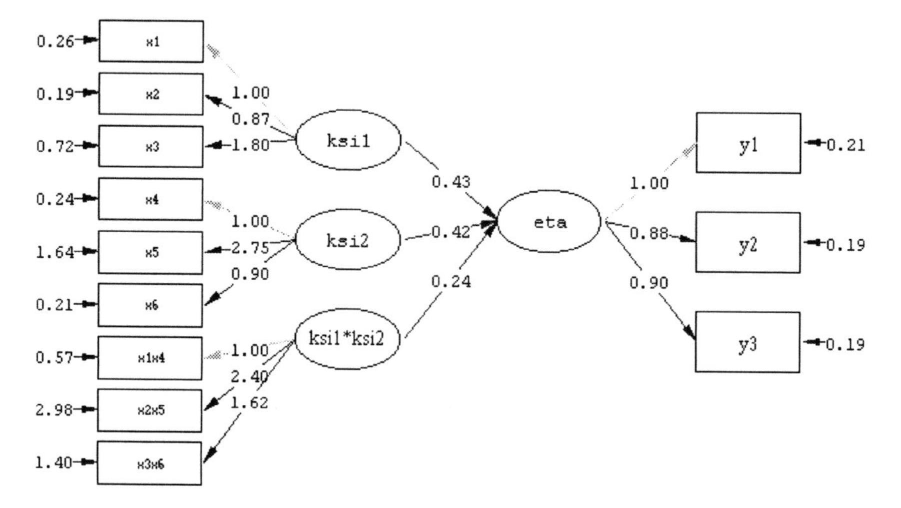

Figure 8.2. Path diagram and estimations of the constrained interaction model. Although not shown in the figure, the correlation between ξ_1 and ξ_2 (i.e., ϕ_{21}) is estimated freely. The correlations between $\xi_1\xi_2$ and its components (i.e., ϕ_{31}, ϕ_{32}) are fixed to be zero in the constrained approach.

Interaction Effects: Partially Constrained Approach (see Appendix 2b)

The syntax of the partially constrained approach is shown in Appendix 2b. In contrast to the constrained method, we allow the covariance between $\xi_1\xi_2$ and ξ_1, $\xi_1\xi_2$ and ξ_2, and the var($\xi_1\xi_2$) to be freely estimated. The estimated main effects are $\gamma_1 = 0.434$ (SE = 0.041, t = 10.584) and $\gamma_2 = 0.418$ (SE = 0.041, t = 10.124). The estimated interaction effect is $\gamma_3 = 0.246$ (SE = 0.045, t = 5.484), while the fit indices are $\chi^2(62) = 63.7$ (p = 0.417), RMSEA = 0.006, SRMR = 0.022, NNFI = 1.000, and CFI = 1.000.

Interaction Effects: Unconstrained Approach (see Appendix 2c)

The syntax of the unconstrained approach is shown in Appendix 2c. In this method, all of the constraints imposed in the constrained and partially constrained approaches are removed. The estimated main effects are $\gamma_1 = 0.425$ (SE = 0.040, t = 10.531) and $\gamma_2 = 0.414$ (SE = 0.041, t = 10.088). The estimated interaction effect is $\gamma_3 = 0.248$ (SE = 0.046, t = 5.356), while the fit indices are $\chi^2(58)=68.6$ (p=0.160), RMSEA=0.018, SRMR=0.044, NNFI=0.997, and CFI=0.998.

ROBUSTNESS TO VIOLATIONS OF NORMALITY ASSUMPTIONS IN CONSTRAINED, PARTIALLY CONSTRAINED, AND UNCONSTRAINED APPROACHES

There are three fundamentally different problems associated with violations in the assumption of normality that must be considered in the evaluation of the constrained, partially constrained, and unconstrained approaches to the estimation of interaction effects. First, the ML estimation typically used with each of these approaches is based on the assumption of the multivariate normality of the indicators of ξ_1 and ξ_2. This problem is common to all CFA and SEM research based on ML estimation, and is not specific to the evaluation of latent interaction and quadratic effects. Second, even when the indicators of ξ_1 and ξ_2 are normally distributed, the distributions resulting from the products of these indicators are known to be non-normal (Jöreskog & Yang, 1996). Hence, this second problem is specific to the evaluation of latent interaction and quadratic effects with ML estimation. Thus, the constrained, partially constrained, and unconstrained approaches all suffer from both of these problems when ML estimation is used. Fortunately, ML estimation tends to be robust to violations of normality such as these in terms of parameter estimates (e.g., Boomsma, 1983; Hau & Marsh, 2004). However, previous research suggests that the ML likelihood ratio test is too large and that ML standard errors are too small under some conditions of non-normality (Hu, Bentler, & Kano, 1992; West, Finch, & Curran, 1995). Although there are alternative estimators that do not assume multivariate normality (e.g., arbitrary distribution function, weighted least squares), simulation studies of latent interaction effects suggest that ML estimation outperforms these alternative estimation procedures under most conditions (Jaccard & Wan, 1995; Wall & Amemiya, 2001; see also Marsh et al., 2004). Nevertheless, it may be appropriate to impose adjustments on the standard errors and χ^2-statistics to correct for this bias (Marsh et al., 2004; Yang-Wallentin & Jöreskog, 2001).

The third problem associated with violations of the assumptions of normality is specific to the constrained approach, but does not apply to the partially constrained and unconstrained approaches. As emphasized by Wall and Amemiya (2001) and by Marsh and colleagues (2004), the nature of the constraints imposed in the constrained approach depends fundamentally on the assumption that ξ_1 and ξ_2 are normally distributed. Importantly, estimates of the interaction effect based on these constraints are not robust in relation to violations of this assumption of normality; neither the size nor even the direction of this bias is predictable a priori, and the size of the bias does not decrease systematically with increasing N. Furthermore, this problem is not a function of the ML estimation procedure, but is inherent to the constrained approach. In contrast to the constrained approach, both the partially constrained and the unconstrained approaches specifically avoid this problem in that they do not impose any constraints based on this assumption of normality.

There is not good evidence about how serious this third problem must be in order to threaten the validity of interpretations of the constrained approach. However, simulation studies by Wall and Amemiya (2002) and by Marsh and colleagues (2004) each provide examples in which the bias is substantial. Because most published evaluations of the constrained approach are based on simulated data that are specifically constructed to be multivariate normal, applied researchers have little research evidence upon which to justify the use of the constrained approach unless their data are multivariate normal. In contrast, both the partially constrained and unconstrained approaches provide relatively unbiased estimates of the latent interaction effects under the widely varying conditions of non-normality considered by Marsh and colleagues. Furthermore, the small biases in non-normally distributed data that were observed became smaller as N increased. On this basis, we recommend that the constrained approach should not be used when assumptions of normality are violated.

TYPES OF PRODUCT INDICATORS

The determination of the optimal number and the best type of product indicators to be used is an important issue in the analysis of latent interactions using the constrained, partially constrained, and unconstrained approaches. Surprisingly, this critical issue has received relatively little attention in previous research. Particularly when there are many multiple indicators of each latent independent variable, there are a variety of different strategies that can be used in the formation of multiple product indicators used to infer latent interaction effects. Existing research, however, is typically based on ad hoc strategies that are, perhaps, idiosyncratic to a particular application, and has not systematically compared alternative strategies.

In their original demonstration of the constrained approach, Kenny and Judd (1984) used all possible cross-products x_1x_3, x_1x_4, x_2x_3, and x_2x_4 as indicators of $\xi_1\xi_2$, the latent interaction factor. In the Jaccard and Wan (1995) study, there were three indicators of each latent variable. Although there were nine possible products, they only used two out of three indicators for each latent factor to form four possible products as indicator variables of $\xi_1\xi_2$. Jöreskog and Yang (1996) included only one product indicator (x_1x_3) in their example and Jöreskog (1998) outlined a simplified approach that also used only one product indicator, but Yang (1998) included two matched pairs of product indicators (x_1x_3 and x_2x_4) in another study. Based on a comparison of models with one and four indicators, Yang reported that the bias and mean squared error were smaller for the model with four products, suggesting that more indicators were better (also see Marsh, Hau, Balla, & Grayson, 1998). However, Marsh and colleagues (2004) expressed concerns about strategies that failed to use all of the information (i.e., some indicators of the independent variables were not used at all) or that reused the same information (i.e., the same indicator was used repeatedly in the construction of more than one of the multiple indicators of the latent interaction effect).

Particularly when the number of multiple indicators of each independent variable is large, increasing the number of product indicators may introduce other problems. Ping (1998) suggested that for large models, the size of the covariance matrix created by the addition of product indicators and the number of additional variables implied by the constrained equations can create model nonconvergence and other model estimation problems. As an alternative to Kenny and Judd's (1984) product indicator specification, Ping (1995) proposed a single-indicator specification in which the product of single indicators represents the average of the indicators of each latent variable (average-parcel). Ping's one-parcel strategy had the advantage of using all information and not using the same measured variable in the construction of more than one product indicator (as is the case with the all-possible strategy). However, Ping's specification required a two-step estimation procedure that has been found to be less efficient in the simulation study by Moulder and Algina (2002; see also Algina & Moulder, 2001) and still necessitated the construction of cumbersome nonlinear constraints.

In a series of simulation studies, Marsh and colleagues (2004) compared three types of product indicators formed from the three indicators of X_1 and the three indicators of X_2: (1) all possible products (9 products in this example): $x_1x_4, x_1x_5, x_1x_6, x_2x_4, x_2x_5, x_2x_6, x_3x_4, x_3x_5, x_3x_6$; (2) matched-pair products (three matched pairs in this example): x_1x_4, x_2x_5, x_3x_6; and (3) one pair: x_1x_4. Different combinations of matched pairs and one pair were also compared. Their results showed that the precision of estimation for matched pairs was systematically better than other types of products. Based on their results, they posited two guidelines in the selection of the most appropriate strategies for the estimation of latent interactions: (1) *Use all of the information.* All of the multiple indicators should be used in the formation of the indicators of the latent variable interaction factor. (2) *Do NOT reuse any of the information.* Each of the multiple indicators should only be used once in the formation of the multiple indicators of the latent variable interaction factor, to avoid creating artificially correlated residuals when the same variable is used in the construction of more than one product term. Importantly, when the indicators are not reused, the variance–covariance matrix of errors will become diagonal. This subsequently decreases the number of constraints in constrained-type analyses. Although intuitive and supported by their simulation research, more research is necessary to endorse their suggestions fully as guiding principles.

For situations where two latent variables have the same number of indicators (three indicators of each latent variable as in our simulations), matched pairs are easily formed with arbitrary combinations (or nonarbitrary combinations if there is a natural pairing of indicators of the two latent constructs). What if the number of indicators differs for the two first-order effect factors? Assume, for example, that there were five indicators for the first factor and 10 for the second. In this situation, the application of the match strategy is not so clear-cut. The "do not reuse" principle dictates that we should only have five product indicators of the latent interaction factor, but our "use all of the information" principle dictates that we should have 10 product indicators. We can suggest several alternative strategies. One approach would be to use 5 indicators and 10 indicators, respectively, to define the two first-order effect factors, but to use the best (e.g., most reliable) five indicators of the second factor to form five matched-product indicators of the latent interaction effect. This suggestion suffers in that it does not use all of the information (i.e., only 5 of the 10 indicators of the second factor are used) and may require the research to select the best indicators based on the same data used to fit the model, but we suspect that it would work reasonably well so long as there were good indicators of both factors. A second approach would be to use the items from the second factor to form item parcels. Thus, for example, the researcher could form five (item pair) parcels by taking the average of the first two items to form the first item parcel, the average of the second two items to form the second parcel, and so on. In this way, the first factor would be defined in terms of five (single-indicator) indicators, the second factor would be defined by five (item-pair parcel) indicators, and the latent interaction factors would be defined in terms of five matched-product indicators. This strategy has the advantage of satisfying both of our principles. However, the use of item parcels instead of single indicators to define a first-order latent factor remains a controversial issue. Whereas suggestions such as those outlined here are consistent with our empirical results and proposed guidelines, further research is needed to evaluate their appropriateness more fully.

QUADRATIC EFFECTS: CONSTRAINED, PARTIALLY CONSTRAINED, AND UNCONSTRAINED APPROACHES

Starting with the early work of Kenny and Judd (1984), quadratic effects were often considered together with interaction effects because of the nonlinearity of latent factors, and this continues to be the case (e.g., Klein & Muthén, 2002; Laplante, Sabourin, Cournoyer, & Wright, 1998; Wall & Amemiya, 2000). Here we briefly introduce a simple latent quadratic model (see earlier discussion of substantive examples whereby this quadratic model might be used) and describe the application of the constrained, partially constrained, and unconstrained approaches to this issue. More generally, we stress that much less research has been conducted to evaluate the generalizability and robustness of approaches used to estimate interaction effects to the problem of polynomial effects.

Suppose the simple quadratic model is

$$\eta = \gamma_1 \xi + \gamma_2 \xi^2 + \zeta, \tag{14}$$

and that both latent variables η and ξ have three indicators, namely, y_1, y_2, y_3 and x_1, x_2, x_3, respectively. Further suppose that all of the independent indicators are centered so that each has a mean of zero. Using x_1^2, x_2^2, and x_3^2 as the multiple indicators of the latent quadratic term ξ^2, the LISREL specification is:

$$\begin{bmatrix} y_1 \\ y_2 \\ y_3 \end{bmatrix} = \begin{bmatrix} \tau_{y1} \\ \tau_{y2} \\ y_{y3} \end{bmatrix} + \begin{bmatrix} 1 \\ \lambda_{y2} \\ \lambda_{y3} \end{bmatrix} \eta + \begin{bmatrix} \varepsilon_1 \\ \varepsilon_2 \\ \varepsilon_3 \end{bmatrix}$$

$$\begin{bmatrix} x_1 \\ x_2 \\ x_3 \\ x_1^2 \\ x_2^2 \\ x_3^2 \end{bmatrix} = \begin{bmatrix} 1 & 0 \\ \lambda_2 & 0 \\ \lambda_3 & 0 \\ 0 & 1 \\ 0 & \lambda_5 \\ 0 & \lambda_6 \end{bmatrix} \begin{bmatrix} \xi \\ \xi^2 \end{bmatrix} + \begin{bmatrix} \delta_1 \\ \delta_2 \\ \delta_3 \\ \delta_4 \\ \delta_5 \\ \delta_6 \end{bmatrix},$$

$$\eta = \gamma_1 \xi + \gamma_2 \xi^2 + \zeta,$$

$$\boldsymbol{\kappa} = \begin{bmatrix} 0 \\ \kappa_2 \end{bmatrix}, \quad \boldsymbol{\Phi} = \begin{bmatrix} \phi_{11} & \\ 0 & \phi_{22} \end{bmatrix}$$

$$\boldsymbol{\Theta}_{\varepsilon} = \text{diag}(\theta_{\varepsilon 1}, \theta_{\varepsilon 2}, \theta_{\varepsilon 3})$$

$$\boldsymbol{\Theta}_{\delta} = \text{diag}(\theta_{\delta 1}, \theta_{\delta 2}, \theta_{\delta 3}, \theta_{\delta 4}, \theta_{\delta 5}, \theta_{\delta 6})$$

The necessary constraints are:

(i) $\lambda_5 = \lambda_2^2$, $\lambda_6 = \lambda_3^2$;

(ii) $\kappa_2 = \phi_{11}$;

(iii) $\phi_2 = 2\phi_{11}^2$;

(iv) $\theta_{\delta 4} = 4\phi_{11}\theta_{\delta 1} + 2\theta_{\delta 1}^2$, $\theta_{\delta 5} = 4\lambda_2^2\phi_{11}\theta_{\delta 2} + 2\theta_{\delta 2}^2$,

$\theta_{\delta 6} = 4\lambda_3^2\phi_{11}\theta_{\delta 3} + 2\theta_{\delta 3}^2$

Similar to interaction models, the intercept terms τ_{y1}, τ_{y2}, and τ_{y3} of the dependent indicators must be included even if y_1, y_2, and y_3 are mean-centered, because η typically does not have a zero mean. The path diagram for the quadratic model is shown in Figure 8.3.

The above constraints can be proved in the same way as those for the interaction model presented earlier. In the partially constrained approach to quadratic effects, similarly to the partially constrained interaction model, the covariance matrix $\mathbf{\Phi}$ should be changed to

$$\mathbf{\Phi} = \begin{bmatrix} \phi_{11} & \\ \phi_{21} & \phi_{22} \end{bmatrix},$$

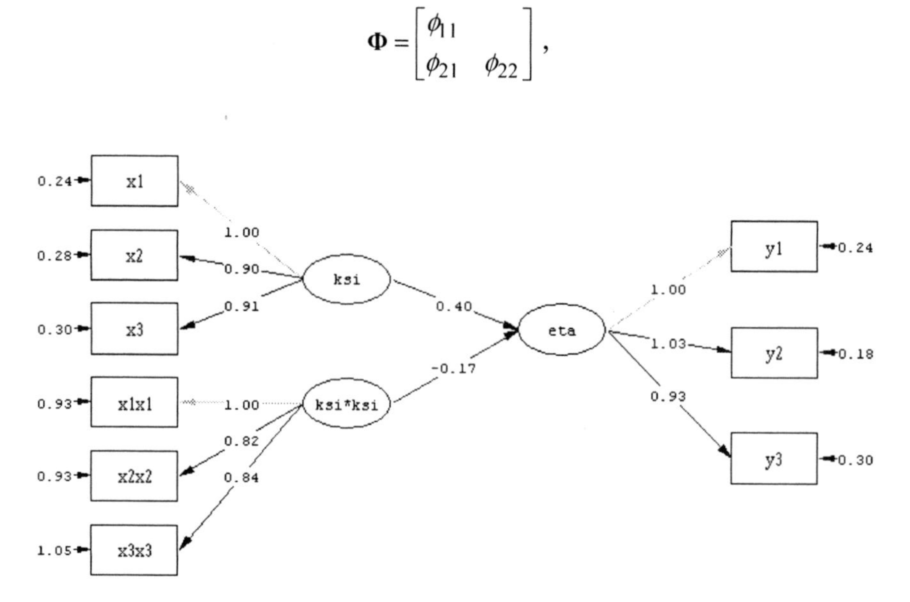

Figure 8.3. Path diagram and the estimations of the constrained quadratic model. The correlation between ξ and ξ^2 (i.e., ϕ_{21}) is fixed to be zero in the constrained approach, but is estimated freely in the partially constrained approach and unconstrained approach.

with all of the elements of $\mathbf{\Phi}$ allowed to be estimated freely. All other constraints are retained.

The unconstrained quadratic model is similar to the constrained and partly constrained models except that the constraints are not included. However, we know of no applications or simulation studies to evaluate the unconstrained approach to estimating quadratic effects, and there is clearly a need for more research.

In order to illustrate the three approaches to quadratic effects (constrained, partially constrained, and unconstrained), consider the substantive issue of the relation between university students' evaluations of teaching effectiveness and student perceptions of course workload/difficulty (see Marsh, 2001). Let y_1, y_2, and y_3 be three rating items completed by students that assess the overall effectiveness of the teacher, whereas x_1, x_2, and x_3 are three indicators of students' perceptions of the course workload/difficulty. The means and SDs of simulated data are y_1 (3.810, 0.833), y_2 (3.844, 0.813), y_3 (3.723, 0.834); x_1 (3.554, 1.038), x_2 (3.462, 0.962), x_3 (3.666, 0.991). Sample size N is 852. Independent indicators x_1, x_2, and x_3 are centered to have zero means.

The covariance matrix and mean vector of six indicators and three squared-indicators are presented as part of the LISREL syntax for the constrained approach (see Appendix 3a) and the path model and parameter estimates are shown in Figure 8.3. The estimated linear effect is $\gamma_1 =$ 0.399 (SE = 0.026, t = 15.25). The estimated quadratic effect is $\gamma_2 =$ −0.169 (SE = 0.020, t = −8.38). Both linear and quadratic effects are significant. The model fits the data very well: $\chi^2(37) = 79.4$, RMSEA = 0.033, SRMR = 0.049, NNFI = 0.988, and CFI = 0.987.

In order to explain the model, we should use the original scales. Noting that in the original scales, we should fix the η to have the same units and mean as y_1 (i.e., $y_1 = \eta + \varepsilon_1$), and ξ as x_1 (i.e., $x_1 = \xi + \delta_1$), then η and ξ in Equation 14 should be replaced by $\eta - 3.810$ and $\xi - 3.554$, respectively. Hence we can write the estimated quadratic equation as:

$$\eta - 3.810 = 0.399(\xi - 3.554) - 0.169(\xi - 3.554)^2$$

That is,

$$\eta = -0.169\xi^2 + 1.600\xi + 0.257$$

The results indicate that both the linear and quadratic components of workload/difficulty are statistically significant. The linear effect is positive whereas the negative quadratic component represents an inverted U-shaped function. The hypothetical results indicate that student ratings of teaching effectiveness improve with increases in workload/difficulty for most of the range of workload/difficulty. However, using derivatives we see that when $\xi = -\left(\frac{1.600}{2 \times (-0.169)}\right) = 4.73$ (near the top of the range of workload/difficulty, which varies along a 1 to 5 scale.) there is an inflection point where the ratings plateau ($\eta = 4.04$), and then decrease slightly for courses that are perceived to be excessively difficult.

The LISREL syntax of the partly constrained approach is shown in Appendix 3b. The main results of the partly constrained model are $\gamma_1 = 0.386$ (SE = 0.025, t = 15.51) and $\gamma_2 = -0.177$ (SE = 0.024, t = –7.50). The fit is very good: $\chi^2(35)=51.9$, RMSEA = 0.022, SRMR = 0.051, NNFI = 0.995, and CFI = 0.995.

The LISREL syntax of the unconstrained approach is shown in Appendix 3c. For this model, the main results are $\gamma_1 = 0.391$ (SE = 0.028, t = 14.18) and $\gamma_2 = -0.177$ (SE = 0.024, t = –7.41). The fit is very good: $\chi^2(29)=25.6$, RMSEA=0.0, SRMR=0.017, NNFI = 1.001, and CFI = 1.000.

Although our main focus has been on the application of the constrained, partially constrained, and unconstrained approaches to estimating interaction effects, some additional concerns exist when these approaches are used to estimate quadratic effects. As is the case with the estimation of quadratic effects with manifest variables (e.g., Cohen & Cohen, 1983), there are substantial problems of multicollinearity in that the same variables are used to infer the linear and quadratic effects. In the latent variable approach, this is likely to result in less precise estimates and negatively affect the convergence behavior of the models used to estimate the quadratic effects. This problem of multicollinearity is countered substantially by mean-centering each of the indicators. Nevertheless, there has been insufficient evaluation of these three approaches to the estimation of quadratic effects with either simulated data in Monte Carlo studies or real data in substantive applications. Hence, the applied researcher must apply appropriate caution in the interpretation of results based on these techniques.

NEW, EVOLVING PROCEDURES FOR THE ANALYSIS OF LATENT INTERACTION AND QUADRATIC EFFECTS

We conclude our chapter with a brief overview of new and evolving procedures for the analysis of latent interaction and quadratic effects. Whereas all of these procedures have many desirable features, they are each very computationally demanding and, at the time of writing of this chapter, have not yet been incorporated into widely accessible SEM packages such as LISREL, EQS, AMOS, or Mplus. For this reason, they are unlikely to be used in current applications of applied research, but are likely to be the basis of important developments in the near future.

Latent Moderated Structural (LMS) Equations and Quasi-Maximum Likelihood (QML)

QML belongs to the new generation of methods for analyzing nonlinear effects. Before introducing QML, we should mention the LMS (latent moderated structural equations) method that was developed as part of the doctoral thesis of Klein (see Klein & Moosbrugger, 2000; Schermelleh-Engel et al., 1998). Without the need of product indicators, LMS is a radically different approach to solving the non-normality problems implied by latent interaction effects—recognizing that these effects are not generally distributed normally even when the indicators of each of the first-order factors are normally distributed. In the LMS method, however, the latent independent variables and the error variables are still assumed to be normally distributed (as they were in the constrained approach). The distribution of the indicator variables is approximated by a finite mixture distribution and the log likelihood function is maximized by use of the EM algorithm. Simulation studies showed that LMS provided efficient parameter estimators and a reliable model difference test, and that standard errors were relatively unbiased (Klein & Moosbrugger, 2000; Schermelleh-Engel et al., 1998). Although models with multiple latent product terms can be analyzed with the LMS method, the method can become computationally too intensive for models with more than one product term in the structural equation and many indicators of latent variables.

The QML method for structural equation models with quadratic forms was developed for the efficient estimation of more complex nonlinear structural equation models, which cannot be analyzed with the LMS method because of the computational burden involved in a more complex nonlinear model structure (Klein & Muthén, 2002). In their QML study, Klein and Muthén (2002) dealt with a structural equation model with a general quadratic form of latent independent variables as

$$\eta = \alpha + \mathbf{\Gamma}\boldsymbol{\xi} + \boldsymbol{\xi}'\mathbf{\Omega}\boldsymbol{\xi} + \zeta, \quad (15)$$

where η is a latent dependent variable, α is an intercept term, $\boldsymbol{\xi}$ is a $(k \times 1)$ vector of latent predictor variables, $\mathbf{\Gamma}$ is a $(1 \times k)$ coefficient matrix, $\mathbf{\Omega}$ is an upper triangular $(k \times k)$ coefficient matrix, and ζ is a disturbance variable. The usual interactions caused by two latent variables and the quadratic effect of one latent variable are the special cases of Equation 15. For example, when

$$\boldsymbol{\Gamma} = (\gamma_1, \gamma_2), \boldsymbol{\xi} = (\xi_1, \xi_2)', \text{ and } \boldsymbol{\Omega} = \begin{bmatrix} 0 & \gamma_3 \\ 0 & 0 \end{bmatrix}$$

then Equation 15 becomes Equation 13.

After some rather complicated transformations and calculations, the QML method was based on an approximation of the non-normal density function of the joint indicator vector by a product of a normal and a conditionally normal density. Simulation studies showed that the QML estimates were very close to those given by the LMS method, and the QML estimators were almost as efficient as the ML estimators. Also, the QML estimator was substantially more efficient than the two-step method of moments (2SMM) estimator discussed below (Klein & Muthén, 2002).

However, the QML approach still relies on the assumption that the first-order effect factors are normally distributed, which is generally violated when the measured variables are not normally distributed. In Marsh and colleagues' (2004) simulation results, the QML estimates were substantially biased when this assumption of normality of the measured variable indicators was severely violated, whereas estimates based on the unconstrained and partially constrained approaches were not substantially biased. However, in all cases, the QML estimates were consistently more precise than those based on the unconstrained and partially constrained approaches. Hence, except in cases where non-normality was extreme, the tradeoff between the small bias in QML estimates appeared to be offset by its greater precision. Although the special software package for QML is in the process of being developed, at least for the time being it cannot be conducted with the commonly used structural equation modeling packages. However, the QML approach is likely to become an attractive choice among the various approaches for analyzing interaction and quadratic effects once a stand-alone statistical package is readily accessible, or the QML approach is integrated into widely available CFA/SEM statistical packages (see adaptation of the Klein & Moosbrugger, 2000, approach implemented in Version 3 of MPlus).

Two-Step Method of Moments (2SMM)

Wall and Amemiya (2000) developed a two-step method of moments (2SMM) approach for a general polynomial structural equation model:

$$\eta = \boldsymbol{\alpha}' h(\boldsymbol{\xi}) + \zeta, \tag{16}$$

where η is a dependent latent variable, $\boldsymbol{\xi}$ is a $(k \times 1)$ vector of latent factors, $h(\boldsymbol{\xi})$ is an $(r \times 1)$ vector with each component being a pure or mixed power of the elements of $\boldsymbol{\xi}$, $\boldsymbol{\alpha}$ is an $(r \times 1)$ vector of unknown coefficients, ζ is a disturbance with zero mean and uncorrelated with each ξ and its measurement errors. Even the rather general structural model in Equation 15 is a special case of Equation 16. In fact, Equation 16 can produce a model including any polynomial effects. For example, when $\boldsymbol{\xi} = (\xi_1, \xi_2)'$, $h(\boldsymbol{\xi}) = (\xi_1, \xi_2, \xi_1\xi_2)'$, and $\boldsymbol{\alpha} = (\gamma_1, \gamma_2, \gamma_3)'$, then Equation 16 becomes Equation 13.

In the first step of 2SMM, the parameters of the measurement model are estimated by using a factor-analytical technique. In the second step, conditional moments of products of latent variables are calculated and a method of moments procedure using these conditional moments is applied to estimate the parameters of the structural equation. In particular, the assumption of normally distributed latent variables can be relaxed when distribution-free factor score estimators are used in the first step. In theory, the 2SMM approach can be generalized to more complex nonlinear latent variable models. However, more research is needed to determine how much the nonsimultaneous, two-step estimation procedure and the choice of a complex nonlinear model affects the efficiency of the 2SMM estimators, particularly in cases in which the reliability of indicator variables is low to medium (Klein & Muthén, 2002). With deep statistical theory and high computational burden, but without a special and user-friendly statistical package, the 2SMM approach is not yet ready to be used for applied researchers, but holds promising potential for the future.

SUMMARY AND LIMITATIONS

In Table 8.1 we briefly summarize the strengths, weaknesses, and assumptions of the various approaches that we have discussed. This leads us to offer the following advice, particularly for novice researchers. For researchers still new to CFA and SEM, it may be useful to begin with ANOVA (if both independent variables are categorical) or the multiple regression approach to ANOVA (if at least one of the independent variables is reasonably continuous) using scale scores instead of latent constructs. This should provide a preliminary indication of the expected results when more sophisticated approaches are applied. It may also be useful to use the factor score approach in which factor scores are used instead of scale scores. However, the use of scale scores and factor scores should give similar results for well-defined measures (i.e., the multiple indicators are all substantially related to the latent construct). When one of the independent variables is a categorical variable that can be repre-

Table 8.1. Procedures/Assumptions, Strengths, and Weaknesses of Various Approaches Introduced in This Chapter

Method	*Procedure/Assumption*	*Strength*	*Weakness*
Nonlatent Approaches			
Factor score analyses	Uses factor scores in regression; assumes error-free measurement	Simple, easy to understand and implement; evaluates the factor structure of independent and dependent variables	Measurement errors are ignored, standard errors not precise; does not analyze measurement and structural models simultaneously
Two-stage least square method	Instrumental variables are formed and used in the subsequent ordinary least square regression; does not assume normality of factors or indicators	Easy to implement because does not involve SEMs (but more difficult than the factor score approach)	Limited to a single-indicator outcome (without multiple indicators); results dependent on how latent variables are scaled; larger bias and lower power than fully latent approaches, which may increase with sample size, particularly when indicator reliability is low
Multiple-group analysis	Multiple-group SEM analysis	Easy to understand and implement by using most SEM packages	Applicable only when one of the independent variables can be represented by a small number of groups based on a single-indicator construct; requires a reasonable sample size for each group; strength of interaction not directly estimated but inferred on the basis of lack of invariance in parameter estimates across the multiple groups

Fully Latent Variable Approaches			
Constrained approach	Measurement and structural models estimated simultaneously, under the large number of constraints; normality assumed for factors, indicators, and constraints	Appropriate consideration and treatment of measurement errors; less bias, better control of Type 1 error rate, higher power	Complicated constraints required; imposed constraints based on multivariate normal indicators so that interaction effects are biased when variables are non-normal
Partially constrained approach	Similar to constrained approach; covariances among interaction effects and latent variables freely estimated; no constraints are based on assumption of normality	Similar to those of constrained approach; one constant less; relax normality assumption; less biased interaction effects than the constrained approach for non-normal variables, particularly for large sample sizes	Complicated constraints required
Unconstrained approach	Similar to the constrained and partially constrained approach, except constraints are not imposed	Similar to those of the partially constrained approach, but without complicated constraints	Somewhat less precise than constrained approach when indicators are normally distributed and small sample sizes
Latent moderated structural equation (LMS)/ quasi-maximum likelihood (QML)	Measurement and structural relation estimated simultaneously using special software; assumes latent variables and their errors are normal	More precise than constrained, partially constrained, and unconstrained approaches	Need special software, but simple to use, computational intensive in LMS; factors/indicators still assumed to be normal so that interaction effects are biased when indicators are not multivariate normal
Two-step method of moments	Measurement model estimated first, then methods of moment are used	Comprehensive models accommodating wide range of nonlinear models; normality assumptions can be relaxed	Has not been incorporated into commonly used SEM packages and no commercial software available

sented by a relatively small number of groups, the multiple-group CFA approach is an attractive alternative.

Before proceeding to true latent variable approaches to assessing interactions, data should be evaluated for violations of univariate and multivariate normality. Particularly if there are violations of normality, then the various constrained approaches should only be used with great caution and researchers should apply the partially constrained and unconstrained approaches. However, because of the simplicity of the unconstrained approach—compared to the constrained and partially constrained approaches—the unconstrained approach is recommended.

Finally, the use of highly sophisticated statistical tools such as structural equation modeling can mislead otherwise knowledgeable researchers into thinking that well-known problems that exist with less complicated approaches are no longer relevant. Although we have focused on important advantages to the use of latent variable approaches to these problems, the reader should not think that the latent variable approaches are a panacea for all of the complications in evaluating interaction and nonlinear effects. Theoretical and empirical research (e.g., Aiken & West, 1991) demonstrates that there are many inherent difficulties in estimating interaction effects, even when manifest (nonlatent) variable approaches are used. Misspecified models may be a problem. Also, the analysis of interaction effects becomes more problematic when the relation between a given independent variable and the dependent variable is nonlinear or when the correlations among the interacting variables become increasingly large. There is no reason to suggest that problems such as these will not also present difficulties for latent variable approaches.

Most of the research conducted with fully latent approaches (e.g., the constrained, partially constrained, and unconstrained approaches) has been based on simulated data with multivariate normal indicators. Whereas there has been some simulation research with non-normal indicators (e.g., Marsh et al., 2004), these approaches have not been widely applied to substantive issues with real data. Hence, a more complete evaluation of the different procedures in actual practice is not possible. Because the robustness and convergence behavior of the constrained, partially constrained, and unconstrained approaches have not been systematically evaluated with "real" data, researchers are encouraged to apply different approaches and to compare the results based on the different approaches. Particular caution is needed in interpreting polynomial effects, because the fully latent approaches discussed here have not been sufficiently evaluated—even with well-behaved, simulated data—for this application.

The estimation of latent interaction and quadratic effects is a rapidly developing area of research. Hence, applied researchers need to be aware that new procedures (e.g., QML) may be recommended over the approaches demonstrated here once efficient, user-friendly software becomes widely available or is incorporated into existing SEM packages such as LISREL, EQS, AMOS, and Mplus.

APPENDIX 1: PROOF OF THE CONSTRAINTS IN THE CONSTRAINED APPROACH

(i) From $x_2 = \lambda_2\xi_1 + \delta_2$, $x_5 = \lambda_5\xi_2 + \delta_5$, we have

$$x_2x_5 = \lambda_2\lambda_5\xi_1\xi_2 + \delta_8,$$

where $\delta_8 = \lambda_2\xi_1\delta_5 + \lambda_5\xi_2\delta_2 + \delta_2\delta_5$. So, the loading of x_2x_5 on $\xi_1\xi_2$ is $\lambda_2\lambda_5$, that is, $\lambda_8 = \lambda_2\lambda_5$. For the same reason, it follows that $\lambda_9 = \lambda_3\lambda_6$. Because λ_1 and λ_4 are fixed to 1, $\lambda_7 = \lambda_1\lambda_4$ must be fixed to 1, accordingly.

(ii) $\kappa_3 = E(\xi_1\xi_2) = \text{cov}(\xi_1,\xi_2) = \phi_{21}$.

(iii) $\phi_{33} = \text{var}(\xi_1\xi_2) = \text{var}(\xi_1)\text{var}(\xi_2) + \text{cov}^2(\xi_1,\xi_2) = \phi_{12}\phi_{22} + \phi_{21}^2$.

This result holds under the supposition that ξ_1 and ξ_2 have multivariate normal distribution with zero means (see, e.g., Kenny & Judd, 1984).

(iv) $\theta_{\delta 8} = \text{var}(\delta_8) = \text{var}(\lambda_2\xi_1\delta_5 + \lambda_5\xi_2\delta_2 + \delta_2\delta_5) =$

$\lambda_2^2\phi_{11}\theta_{\delta 5} + \lambda_5^2\phi_{22}\theta_{\delta 2} + \theta_{\delta 2}\theta_{\delta 5}$.

Similar derivations can be conducted with the other two constraints for $\theta_{\delta 7}$ and $\theta_{\delta 9}$. These results hold when ξ_1, ξ_2, and all δ terms are multivariate normal with mean zero, and all of them are uncorrelated with each other (except that ξ_1 and ξ_2 are allowed to be correlated). Syntax for estimating this model appears in Appendix 2a.

APPENDIX 2a: INTERACTION MODEL: LISREL SYNTAX FOR THE CONSTRAINED APPROACH

```
Constrained Approach
Independent indicators were centered
ML ESTIMATION
DA NI=12 NO=500
CM
```

```
1.221
0.891 0.968
0.907 0.800 1.007
0.493 0.401 0.449 1.285
0.458 0.385 0.413 0.885 0.955
0.834 0.697 0.733 1.838 1.581 3.999
0.470 0.411 0.444 0.200 0.174 0.436 1.217
1.277 1.122 1.247 0.366 0.339 0.777 2.691 8.963
0.416 0.379 0.390 0.177 0.143 0.331 0.874 2.381 0.989
0.157 0.107 0.062 -0.183 -0.174 -0.332 -0.027 -0.154 -0.048 1.407
0.406 0.266 0.266 -0.348 -0.220 -0.487 -0.052 -0.106 -0.057 2.021 8.104
0.348 0.220 0.204 -0.193 -0.092 -0.182 -0.003 -0.111 -0.110 1.392 3.522 3.711
ME
1.057 2.075 3.057 0.000 0.000 0.000 0.000 0.000 0.000 0.199 0.338 0.330
LA
y1 y2 y3 x1 x2 x3 x4 x5 x6 x1x4 x2x5 x3x6
MO NY=3 NE=1 NX=9 NK=3 LY=FU,FR PS=SY,FR   C
  TE=DI,FR TD=DI,FR TY=FR KA=FR
LE
eta
LK
ksi1 ksi2 ksi1*ksi2
PA PH
1
1 1
0 0 1
! fixed PH 3 1 and PH 3 2 to 0, see Equation 12
FI LY 1 1
FR LX 2 1 LX 3 1 LX 5 2 LX 6 2 LX 8 3 LX 9 3
VA 1 LY 1 1 LX 1 1 LX 4 2 LX 7 3
VA 0 LX 7 1 LX 7 2 LX 8 1 LX 8 2 LX 9 1 LX 9 2
FR GA 1 1 GA 1 2 GA 1 3
FI KA(1) KA(2)
VA 0 KA(1) KA(2)
CO LX 8 3 = LX(2,1) *LX(5,2)
CO LX 9 3 = LX(3,1) * LX(6,2)
! see Constraint (i) under "Constrained Approach"
CO KA(3)=PH(2,1)
! see Constraint (ii) under "Constrained Approach"
CO PH 3 3 = PH(1,1)*PH(2,2) + PH(2,1)** 2
! see Constraint (iii) under "Constrained Approach"
CO TD 7 7 = TD(4,4) + TD(1,1) +TD(1,1)*TD(4,4)
```

```
CO TD 8 8 = TD(5,5)*LX(2,1)**2 + TD(2,2)*LX(5,2)**2 +   C
  TD(2,2)*TD(5,5)
CO TD 9 9 = TD(6,6)*LX(3,1)**2 + TD(3,3)*LX(6,2)**2 +   C
  TD(3,3)*TD(6,6)
! see Constraint (iv) under "Constrained Approach"
PD
OU ND=3 AD=off EP=0.0001 IT=500 XM
```

APPENDIX 2b: INTERACTION MODEL: LISREL SYNTAX FOR THE PARTIALLY CONSTRAINED APPROACH

```
Partially Constrained Approach
Independent indicators were centered
ML ESTIMATION
DA NI=12 NO=500
CM <<insert covariance matrix from Appendix 2a>>
ME <<insert vector of means from Appendix 2a>>
LA
y1 y2 y3 x1 x2 x3 x4 x5 x6 x1x4 x2x5 x3x6
MO NY=3 NE=1 NX=9 NK=3 LY=FU,FR PS=SY,FR TE=DI,FR   C
  TD=DI,FR TY=FR KA=FR
LE
eta
LK
ksi1 ksi2 ksi1*ksi2
PA PH
1
1 1
1 1 1
! freely estimate PH 3 1 and PH 3 2
FI LY 1 1
FR LX 2 1 LX 3 1 LX 5 2 LX 6 2 LX 8 3 LX 9 3
VA 1 LY 1 1 LX 1 1 LX 4 2 LX 7 3
VA 0 LX 7 1 LX 7 2 LX 8 1 LX 8 2 LX 9 1 LX 9 2
FR GA 1 1 GA 1 2 GA 1 3
FI KA(1) KA(2)
VA 0 KA(1) KA(2)
CO LX 8 3 = LX(2,1) *LX(5,2)
CO LX 9 3 = LX(3,1) * LX(6,2)
! see Constraint (i) under "Constrained Approach"
CO KA(3)=PH(2,1)
! see Constraint (ii) under "Constrained Approach"
```

```
CO TD 7 7 = TD(4,4) + TD(1,1) + TD(1,1)*TD(4,4)
CO TD 8 8 = TD(5,5)*LX(2,1)**2 + TD(2,2)*LX(5,2)**2 +   C
   TD(2,2)*TD(5,5)
CO TD 9 9 = TD(6,6)*LX(3,1)**2 + TD(3,3)*LX(6,2)**2 +   C
   TD(3,3)*TD(6,6)
! see Constraint (iv) under "Constrained Approach"
PD
OU ND=3 AD=off EP=0.0001 IT=500 XM
```

APPENDIX 2c: INTERACTION MODEL: LISREL SYNTAX FOR THE UNCONSTRAINED APPROACH

```
Unconstrained Approach
Independent indicators were centered
ML ESTIMATION
DA NI=12 NO=500
CM <<insert covariance matrix from Appendix 2a>>
ME <<insert vector of means from Appendix 2a>>
LA
y1 y2 y3 x1 x2 x3 x4 x5 x6 x1x4 x2x5 x3x6
MO NY=3 NE=1 NX=9 NK=3 LY=FU,FR PS=SY,FR   C
   TE=DI,FR TD=DI,FR TY=FR KA=FR
LE
eta
LK
ksi1 ksi2 ksi1*ksi2
PA PH
1
1 1
1 1 1
! PH 3 1 PH 3 2 should be fixed to zero for normal data
FI LY 1 1
FR LX 2 1 LX 3 1 LX 5 2 LX 6 2 LX 8 3 LX 9 3
VA 1 LY 1 1 LX 1 1 LX 4 2 LX 7 3
VA 0 LX 7 1 LX 7 2 LX 8 1 LX 8 2 LX 9 1 LX 9 2
FR GA 1 1 GA 1 2 GA 1 3
FI KA(1) KA(2)
VA 0 KA(1) KA(2)
PD
OU ND=3 AD=off EP=0.0001 IT=500 XM
```

APPENDIX 3a: QUADRATIC MODEL: LISREL SYNTAX FOR THE CONSTRAINED APPROACH

```
Constrained Approach
Independent indicators were centered
ML ESTIMATION
DA NI=9 NO=852
CM
 0.693
 0.472 0.661
 0.418 0.430 0.695
 0.335 0.352 0.339 1.075
 0.299 0.328 0.318 0.740 0.925
 0.301 0.313 0.311 0.749 0.662 0.980
-0.269 -0.265 -0.275 -0.117 -0.125 -0.131 2.187
-0.181 -0.191 -0.186 -0.109 -0.100 -0.066 1.002 1.729
-0.224 -0.240 -0.196 -0.122 -0.069 -0.123 0.993 0.740 1.805
ME
3.810 3.844 3.722 0.000 0.000 0.000 1.073 0.924 0.979
LA
y1 y2 y3 x1 x2 x3 x1x1 x2x2 x3x3
MO NY=3 NE=1 NX=6 NK=2 LY=FU,FR PS=SY,FR PH=SY,FR   C
   TE=DI,FR TD=DI,FR KA=FR TY=FR
LE
eta
LK
ksi ksi*ksi
FI PH 2 1
VA 0 PH 2 1
FI LY 1 1
VA 1 LY 1 1 LX 1 1 LX 4 2
FR LX 2 1 LX 3 1 LX 5 2 LX 6 2
FR GA 1 1 GA 1 2
FI KA(1)
VA 0 KA(1)
CO PH 2 2=2*PH(1,1)**2
CO LX 5 2=LX(2,1)**2
CO LX 6 2=LX(3,1)**2
CO TD 4 4=4*PH(1,1)*TD(1,1) + 2*TD(1,1)**2
CO TD 5 5=4*LX(2,1)**2*PH(1,1)*TD(2,2) + 2*TD(2,2)**2
CO TD 6 6=4*LX(3,1)**2*PH(1,1)*TD(3,3) + 2*TD(3,3)**2
CO KA(2)=PH(1,1)
```

```
PD
OU ND=3 AD=off EP=0.0001 IT=500 XM
```

APPENDIX 3b: QUADRATIC MODEL: LISREL SYNTAX FOR THE PARTIALLY CONSTRAINED APPROACH

```
Partially Constrained Approach
Independent indicators were centered
ML ESTIMATION
DA NI=9 NO=852
CM <<insert covariance matrix from Appendix 3a>>
ME <<insert vector of means from Appendix 3a>>
LA
y1 y2 y3 x1 x2 x3 x1x1 x2x2 x3x3
MO NY=3 NE=1 NX=6 NK=2 LY=FU,FR PS=SY,FR PH=SY,FR   C
   TE=DI,FR TD=DI,FR KA=FR TY=FR
LE
eta
LK
ksi ksi*ksi
FI LY 1 1
VA 1 LY 1 1 LX 1 1 LX 4 2
FR LX 2 1 LX 3 1 LX 5 2 LX 6 2
FR GA 1 1 GA 1 2
FI KA(1)
VA 0 KA(1)
CO LX 5 2=LX(2,1)**2
CO LX 6 2=LX(3,1)**2
CO TD 4 4=4*PH(1,1)*TD(1,1) + 2*TD(1,1)**2
CO TD 5 5=4*LX(2,1)**2*PH(1,1)*TD(2,2) + 2*TD(2,2)**2
CO TD 6 6=4*LX(3,1)**2*PH(1,1)*TD(3,3) + 2*TD(3,3)**2
CO KA(2)=PH(1,1)
PD
OU ND=3 AD=off EP=0.0001 IT=500 XM
```

APPENDIX 3c: QUADRATIC MODEL: LISREL SYNTAX FOR THE UNCONSTRAINED APPROACH

```
Unconstrained Approach
Independent indicators were centered
ML ESTIMATION
```

```
DA NI=9 NO=852
CM <<insert covariance matrix from Appendix 3a>>
ME <<insert vector of means from Appendix 3a>>
LA
y1 y2 y3 x1 x2 x3 x1x1 x2x2 x3x3
MO NY=3 NE=1 NX=6 NK=2 LY=FU,FR PS=SY,FR PH=SY,FR   C
   TE=DI,FR TD=DI,FR KA=FR TY=FR
LE
eta
LK
ksi ksi*ksi
FI LY 1 1
VA 1 LY 1 1 LX 1 1 LX 4 2
FR LX 2 1 LX 3 1 LX 5 2 LX 6 2
FR GA 1 1 GA 1 2
FI KA(1)
PD
OU ND=3 AD=off EP=0.0001 IT=500 XM
```

REFERENCES

Aiken, L. S., & West, S. G. (1991). *Multiple regression: Testing and interpreting interactions*. Newbury Park, CA: Sage.

Algina, J., & Moulder, B. C. (2001). A note on estimating the Jöreskog-Yang model for latent variable interaction using LISREL 8.3. *Structural Equation Modeling: A Multidisciplinary Journal, 8*, 40–52.

Arminger, G., & Muthén, B.O. (1998). A Bayesian approach to nonlinear latent variable models using the Gibbs sampler and the Metropolis-Hastings algorithm. *Psychometrika, 63*, 271–300.

Bagozzi, R. P., & Yi, Y. (1989). On the use of structural equation models in experimental designs. *Journal of Marketing Research, 26*, 271–284.

Baron, R. M., & Kenny, D. A. (1986). The moderator-mediator variable distinction in social psychological research: Conceptual, strategic, and statistical considerations. *Journal of Personality and Social Psychology, 51*, 1173–1182.

Blom, P., & Christoffersson, A. (2001). Estimation of nonlinear structural equation models using empirical characteristic functions. In R. Cudeck, S. Du Toit, & D. Sörbom (Eds.), *Structural equation modeling: Present and future* (pp. 443–460). Lincolnwood, IL: Scientific Software.

Bollen, K. A. (1989). *Structural equations with latent variables*. New York: Wiley.

Bollen, K. A. (1995). Structural equation models that are nonlinear in latent variables: A least squares estimator. In P. Marsden (Ed.), *Sociological methodology 1995* (pp. 223–251). Cambridge, MA: Blackwell.

Bollen, K. A., & Paxton, P. (1998). Two-stage least squares estimation of interaction effects. In R. E. Schumacker & G. A. Marcoulides (Eds.), *Interaction and nonlinear effects in structural equation modeling* (pp. 125–151). Mahwah, NJ: Erlbaum.

Boomsma, A. (1983). *On the robustness of LISREL against small sample size and non-normality.* Doctoral dissertation, University of Groningen.

Cohen, J., & Cohen, P. (1983). *Applied multiple regression /correlational analysis for the behavioral sciences.* Hillsdale, NJ: Erlbaum.

Fraser, C. (1980). *COSAN user's suide.* Toronto: Ontario Institute for Studies in Education.

Hayduk, L. A. (1987). *Structural equation modeling with LISREL: Essentials and advances.* Baltimore: Johns Hopkins University.

Hau, K. T., & Marsh, H. W. (2004). The use of item parcels in structural equation modeling: Non-normal data and small sample sizes. *British Journal of Mathematical and Statistical Psychology, 57*, 327–351.

Hu, L., Bentler, P. M., & Kano, Y. (1992). Can test statistics in covariance structure analysis be trusted? *Psychological Bulletin, 112*, 351–362.

Jaccard, J., & Wan, C. K. (1995). Measurement error in the analysis of interaction effects between continuous predictors using multiple regression: Multiple indicator and structural equation approaches. *Psychological Bulletin, 117*, 348–357.

Jöreskog, K. G. (1998). Interaction and nonlinear modeling: issue and approaches. In R. E. Schumacker & G. A. Marcoulides (Eds.), *Interaction and nonlinear effects in structural equation modeling* (pp. 239–250). Mahwah, NJ: Erlbaum.

Jöreskog, K. G. & Yang, F. (1996). Nonlinear structural equation models: The Kenny-Judd model with interaction effects. In G. A. Marcoulides & R. E. Schumacker (Eds.), *Advanced structural equation modeling: Issues and techniques* (pp. 57–88). Mahwah, NJ: Erlbaum.

Kaplan, D. (2000). *Structural equation modeling: Foundations and extensions.* Newbury Park, CA: Sage.

Kenny, D. A., & Judd, C. M. (1984). Estimating the nonlinear and interactive effects of latent variables. *Psychological Bulletin, 96*, 201–210.

Klein, A., & Moosbrugger, H. (2000). Maximum likelihood estimation of latent interaction effects with the LMS method. *Psychometrika, 65*, 457–474.

Klein, A. G., & Muthén, B. O. (2002). *Quasi maximum likelihood estimation of structural equation models with multiple interaction and quadratic effects.* Unpublished manuscript, Graduate School of Education, University of California Los Angeles.

Laplante, B., Sabourin, S., Cournoyer, L. G., & Wright, J. (1998). Estimating nonlinear effects using a structured means intercept approach. In R. E. Schumacker & G. A. Marcoulides (Eds,), *Interaction and nonlinear effects in structural equation modeling* (pp. 183–202). Mahwah, NJ: Erlbaum.

Lubinski, D., & Humphreys, L. G. (1990). Assessing spurious "moderator effects": Illustrated substantively with the hypothesized ("synergistic") relation between spatial and mathematical ability. *Psychological Bulletin, 107*, 385–393.

MacCallum, R. C., Zhang, S., Preacher, K. J., & Rucker, D. D. (2002). On the practice of dichotomization of quantitative variables. *Psychological Methods, 7*, 19–40.

Marsh, H. W. (2001). Distinguishing between good (useful) and bad workload on students' evaluations of teaching. *American Educational Research Journal, 38,* 183–212.

Marsh, H. W., Hau, K. T., Balla, J. R., & Grayson, D. (1998). Is more ever too much? The number of indicators per factor in confirmatory factor analysis. *Multivariate Behavioral Research, 33,* 181–220.

Marsh, H. W., Wen, Z., & Hau, K. T. (2004). Structural equation models of latent interactions: Evaluation of alternative estimation strategies and indicator construction. *Psychological Methods, 9,* 275–300.

Moulder B. C., & Algina, J. (2002). Comparison of methods for estimating and testing latent variable interactions. *Structural Equation Modeling: A Multidisciplinary Journal, 9,* 1–19.

Ping, R. A., Jr. (1995). A parsimonious estimating technique for interaction and quadratic latent variables. *Journal of Marketing Research, 32,* 336–347.

Ping, R. A., Jr. (1996). Latent variable interaction and quadratic effect estimation: A two-step technique using structural equation analysis. *Psychological Bulletin, 119,* 166–175.

Ping, R. A., Jr. (1998). EQS and LISREL examples using survey data. In R. E. Schumacker & G. A. Marcoulides (Eds.), *Interaction and nonlinear effects in structural equation modeling* (pp. 63–100). Mahwah, NJ: Erlbaum.

Rigdon, E. E., Schumacker, R. E., & Wothke, W. (1998). A comparative review of interaction and nonlinear modeling. In R. E. Schumacker & G. A. Marcoulides (Eds.), *Interaction and nonlinear effects in structural equation modeling* (pp. 1–16). Mahwah, NJ: Erlbaum.

Schermelleh-Engel, K., Klein, A., & Moosbrugger, H. (1998). Estimating nonlinear effects using a latent moderated structural equations approach. In R. E. Schumacker & G. A. Marcoulides (Eds.), *Interaction and nonlinear effects in structural equation modeling* (pp. 203–238). Mahwah, NJ: Erlbaum.

Schumacker, R. E., & Marcoulides, G. A. (Eds.). (1998). *Interaction and nonlinear effects in* structural equation modeling. Mahwah, NJ: Erlbaum.

Wall, M. M., & Amemiya, Y. (2000). Estimation for polynomial structural equation models. *Journal of the American Statistical Association, 95,* 929–940.

Wall, M. M., & Amemiya, Y. (2001). Generalized appended product indicator procedure for nonlinear structural equation analysis. *Journal of Educational and Behavioral Statistics, 26,* 1–29.

West, S. G., Finch, J. F., & Curran, P. J. (1995). Structural equation models with nonnormal variables: Problems and remedies. In R. H. Hoyle (Ed.), *Structural equation modeling: Concepts, issues and applications* (pp. 56–75). Newbury Park, CA: Sage.

Yang, F. (1998). Modeling interaction and nonlinear effects: a step-by-step LISREL example. In R. E. Schumacker & G. A. Marcoulides (Eds.), *Interaction and nonlinear effects in structural equation modeling* (pp. 17–42). Mahwah, NJ: Erlbaum.

Yang-Wallentin, F., & Jöreskog, K. G. (2001). Robust standard errors and chi-squares for interaction models. In G.A. Marcoulides & R.E. Schumacker (Eds.), *New developments and techniques in structural equation modeling* (pp. 159–171). Mahwah, NJ: Erlbaum.

PART III

ASSUMPTIONS

CHAPTER 9

NON-NORMAL AND CATEGORICAL DATA IN STRUCTURAL EQUATION MODELING

Sara J. Finney and Christine DiStefano

Structural equation modeling (SEM) has become an extremely popular data analytic technique in education, psychology, business, and other disciplines (Austin & Calderón, 1996; MacCallum & Austin, 2000; Tremblay & Gardner, 1996). Given the frequency of its use, it is important to recognize the assumptions associated with different estimation methods, demonstrate the conditions under which results are robust to violations of these assumptions, and specify the procedures that should be employed when assumptions are not met. The importance of attending to assumptions and, consequently, selecting appropriate analysis strategies based on the characteristics of the data and the study's design cannot be understated. Put simply, violating assumptions can produce biased results in terms of model fit as well as parameter estimates and their associated significance tests. Biased results may, in turn, result in incorrect decisions about the theory being tested.

While there are several assumptions underlying the popular normal theory (NT) estimators used in SEM, the two assumptions that we focus

Structural Equation Modeling: A Second Course, 269–314

on in this chapter concern the metric and distribution of the data. Specifically, the data required by NT estimators are assumed to be continuous and multivariate normally distributed in the population. We focus on these two assumptions because often the data modeled in the social sciences do not follow a multivariate normal distribution. For example, Micceri (1989) noted that much of the data gathered from achievement and other measures are not normally distributed. This is disconcerting given that Gierl and Mulvenon (1995) found that most researchers do not examine the distribution of their data, but instead simply assume normality. In addition to the pervasiveness of non-normal data, the applied literature is thick with examples of categorical data collected using ordinal measures (e.g., Likert-type scales).

Because of the prevalence of both non-normal and categorical data in empirical research, this chapter focuses on issues surrounding modeling data with these characteristics using SEM. First, we review the assumptions underlying NT estimators. We next describe non-normal and categorical data and review robustness studies of the most popular NT estimator, maximum likelihood (ML), in order to understand the consequences of violating these assumptions. We then discuss four popular strategies that have been used to accommodate non-normal and/or categorical data:

1. Asymptotically distribution-free (ADF) estimation
2. Satorra-Bentler scaled χ^2 and standard errors
3. Robust weighted least squares (WLS) estimation methods implemented in the software program Mplus (e.g, WLSM, WLSMV)
4. Bootstrapping.

For each strategy we present the following: (a) a description of the strategy; (b) a summary of research concerning the robustness of the χ^2 statistic, fit indices, parameter estimates, and standard errors; and (c) a description of implementation across three software programs.

NORMAL THEORY ESTIMATORS

Assumptions of Normal Theory Estimators

As with most statistical techniques, SEM is based on assumptions that should be met in order for researchers to trust the obtained results. Central to SEM is the choice of an estimation method that is used to obtain parameter values, standard errors, and fit indices. The two common NT estimators are maximum likelihood (ML) and generalized least squares (GLS), and require the following set of assumptions (e.g., Bentler & Dudgeon, 1996; Bollen, 1989):

- Independent observations: Observations for different subjects are independent. This can be achieved through simple random sampling.
- Large sample size: All statistics estimated in SEM are based on an assumption that the sample is sufficiently large.
- Correctly specified model is estimated: The model being estimated reflects the true structure in the population.
- Multivariate normal data: The observed scores have a (conditionally) multivariate normal distribution.
- Continuous data: The assumption of multivariate normality implies that the data are continuous in nature. Categorical data, such as dichotomies or even Likert-type data, cannot by definition be normally distributed because they are discrete in nature (Kaplan, 2000). Therefore, it is often noted that NT estimators require *continuous normally distributed* endogenous variables.

If NT estimators are applied when the above conditions are satisfied, the parameter estimates have three desirable properties: asymptotic unbiasedness (they neither over- nor underestimate the true population parameters in large samples), asymptotic efficiency (variability of the parameter estimate is at a minimum in large samples), and consistency (parameter estimates converge to population parameters as sample size increases).

Defining Normal Theory Estimators

For both ML and GLS estimation methods, model parameters are estimated using an iterative process. The final set of parameters minimizes the discrepancy between the observed sample covariance matrix ($\mathbf{S}$) and the model-implied covariance matrix calculated from the estimated model parameters $[\mathbf{\Sigma}(\hat{\boldsymbol{\theta}})]$. The fit function that is minimized, $F = F[\mathbf{S}, \mathbf{\Sigma}(\hat{\boldsymbol{\theta}})]$, will equal zero if the model perfectly predicts the elements in the sample covariance matrix. If the assumptions noted above are met, the overall fit between the model and the data can be expressed as $T = F(N - 1)$, which follows a central χ^2 distribution.

The fit function for both ML and GLS estimators can be written in the same general form:

$$F = \frac{1}{2} tr[([\mathbf{S} - \boldsymbol{\Sigma}(\hat{\theta})]\mathbf{W}^{-1})^2] \quad (1)$$

where *tr* is the trace of a matrix (i.e., the sum of the diagonal elements), and $[\mathbf{S} - \boldsymbol{\Sigma}(\hat{\theta})]$ represents the discrepancy between the elements in the sample covariance matrix and the elements in the model-implied covariance matrix. These residuals $[\mathbf{S} - \boldsymbol{\Sigma}(\hat{\theta})]$ are weighted by a weight matrix, **W**. The weight matrix differs between the two NT procedures; GLS employs the observed sample covariance matrix **S** as the weight matrix, whereas ML employs the model-implied covariance matrix $\boldsymbol{\Sigma}(\hat{\theta})$.[1] If all assumptions are met, the two weight matrices will be equivalent at the last iteration and the estimators will produce convergent results (Olsson, Troye, & Howell, 1999). However, if the model is misspecified, **W** at the last iteration will differ between the two techniques, even if all other assumptions are met. This difference in **W** results in different parameter estimates and fit indices across the estimators. Specifically, GLS has been found to produce overly optimistic fit indices and more biased parameter estimates than ML if the estimated model is misspecified. Seeing that most applied researchers are interested in the plausibility of a specified model and would, therefore, prefer fit indices sensitive to model misspecification, ML has been recommended over GLS (Olsson et al., 1999; Olsson, Foss, Troye, & Howell, 2000). We, therefore, limit subsequent discussion of NT estimators to ML.

NON-NORMAL DATA

Assessing Non-Normality

In general, the effects of non-normality on ML-based results depend on its extent; the greater the non-normality, the greater the impact on results. Therefore, researchers should assess the distribution of the observed variables prior to analyses in order to make an informed decision concerning estimation method. Three indices of non-normality are typically used to evaluate the distribution: univariate skew, univariate kurtosis, and multivariate kurtosis. Unfortunately, there is no clear consensus regarding an "acceptable" degree of non-normality. Studies examining the impact of univariate normality on ML-based results suggest that problems may occur when univariate skewness and univariate kurtosis approach values of 2 and 7, respectively (e.g., Chou & Bentler, 1995; Curran, West, & Finch, 1996; Muthén & Kaplan, 1985). In addition, there is no generally accepted cutoff value of multivariate kurtosis that indicates non-normality. A guideline offered through the EQS software program

(Bentler, 2004) suggests that data associated with a value of Mardia's normalized multivariate kurtosis (see Bollen, 1989, p. 424, equation 4, for formula) greater than 3 could produce inaccurate results when used with ML estimation (Bentler & Wu, 2002). This guideline is consistent with discussions by many applied and methodological researchers regarding this issue found on SEMNET (structural equation modeling listserv). Future research should investigate the utility of such a cutoff value and the conditions under which it is relevant (e.g., size of model).

Effects of Analyzing Non-Normal Continuous Data: Empirical Results

Given the abundance of non-normal and categorical data analyzed in the social sciences, a question of significant interest concerns the robustness of ML to these conditions. Research examining the effects of nonnormality has typically focused on (a) the χ^2 statistic, (b) other model fit indices, (c) parameter estimates, and (d) standard errors. As detailed below, ML has been found to produce relatively accurate parameter estimates under conditions of non-normality (e.g., Finch, West, & MacKinnon, 1997); however, both the χ^2 statistic and standard errors of the parameter estimates tend to exhibit bias as non-normality increases (e.g., Bollen, 1989; Chou, Bentler, & Satorra, 1991; Finch et al., 1997).

Chi-Square Statistic and Fit Indices

When estimating a correctly specified model, the ML-based χ^2 does not follow the expected central χ^2 distribution if the multivariate normality assumption is violated. More specifically, research has shown that χ^2 is inflated under conditions of moderate non-normality with values becoming more inflated as non-normality increased (e.g., Chou et al., 1991; Curran et al., 1996; Hu, Bentler, & Kano, 1992; Yu & Muthén, 2002). Kurtotic distributions, especially leptokurtic distributions (positive kurtosis), seem to have the greatest effect on χ^2 (e.g., Browne, 1984; Chou et al., 1991). The inflation of the χ^2 statistic may lead to an increased Type I error rate, which is a greater rate of rejecting a correctly specified model than expected by chance.

In addition to the χ^2 statistic, the performance of other fit indices is important to understand given that most researchers are interested in the approximate fit of the model to the data instead of an exact fit evaluation determined solely by the χ^2 test (Bentler, 1990). As Hu and Bentler (1998) explained, "A fit index will perform better when its corresponding chi-square test performs well" (p. 427), meaning that because many fit indices (e.g., comparative fit index [CFI]) are a function of the obtained

χ^2, these too can be affected by the same factors that influence χ^2. Research has shown that if moderately to severely non-normal data are coupled with a small sample size ($N \leq 250$), the ML-based Tucker-Lewis Index (TLI), CFI, and root mean square error of approximation (RMSEA) tend to overreject correctly specified models (Hu & Bentler, 1999; Yu & Muthén, 2002).

Parameter Estimates and Standard Errors

Whereas parameter estimates are unaffected by non-normality, their associated significance tests are incorrect if ML estimation is applied to non-normal data. Specifically, the ML-based standard errors underestimate the true variation of the parameter estimates (e.g., Chou et al., 1991; Finch et al., 1997; Olsson et al., 2000), which results in increased Type I error rates associated with statistical significance tests of the parameter estimates. This would imply that estimates of truly zero parameters could be deemed significantly different than zero, and thus, important, to include in the model. Similar to the χ^2, it appears that kurtotic distributions, specifically leptokurtic distributions, have the greatest impact on standard errors (Hoogland & Boomsma, 1998).

ORDERED CATEGORICAL DATA

Defining Ordered Categorical Data

As stated previously, ML estimation assumes that the observed data are a sample drawn from a continuous and multivariate normally distributed population. In the social sciences, data with these characteristics are not always collected. Frequently, researchers collect and analyze ordinal data, such as data obtained from the use of a Likert scale. While researchers often treat ordinal data as continuous, the ordinal measurements are, as Bollen (1989, p. 433) noted, "coarse" and "crude." Even if the data appear to be approximately normally distributed (e.g., indices of skewness and kurtosis approach zero or plots of the observed data appear to be normal), ordered categorical data are discrete in nature and, therefore, cannot be normally distributed by definition. The crude nature of the measurement will induce some level of non-normality in the data (Kaplan, 2000).

One way in which observed ordered categorical data are thought to occur is when a continuous latent response variable (y^*) is divided into distinct categories (e.g., Bollen, 1989; Muthén, 1993). This has been referred to as the latent response variable formulation (e.g., Muthén & Muthén, 2001). The points that divide the continuous latent response variable (y^*) into a set number of categories (c) are termed thresholds (τ), where the total number of thresholds is equal to the number of categories less one ($c - 1$). For example, if a Likert scale has five response choices, four threshold values are needed to divide y^* into five ordered categories. The observed ordinal data (y) are thought to be produced as follows:

$$y = \begin{Bmatrix} 1 & \text{if } y^* \leq \tau_1 \\ 2 & \text{if } \tau_1 < y^* \leq \tau_2 \\ 3 & \text{if } \tau_2 < y^* \leq \tau_3 \\ 4 & \text{if } \tau_3 < y^* \leq \tau_4 \\ 5 & y^* > \tau_4 \end{Bmatrix} \quad (2)$$

As a result, $y^* \neq y$. Specifically, because subjects respond to the five-point Likert scale, the observed ordinal-level data can only be reported as discrete values from 1 to 5. However, subjects' "true" levels of the latent response variable (y^*) are much more precise than allowed by the five-point response scale. Figure 9.1 illustrates the relation between the continuous latent response variable (y^*), observed level data (y), and the four ($c - 1$) threshold values for a variable with five ordered categories. This figure illustrates how the observed ordinal data provide an approximation of the underlying continuous latent response variable.

This difference between y and y^* has two important consequences when modeling the data. First, unlike y^*, the standard linear measurement model ($y^* = b\text{F} + \text{E}$) does not hold when modeling y ($y \neq b\text{F} + \text{E}$).[2] Second, the assumption that the model estimated reflects the true structure in the population ($\boldsymbol{\Sigma} = \boldsymbol{\Sigma}(\boldsymbol{\theta})$) does not hold when ordinal data are present

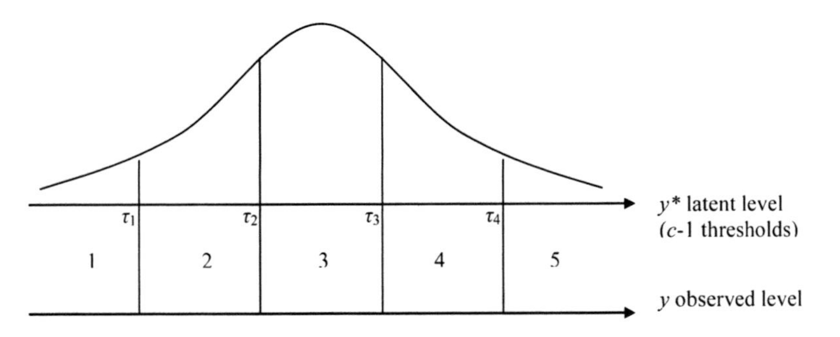

Figure 9.1. Relation between y^*, y, and thresholds.

present (Bollen, 1989). Therefore, many studies have been conducted to examine the extent of bias present when applying the standard linear measurement model to ordinal data.

Effects of Analyzing Approximately Normally Distributed Ordered Categorical Data: Empirical Results

When modeling ordinal data, researchers often ignore the categorical nature of the data and apply ML estimation. By creating a covariance matrix based on Pearson product–moment (PPM) correlational techniques and estimating relations at the observed level (y), one is treating the ordinal data as if they were continuous. As the number of ordered categories increases, data more closely *approximate* continuous-level data and, in turn, the obtained correlations are closer to their true values (Bollen, 1989). The fewer categories present, the more severe the attenuation in the PPM correlations and the greater the discrepancy between true and obtained values. As discussed below, if few categories are used and ML estimation is employed, the model fit indices, parameter estimates, and standard errors can be biased.

Chi-Square and Fit Indices

In general, fit indices have been found to perform well if approximately normally distributed five-category ordinal data are treated as continuous (Babakus, Ferguson, & Jöreskog, 1987; Hutchinson & Olmos, 1998). While the χ^2 was found to be robust when modeling ordinal data was collected using four ordered categories, inflation was present if fewer than four categories were used (Green, Akey, Fleming, Hershberger, & Marquis, 1997). In addition, slight underestimation of the goodness-of-fit index (GFI), adjusted GFI (AGFI), and root mean square residual (RMR) has been found if sample sizes are small and ordered categorical data with five categories are analyzed as continuous (Babakus et al., 1987). Researchers generally agree that when ordinal data are approximately normal and have at least five ordered categories that the ordered categorical data may be treated as if they were continuous without great distortion in the fit indices (e.g., Bollen, 1989; Dolan, 1994; Muthén & Kaplan, 1985).

Parameter Estimates and Standard Errors

Previous research has shown that when ordinal data have at least five categories and are approximately normal, treating data as continuous and applying the ML estimator produces slight underestimation in parameter estimates and factor correlations (Babakus et al., 1987;

Muthén & Kaplan, 1985). Standard errors have shown a greater sensitivity to categorization than the parameter estimates, exhibiting negative bias (Babakus et al., 1987; Muthén & Kaplan, 1985; West, Finch, & Curran, 1995). If standard errors are too small, tests of parameter significance may be inflated, resulting in Type I errors. As the number of ordered categories decreases, the underestimation in both the parameter estimates and standard errors becomes more severe, even if ordinal data are symmetric.

Effects of Analyzing Non-Normal Ordered Categorical Data: Empirical Results

Ordinal data are considered by some researchers as inherently non-normal (e.g., Muthén & Kaplan, 1985). However, as just described, if the observed data have many categories (e.g., at least five ordered categories) and are approximately normal, use of ML estimation techniques does not result in severe levels of bias in fit indices, parameter estimates, or standard errors. Problems begin to emerge as the number of response options decreases or the observed item distributions diverge widely from a normal distribution. As the number of ordered categories is reduced, there are fewer response choices available for subjects to choose. The fewer the number of categories, the greater the amount of attenuation in PPM estimates. Also, as the number of categories decreases, it becomes less likely that observed data could approximate a normal distribution.

Chi-Square and Fit Indices

Similar to results found when modeling continuous non-normal data, fit indices are adversely affected when ordinal data follow non-normal distributions. When modeling non-normal ordered categorical data, ML-based χ^2 values (Green et al., 1997: West et al., 1995) and RMR values were inflated, and values of the non-normed fit index (NNFI), GFI, and CFI were underestimated (Babakus et al, 1987; Hutchinson & Olmos, 1998). This may suggest that a correctly specified model does not fit the data well and could lead a researcher to discard a plausible model.

Parameters and Standard Errors

As univariate skewness and univariate kurtosis levels of the observed ordered categorical data increase, the negative bias observed with the ML-based parameter estimates and standard errors becomes more pronounced (Babakus et al., 1987; Muthén & Kaplan, 1985). Bias levels increased with lower sample sizes, fewer categories, weaker relations between factors and indicators, or higher levels of non-normality (e.g., Babakus et al., 1987; Bollen, 1989; Dolan, 1994).

TECHNIQUES TO ADDRESS NON-NORMAL AND ORDERED CATEGORICAL DATA

In this section we describe four methods that have been developed to address problems encountered when modeling non-normal and/or categorical data. The first method involves an alternative method of estimation that does not make the distributional assumptions of ML estimation (ADF; Browne, 1984). The second method involves adjusting the ML-based χ^2 and standard errors by a factor based on the level of multivariate kurtosis displayed in the observed data (Satorra-Bentler scaled χ^2 and standard errors). This method also involves adjusting any fit indices used to assess model fit (e.g., CFI, RMSEA). The third method involves employing robust WLS estimation methods (e.g., WLSM, WLSMV) available in the software package Mplus (Muthén & Muthén, 2004). These estimators can be conceptualized as combining an alternative estimation method with an adjustment method. Finally, a fourth method involves bootstrapping empirical distributions of each parameter estimate and the χ^2 statistic in order to produce more accurate standard errors and probability values associated with χ^2.

Asymptotically Distribution-Free (ADF) Estimator

Given the unrealistic assumption of multivariate normality and the lack of robustness of ML to non-normal data, Browne (1984) developed the ADF estimator. Unlike ML, ADF makes no assumption of normality; therefore, variables that are kurtotic have no detrimental effect on the ADF χ^2 or standard errors. In addition, input matrices that take the metric of the variables into consideration can be employed to handle the problems of parameter estimate attenuation associated with a small number of ordered categories (Muthén, 1984). For these reasons, it would seem as though non-normally distributed and/or ordered categorical data could be accommodated by the ADF estimation technique, thus avoiding the problems encountered by NT estimators.

ADF Estimator with Non-normal Continuous Data: Description

In order to understand the ADF estimator and some of its practical limitations, it is important to understand the form of the fit function, which is typically written as

$$F = (\mathbf{s} - \hat{\boldsymbol{\sigma}})' \mathbf{W}^{-1} (\mathbf{s} - \hat{\boldsymbol{\sigma}}) \,, \tag{3}$$

where **s** represents a vector of the nonduplicated elements in the sample covariance matrix (**S**), $\hat{\boldsymbol{\sigma}}$ represents a vector of the nonduplicated elements in the model-implied covariance matrix $[\boldsymbol{\Sigma}(\hat{\boldsymbol{\theta}})]$, and $(\mathbf{s} - \hat{\boldsymbol{\sigma}})$ represents the discrepancy between the sample values and the model-implied values. These residuals $(\mathbf{s} - \hat{\boldsymbol{\sigma}})$ are weighted by a weight matrix, **W**. The weight matrix utilized with the ADF estimator is the asymptotic covariance matrix, a matrix of the covariances of the observed sample variances and covariances (Bollen, 1989). Elements of the asymptotic covariance matrix, $W_{ij,\,kl}$, are calculated using the covariances among the elements in the sample covariance matrix along with the fourth-order moments (Bentler & Dudgeon, 1996),

$$W_{ij,\,kl} = s_{ijkl} - s_{ij} s_{kl}, \tag{4}$$

where s_{ijkl}, a quantity related to multivariate kurtosis, is defined as

$$s_{ijkl} = \frac{\sum_{a=1}^{N} (x_{ai} - \bar{x}_i)(x_{aj} - \bar{x}_j)(x_{ak} - \bar{x}_k)(x_{al} - \bar{x}_l)}{N} \,, \tag{5}$$

and s_{ij} and s_{kl} are the covariances of observed variables x_i with x_j and x_k with x_l, respectively. This estimator is often called weighted least squares (WLS) or ADF (Bollen, 1989).

There are practical problems in implementing ADF estimation that are related to the weight matrix. Specifically, because the inverse of the weight matrix ($\mathbf{W}^{-1}$) needs to be calculated, a large weight matrix can make ADF estimation computationally intensive. Dimensions of **W** matrix can be calculated as ½ $(p + q)(p + q + 1)$, where p is the number of observed exogenous variables and q is the number of observed endogenous variables (Bollen, 1989). For example, if a researcher has responses from a 10-item scale and wishes to employ confirmatory factor analysis, the dimensions of **W** would be 55 × 55, resulting in 3,025 elements in **W**. As the number of observed variables increases, the number of elements in **W** increases rapidly. For example, if 10 items were added to the original 10-item measure, the dimensions of the weight matrix would be 210 × 210, resulting in 44,100 elements in **W**. Due to the computational intensity of the ADF technique, it requires very large sample sizes for results to converge to stable estimates. A minimum sample size of $1.5(p + q)(p + q + 1)$ has been suggested (Jöreskog & Sörbom, 1996), but much larger sample sizes may be needed to alleviate estimation and convergence problems.

ADF Estimator with Continuous Non-normal Data: Empirical Results

When modeling non-normal data, theoretically, the ADF estimator should produce parameter estimates with desirable properties and fit statistics that perform as expected (Browne, 1984). However, empirical research has shown otherwise. ADF tends to break down under common situations of moderate to large models (more than two factors, eight items) and/or small to moderate sample sizes ($N < 500$). As discussed in detail below, the poor performance of the ADF estimator under many conditions makes it an unattractive option when modeling non-normal continuous data.

Chi-square and fit indices. With respect to model fit, ADF yields misleading results unless sample size is extremely large (e.g., Olsson et al., 2000). For example, when modeling non-normal continuous data, Hu and colleagues (1992) found that ADF estimation produced acceptable Type I error rates only when sample sizes reached 5,000. Similarly, Curran and colleagues (1996) found that the ADF-based χ^2 increased as sample size decreased and/or non-normality increased, resulting in correctly specified models being rejected too frequently. Even more problematic is the ADF estimator's lack of sensitivity to model misspecification. Research has shown that ADF estimation produces overly optimistic fit values when models are misspecified, which in turn could lead researchers to fail to reject an incorrectly specified model. The lack of sensitivity to specification errors becomes worse with increasing departures from normality (e.g., Curran et al., 1996; Olsson et al., 2000).

Parameter estimates and standard errors. Empirical results concerning ADF-based parameter estimates and standard errors are also discouraging. Parameters tend to be negatively biased unless the sample is large, with bias levels becoming more pronounced as kurtosis increases. In addition, ADF-based standard errors estimated from a correctly specified model under conditions of non-normality have been found to be superior to ML-based standard errors only when the observed variables have an average univariate kurtosis larger than three and the sample size is greater than 400 (Hoogland & Boomsma, 1998).

ADF Estimator with Continuous Data: Software Implementation

Researchers who wish to utilize ADF as an estimator will find it easy to employ using LISREL, EQS, or Mplus. In LISREL (Jöreskog & Sörbom, 2004), ADF estimation is called WLS and implementation requires the use of two programs (LISREL and PRELIS). As noted, WLS/ADF employs the asymptotic covariance matrix as the weight matrix. PRELIS (preprocessor for LISREL; Jöreskog & Sörbom, 1996) is used to produce both the asymptotic and observed covariance matrices from the raw data. These matrices are input into the LISREL program to estimate the model.

Appendix A provides an example of the SIMPLIS command language (user-friendly language employed in LISREL program) that specifies WLS/ADF estimation. Notice that WLS must be specified on the options line or else ML estimation will be employed by default. The combination of ML estimation and an asymptotic covariance matrix will produce the Satorra-Bentler scaling procedure (discussed below).

Similar to LISREL, the raw data file is necessary in order to construct the asymptotic covariance matrix in EQS (Bentler, 2004) and Mplus (Muthén & Muthén, 2004). Unlike LISREL, a preprocessor is not needed to construct this matrix in either EQS or Mplus; it is constructed and employed by specifying the estimator. Arbitrary GLS (AGLS) estimation is requested as the estimation method for EQS while WLS is requested for Mplus. AGLS follows the same general form as GLS, with the choice of weight matrix based upon fourth-order moments to allow distribution-free requirements of the variables (Bentler, 1995). While a different name is used, it is equivalent to ADF/WLS. All three programs tend to produce similar parameter estimates, standard errors, and fit indices.

ADF Estimator with Ordered Categorical Data: Description of Categorical Variable Methodology (CVM)

As previously discussed, if researchers ignore the ordinal metric of the data (i.e., treating ordinal data as if continuous and employing ML estimation), data–model fit and parameter estimates may be underestimated. Alternative strategies consider the metric of the ordinal data by including this information in the estimation procedures. Specifically, categorical variable methodology (CVM) incorporates the metric of the data into analyses by considering two components: (1) input for analyses that recognizes the ordered categorical indicators and (2) the use of the correct weight matrix when employing ADF/WLS estimation (Muthén, 1984; Muthén & Kaplan, 1985). Therefore, CVM is basically ADF/WLS estimation with specific input to accommodate ordered categorical variables.

Employing CVM. With regard to metric, if data are continuous then data at the observed level are considered equivalent to the underlying latent response variable, that is, $y = y^*$. On the other hand, if data are ordinal, $y \neq y^*$ (e.g., Jöreskog & Sörbom, 1996; Muthén & Kaplan, 1985; Muthén & Muthén, 2001). In order to avoid the consequences of modeling y using the standard linear confirmatory factor analysis model (e.g., $y \neq b\text{F} + \text{E}$ and $\boldsymbol{\Sigma} \neq \boldsymbol{\Sigma}(\boldsymbol{\theta})$), y^* can be modeled (Bollen, 1989). Applying the linear factor model to the underlying latent response variable is illustrated in Figure 9.2. Notice that the factor does not directly affect y, but instead, directly affects y^*. Because y^* is a continuous variable, the standard linear model ($y^* = b\text{F} + \text{E}$) can be used to estimate the relation between y^* and the factor.

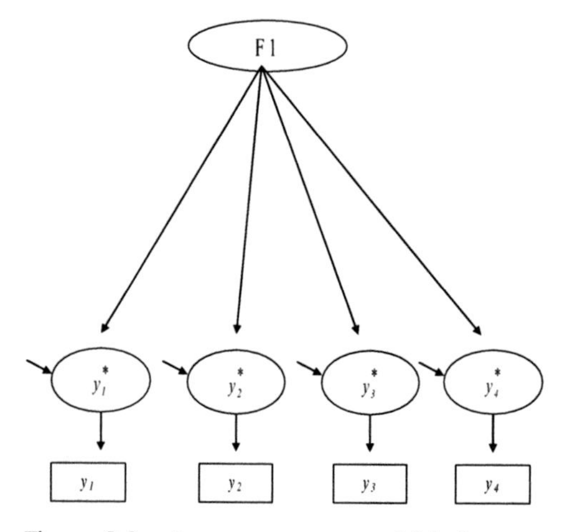

Figure 9.2. Latent response variable formulation.

Modeling the relation between F and y^* entails computing thresholds and latent correlations. Because thresholds are used in the computation of latent correlations, these will be discussed first. Threshold values, which cut the underlying continuum into ordered categories (see Figure 9.1), may be estimated if the number of subjects who chose a certain category is known (Bollen, 1989). Threshold values are important to estimate not only because they are thought of as critical points that "move" a subject from one category to another but also because they are used to create marginal distributions of ordinal variables that assist with estimation procedures. Because the metric of ordered categorical data is arbitrary, the mean and standard deviation are often set to 0 and 1, respectively. Using a mean of 0 and standard deviation of 1, item thresholds may be estimated by considering the cumulative area under the normal curve up to a given point (Bollen, 1989; Jöreskog & Sörbom, 1996) by:

$$\tau_i = \Phi^{-1}\left[\sum_{k=1}^{i} \frac{N_k}{N}\right], \qquad i = 1, 2, \ldots, c - 1 \tag{6}$$

where τ_i is a particular threshold, Φ^{-1} is the inverse of the normal distribution function, N_k is the number of subjects who selected category k, N is the total sample size, and c is the total number of categories. Equation 6 shows that thresholds are calculated using the proportions of subjects within each ordered category (N_k/N). The resulting threshold values (z-values) divide the underlying normal distribution into the c categories and relate the y values to the y^* values. By incorporating thresholds into the standard measurement model, the linear model estimates the relation between the factor and the continuous underlying latent response variable, thus avoiding the problems associated with modeling y (Bollen, 1989).

Recall that in addition to the linear model not applying to ordinal data, the assumption of a correctly specified model does not hold when ordinal variables are modeled. In brief, the population covariance matrix of the ordinal variables ($\boldsymbol{\Sigma}$) will not equal the population covariance matrix of the continuous underlying latent response variable ($\boldsymbol{\Sigma}^*$), for which the model does hold ($\boldsymbol{\Sigma}^* = \boldsymbol{\Sigma}(\boldsymbol{\theta})$). In order to avoid this problem, correlations representing the relations among the y^* variables can be estimated (Bollen, 1989).

The latent correlations represent the theoretical relations between the underlying continuous latent response variables (y^*). For each pair of variables, a latent correlation can be estimated. If both variables are dichotomous, a tetrachoric correlation represents the relation between the y^* variables. If both variables are ordinal, a polychoric correlation represents the relation between the y^* variables. If one variable is ordinal and the other variable is continuous, a polyserial correlation represents the relation between the y^* variables.

The WLS estimator can then be employed using the thresholds and latent correlations:

$$F_{\text{WLS}} = (\mathbf{r} - \hat{\boldsymbol{\rho}})'\mathbf{W}^{-1}(\mathbf{r} - \hat{\boldsymbol{\rho}})\,, \qquad (7)$$

where $\mathbf{r}$ is a vector containing the sample latent correlations and thresholds, $\hat{\boldsymbol{\rho}}$ is the corresponding vector from the implied matrix, and $\mathbf{W}$ is the asymptotic covariance matrix of $\mathbf{r}$ (e.g., Muthén, 1984). This asymptotic covariance matrix, along with the appropriate correlational input, is required to correctly implement CVM. It is important to realize that both pieces (appropriate correlations and correct $\mathbf{W}$) are necessary. If either piece is missing, the estimation technique is not CVM.

An example to illustrate the calculation of a polychoric correlation. To illustrate how ordinal variables may be accommodated, consider an example employing Rosenberg's (1989) self-esteem scale. A sample of 120 college freshmen responded to questions concerning their self-esteem using a scale with four ordered categories anchored at "Strongly Disagree" to "Strongly Agree." Responses to two of the questions, "On the whole, I am satisfied with myself" (SATISFIED) and "I feel that I have a number of good qualities" (QUALITIES), are provided in Table 9.1.

To illustrate how threshold values are obtained, we can compute these values directly from the sample data using two pieces of information: (1)

Table 9.1. Frequency of Responses to Self-Esteem Items

SATISFIED		*QUALITIES*	
Category	*Frequency*	*Category*	*Frequency*
1 (SD)	4	1 (SD)	4
2 (D)	16	2 (D)	27
3 (A)	50	3 (A)	42
4 (SA)	50	4 (SA)	47
N	120	*N*	120
Skew	–.812	Skew	–.514
Kurtosis	.154	Kurtosis	–.726

Note: SD = Strongly Disagree; D = Disagree; A = Agree; SA = Strongly Agree.

the proportion of students who selected a certain category and (2) the area under the normal curve. Consider the first threshold for the variable, SATISFIED. This threshold divides the underlying latent response variable into the two categories of "Strongly Disagree" (category 1) and "Disagree" (category 2). Students below this threshold will have responded "Strongly Disagree" to the statement. There are four students responding "Strongly Disagree" to the SATISFIED item. Using Equation 6, the cumulative probability of cases through category 1 is (4/120), or .033. Considering this as representative of cumulative area under the normal curve, the threshold value is the z-value associated with .033, or a z-value of –1.83. The remaining thresholds may be calculated in a similar manner, as shown in Table 9.2.

To determine the polychoric correlation between two ordinal variables, a contingency table of students' responses to each pair of the variables is needed. The frequency of responses to each option can be tabulated across the pair of variables. Table 9.3 shows the contingency table for the responses to the SATISFIED and QUALITIES variables. From the table, one can see a relation between the responses to the items. For example, those students who agreed or strongly agreed that they were satisfied with themselves generally agreed that they had a number of good qualities. However, some discrepancies are noticed. For example, six students who

Table 9.2. Threshold Values and Cumulative Area

	SATISFIED			*QUALITIES*		
Threshold	1	2	3	1	2	3
Cumulative *N*	4	20	70	4	31	73
Cumulative Area	.033	.167	.583	.033	.258	.608
Threshold Value	–1.834	–0.967	0.210	–1.834	–0.648	0.275

Table 9.3. Contingency Table Between SATISFIED and QUALITIES Variables

	QUALITIES				
SATISFIED	*1* (*SD*)	*2* (*D*)	*3* (*A*)	*4* (*SA*)	*Total*
1 (SD)	2	1	1	0	4
2 (D)	0	8	6	2	16
3 (A)	1	15	26	8	50
4 (SA)	1	3	9	37	50
Total	4	27	42	47	120

Note: SD = Strongly Disagree; D = Disagree; A = Agree; SA = Strongly Agree.

agreed that they possessed good qualities disagreed with the statement about having self-satisfaction.

The contingency table information reported in Table 9.3 and item thresholds are used to calculate latent correlations among the ordinal variables (see Olsson, 1979, for formula). The estimated polychoric correlation represents the value with the greatest likelihood of yielding the observed contingency table, given the estimated thresholds. Here, the polychoric correlation is .649, reflecting the positive relation, while acknowledging some inconsistency in student responses. Note that the polychoric correlation estimate is higher than the PPM correlation estimate (.551) because the polychoric correlation is disattenuated for the error associated with the ordinal variable's coarse categorization of the underlying latent variable's continuum. To illustrate the polychoric correlation graphically, the relation between the observed ordinal variables and the underlying latent responses variables are plotted in Figure 9.3.

ADF/CVM with Ordered Categorical Data: Empirical Results

Although differences may exist in how software programs employ CVM, the information presented here is made without specific reference to software packages (details concerning the implementation of CVM using LISREL, EQS, and Mplus are described below). In general, empirical studies have found that CVM has both desirable and undesirable characteristics.

Chi-square and fit indices. When approximately normally distributed ordinal data were analyzed using CVM, values of χ^2 were close to expected values when small to moderate models were specified (15 parameters or less) or sample sizes were large (N =1,000; Muthén & Kaplan, 1985; Potthast, 1993). The amount of inflation of the CVM-based χ^2 increased as sample size decreased, model size increased, or non-nor-

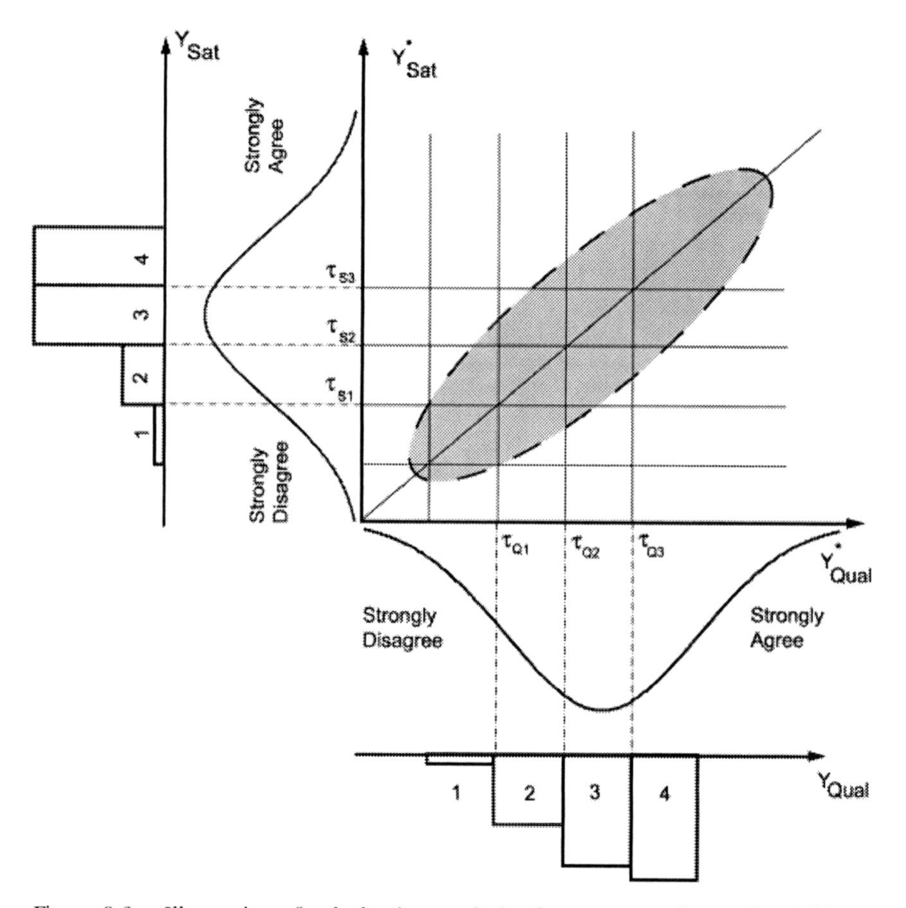

Figure 9.3. Illustration of polychoric correlation between two observed variables. Notes: Sat = SATISFIED; Qual = QUALITIES. Threshold values are specific to the variable in question. For example, S1, S2, S3, respectively, refer to thresholds 1 through 3 for the variable SATISFIED. Similarly, thresholds Q1, Q2, Q3, respectively, refer to thresholds 1 through 3 for the variable QUALITIES.

mality of the data increased (DiStefano, 2002; Muthén & Kaplan, 1992; Potthast, 1993). RMSEA has been found to be somewhat robust to ordinal data analyzed by CVM, and is not sensitive to sample size or model size when correctly specified models were estimated (Hutchinson & Olmos, 1998).

Parameter estimates and standard errors. A major strength of CVM is that parameter estimates appear unbiased when modeling non-normally distributed ordered categorical data. When estimating correctly specified models, parameter estimates were found to have little bias, regardless of whether dichotomous variables (Dolan, 1994; Muthén & Kaplan, 1985) or non-normal ordered categorical data (approximate values: skewness = 2.5, kurtosis = 6) were analyzed (DiStefano, 2002; Dolan, 1994; Potthast, 1993). The unbiased nature of the estimates is attributed to the analyses being conducted at the latent level (e.g., using polychoric correlations), rather than the observed level (Muthén & Kaplan, 1985). Unlike parameter estimates, research suggests that estimated standard errors are too low. This negative bias becomes more pronounced as model size and non-normality increased or sample sizes decreased (DiStefano, 2002; Potthast, 1993).

ADF/CVM with Ordered Categorical Data: Software Implementation

Software packages that conduct CVM estimation techniques typically use a multistep procedure to obtain results. CVM procedures are described for three software programs, LISREL, EQS, and Mplus. The syntax specific to each software package can be found in Appendix A.

The LISREL approach to CVM requires that both LISREL and PRELIS programs be used. PRELIS classifies variables with less than 16 distinct categories as ordinal unless otherwise specified by the researcher (Jöreskog & Sörbom, 1996). First, item threshold values are calculated from the sample data (Jöreskog & Sörbom, 1996). The thresholds are assumed to come from a standard normal distribution and are estimated by using the inverse of the normal distribution function (Jöreskog & Sörbom, 1996). Next, the threshold information and contingency tables between pairs of variables are used to compute the polychoric correlations. PRELIS then estimates the full asymptotic covariance matrix using the correlational and threshold information. Both the correlation and asymptotic covariance matrices are input into LISREL for WLS estimation of model parameters and model fit.

For CVM, a preprocessor is not needed with either EQS or Mplus. All of the necessary components are constructed and employed within the program. EQS uses a two-stage process to analyze categorical variables. In the first stage, partitioned maximum likelihood (PML) is used to estimate the thresholds and the appropriate correlations simultaneously. PML partitions the model into submodels and estimates the relations among smaller parts of the model rather than the entire model at once. If a mixture of continuous and categorical variables is included in the same model, EQS estimates relations between continuous variables and categorical variables first (called the polyserial partition). Following the polyserial partition, relations among ordinal variables are estimated (called the polychoric partition). In the second stage, the asymptotic covariance matrix is computed for use with the AGLS estimator, and estimates of model parameters and fit indices are provided.

Mplus has the capability to employ two different estimation procedures when undertaking CVM. The choice of procedure is dictated by the presence of an exogenous variable (x) influencing the factor. In case A there is no exogenous variable and CVM is conceptualized as discussed above (latent response variable formulation). Specifically, it is assumed that continuous normally distributed latent response variables (y^*) underlie the ordinal variables (y). The thresholds, latent correlations (e.g., polychorics), and the asymptotic covariance matrix are estimated. WLS can then be employed as the estimator.

In case B, there is an exogenous variable influencing the factor (e.g., MIMIC models). In this situation, a different estimation process is employed. As an alternative to the latent response variable formulation there exists an equivalent formulation, termed the conditional probability curve formulation (e.g., Muthén & Asparouhov, 2002; Muthén & Muthén, 2001). Instead of estimating the linear relation between y^* and the factor (as with the latent response variable formulation), the nonlinear relation between y and the factor may be modeled. Specifically, a probit model can be used to estimate the probability (ranging from 0 to 1) that a specific category (k) is selected or exceeded by modeling the nonlinear relation between y and the factor (F):

$$P(y \geq k \,|\, \mathrm{F}) = \Phi\,(\alpha_k + \beta \mathrm{F}), \tag{8}$$

where Φ is the standard normal distribution function.[3] The α (intercept) and β (slope) parameters from this conditional probability formulation can be derived from the parameters estimated using the latent response variable formulation, which illustrates the similarity of the two formulations (see Muthén & Asparouhov, 2002, for formula). In fact, the two formulations produce equivalent results in terms of the probability of being in or exceeding a category given the factor, $P(y \leq k \,|\, \mathrm{F})$ (Muthén & Asparouhov, 2002). Because these two formulations produce equivalent results, Muthén and Asparouhov explained that the assumption of an underlying continuous latent response variable is not necessary but rather a convenience: "It is shown that the two formulations give equivalent results. The discussion clarifies that the latent response variables are a convenient conceptualization, but that it is not necessary that the data have been generated by categorizing latent response variables" (Muthén & Asparouhov, 2002, p. 2).

When the conditional probability formulation is employed, the first step involves computing the sample statistics: probit thresholds, probit regression coefficients, and probit residual correlations. In the second step, the asymptotic covariance matrix of these sample statistics is constructed. In the final step, estimates of model parameters, standard errors, and model fit information can be computed using a WLS estimator (Muthén & Muthén, 2001).

Satorra-Bentler Scaled Chi-Squares and Standard Errors

Description of Method

Another strategy employed to accommodate non-normal continuous and/or categorical data that has become popular in the last several years involves adjusting the χ^2, fit indices, and standard errors by a factor based on the amount of non-normality in the data. With normally distributed data the expected value of the χ^2 is equal to the model's degrees of freedom when the model is correctly specified. Therefore, if a correctly specified model with 60 degrees of freedom was estimated using multivariate normal data, a χ^2 value of approximately 60 would be expected. However, if data are moderately non-normal, the ML-based χ^2 will be biased even though the model is correctly specified. A correction, typically called the Satorra-Bentler (S-B) scaling procedure, uses the observed data's distributional characteristics to adjust the ML-χ^2 in order to better approximate the theoretical χ^2 reference distribution:

$$\text{S-B } \chi^2 = d^{-1}(\text{ML-based } \chi^2), \tag{9}$$

where d is a scaling factor that incorporates the kurtosis of the variables (Chou & Bentler, 1995; Satorra & Bentler, 1994). If no multivariate kurtosis exists, then ML-based χ^2 = S-B χ^2. However, as the level of multivariate kurtosis increases, the S-B χ^2 becomes more discrepant from the ML-based χ^2.

The S-B scaling method is typically applied with ML estimation. Because ML is employed, computation problems experienced with the ADF estimator are avoided. Recall that ADF requires the inversion of a large asymptotic covariance matrix (**W**). The S-B scaling method does not require inversion of this **W** matrix. Instead, the matrices to be inverted in order to compute the scaling factor are of the smaller dimensions of $df \times df$, where df represents the degrees of freedom associated with the model (Satorra & Bentler, 1994).

A similar scaling process is used to correct the standard errors, alleviating some of the attenuation present when modeling non-normal data using ML estimation. Specifically, the scaled standard errors are adjusted upward to approximate those that would have been obtained if the data were normally distributed.[4] Recall that when non-normal data are analyzed using ML, the parameter estimates are not affected. Therefore, the

ML-based parameter estimates are not adjusted in any way when this method is employed.

It must be noted that the typical χ^2 difference test employed for nested model comparisons should not be calculated using the S-B scaled χ^2 values (i.e., simply subtracting χ^2 of the less parsimonious model from the χ^2 of the more parsimonious model). The difference between two S-B χ^2 values is not distributed as a χ^2. Fortunately, fairly simple calculations can be employed to correct the difference test in order to make nested model comparisons using the S-B χ^2 values (see Satorra & Bentler, 2001, for calculations).

S-B Scaling Methods with Continuous Non-normal Data: Empirical Results

Chi-square and fit indices. Studies using correctly specified models and continuous non-normal data have shown that the S-B χ^2 outperforms the ML-based χ^2, particularly as the degree of non-normality increases (e.g., Chou et al., 1991; Curran et al., 1996; Hu et al., 1992; Yu & Muthén, 2002). In addition, it performs better than the ADF-based χ^2 when estimating correctly specified models at all but the largest sample sizes (e.g., N = 1,000, Curran et al., 1996; N = 5,000, Hu et al., 1992).

If the S-B χ^2 is employed to handle non-normal data, it follows that the S-B χ^2 would be incorporated into the calculation of fit indices (e.g., TLI, CFI) in order to gain benefits of the scaling procedure and, in turn, provide more accurate reflections of model–data fit. Few studies have investigated the performance of these S-B scaled indices. Nevitt and Hancock (2000) found that the S-B scaled RMSEA outperformed the unadjusted index. Yu and Muthén (2002) also examined this index in addition to the S-B scaled TLI and CFI and found that, under conditions of moderate to severe non-normality coupled with small sample size ($N \leq 250$), the S-B scaled versions of these three indices are preferred over the ML-based estimates. Yu and Muthén suggested that values at or below .05 for the S-B scaled RMSEA and at or above .95 for the S-B scaled CFI indicate adequate fit, which is quite similar to the cutoff values recommended by Hu and Bentler (1999) for the unadjusted indices.

Parameter estimates and standard errors. The scaled standard errors have also been found to outperform ML-based and ADF-based standard errors under conditions of non-normality (Chou & Bentler, 1995; Chou et al., 1991). Similar to the ML-based and ADF-based standard errors, the scaled standard errors showed some negative bias. However, they tended to be much closer to the expected values of the standard errors than those obtained from either ADF or ML methods. Recall that the ML-based parameter estimates are not adjusted as they are not affected by non-normality.

S-B Scaling Methods with Non-Normal Ordered Categorical Data: Empirical Results

While the functioning of the S-B correction has been generally examined with continuous non-normal data, a few studies have evaluated how the correction performs with ordered categorical data. It is recognized that this approach treats the categorical data as continuous, ignoring the metric level of the data. A caveat to this method is that ML estimation is known to be sensitive to the number of ordered response categories in addition to non-normal distributions.

Chi-square and fit indices. With regard to model fit, Green and colleagues (1997) found that S-B scaling produced χ^2 values very close to the expected χ^2 values when modeling two-, four- or six-ordered category data that displayed symmetric, uniform, and negatively skewed distributions. The S-B χ^2 did show positive bias when modeling data that exhibited differential skew, with bias being the greatest in the two-category condition. In all conditions, the S-B χ^2 outperformed the ML-based χ^2.

Parameter estimates and standard errors. As noted above, the S-B correction simply scales the standard errors and the χ^2. Thus, S-B–based parameter estimates will be equivalent to ML-based estimates. This implies no correction for the attenuation of the parameter estimates due to the categorical nature of the data. With respect to standard errors, research has found that scaled standard errors exhibited greater precision than ML-based standard errors when non-normally distributed ordered categorical data were analyzed (skewness ≈ 2, kurtosis ≈ 6; DiStefano, 2002). The benefit of the S-B scaling method was present even with non-normally distributed data having as few as three ordered categories (DiStefano, 2003).

S-B Scaling Methods: Software Implementation

As discussed above, the S-B scaling method has been applied to ordered categorical data by treating them as continuous (i.e., calculation of PPM covariances instead of latent correlations for ordered categorical data). This implies that the implementation of this method is the same across the two data types. All three software programs calculate the S-B–scaled χ^2 and standard errors. Appendix B provides the syntax needed to employ the S-B scaling method.

Similar to ADF estimation with continuous data, LISREL calculates the S-B χ^2 and scaled standard errors in two steps. The first step involves computing the asymptotic and observed covariance matrix from the raw data file using PRELIS. The second step involves specifying the model and estimation technique in the SIMPLIS program file. Recall that ML is used as the estimator and the obtained χ^2 and standard errors are adjusted for the level of non-normality. It is important to note that LIS-

REL 8.54 does not adjust all fit indices for non-normality. In fact, no incremental indices are adjusted (e.g., CFI, NNFI). Adjustments to these indices must be calculated by hand. This is done by first specifying and estimating the independence model and then using the independence model's S-B χ^2 along with the hypothesized model's S-B χ^2 in the corresponding fit index formula (see, e.g., Hu & Bentler, 1998, for fit index formulas).

Both EQS and Mplus read the raw data file directly into the program in order to construct the necessary matrices used to scale the χ^2 and standard errors. In addition to specifying ML as the estimator in EQS, the word "robust" is included to request the scaling method. In Mplus one simply requests "MLM" as the estimator, which refers to the ML-mean-adjusted χ^2 (equivalent to the S-B χ^2) and scaled standard errors. Unlike LISREL 8.54, both EQS 6.1 and Mplus 3.01 adjust all the fit indices reported when the S-B scaling method is used.

It must be noted that in addition to the MLM estimator, Mplus also provides the MLMV estimator, which produces a mean- and variance-adjusted chi-square. The scaled standard errors are equivalent across the two estimators. Research has shown that the MLM and MLMV chi-squares perform similarly, indicating that the additional adjustment provided by MLMV may not be needed (Muthén, 1999).

Robust WLS Estimation Procedures

Description of Estimators

Because the S-B scaling method does not adjust parameters for the metric of the data when modeling ordered categorical data, it may seem as though CVM is a more attractive option for data of this type. However, the computational demands of the ADF/WLS estimator make CVM an implausible option for dealing with ordered categorical data unless an extremely large sample size is available. Muthén (1993) developed and implemented two robust WLS estimators (WLSM and WLSMV) that avoid the necessity of a large sample size by decreasing the computational intensity found with the traditional ADF/WLS estimator. In addition, these estimators incorporate scaling similar to the S-B scaling methods.

Concerning the details of the estimators, WLSM and WLSMV differ from the conventional ADF/WLS estimator in the use of the asymptotic covariance matrix. Although WLS, WLSM, and WLSMV all use the same asymptotic covariance matrix, they differ in what elements of the weight matrix are used and how they are used. ADF/WLS employs and inverts the full weight matrix in order to estimate parameters, standard errors, and χ^2. Instead of using the full matrix when estimating parameters,

WLSM and WLSMV use only the diagonal elements of the weight matrix (e.g., asymptotic variances of the thresholds and latent correlation estimates). While WLSM and WLSMV do utilize the entire weight matrix to compute standard errors for the parameters, they employ a method that avoids its inversion (Muthén, 1993; Muthén, du Toit, & Spisic, 1997).

Like the standard errors, the χ^2 is calculated using the full weight matrix but avoids its inversion. In addition, a scaling factor, akin to the one employed in the Satorra-Bentler scaled χ^2, is employed to more thoroughly adjust the χ^2. Specifically, WLSM produces a mean-adjusted chi-square. WLSMV differs from WLSM in that the χ^2 is both mean- and variance-adjusted. The standard errors and parameter estimates from WLSM and WLSMV are equivalent. An additional distinction exists between WLSM and WLSMV in that WLSMV does not calculate model degrees of freedom in the standard way. Instead, degrees of freedom are "estimated" to approximate a χ^2 distribution and are lower in value than standard degrees of freedom (Muthén & Muthén, 2001). Therefore, fit indices (e.g., CFI, NNFI) will differ due to both the different degrees of freedom and different χ^2 values used in the calculations. Table 9.4 provides a brief outline of the differences and similarities between these three estimation techniques available in Mplus.

With the robust WLS estimators comes a new index for use with categorical data, the weighted root mean square residual (WRMR). This

Table 9.4. Mplus Estimation Techniques for Ordered Categorical Data

	Description	*Chi-square Estimation*	*Parameter Estimates*	*Standard Errors*	*Applied When?*
WLS	• Weighted least squares parameter estimates • Conventional χ^2 • Conventional standard errors	Full weight matrix used and inverted	Full weight matrix used	Full weight matrix used and inverted	Categorical or continuous endogenous variables
WLSM	• Weighted least squares parameter estimates • Mean-adjusted χ^2 • Scaled standard errors	Full weight matrix used but not inverted	Diagonal weight matrix used	Full weight matrix used but not inverted	At least one categorical endogenous variable
WLSMV	• Weighted least squares parameter estimates • Mean- and variance-adjusted χ^2 • Scaled standard errors	Full weight matrix used but not inverted	Diagonal weight matrix used	Full weight matrix used but not inverted	At least one categorical endogenous variable

index is well suited for categorical data because it incorporates the asymptotic variances into the computation. The WRMR is also appropriate to employ with non-normal continuous data, or if variables have large variances (Muthén & Muthén, 2001). WRMR values under 1.0 have been recommended to represent good fit with continuous or categorical data, with smaller values indicating better fit (Yu & Muthén, 2002).

WLSM and WLSMV: Empirical Results

There is limited research examining the functioning of the robust WLS estimators. Results have shown that these estimators perform much better than the conventional WLS when the tested model is large (15 variables) or sample size is small (N =1,000), yielding less biased χ^2 and standard errors (Muthén, 1993). Tentative results suggest that WLSMV outperforms WLSM, with WLSM showing higher Type I error rates (Muthén, 1999, 2003; Muthén et al., 1997). WLSMV tends to perform well except under conditions of small sample sizes (N = 200) and markedly skewed variables (Muthén et al., 1997).

WLSM and WLSMV: Software Implementation

Neither LISREL 8.54 nor EQS 6.1 have the capabilities to employ WLSM or WLSMV estimation techniques. Therefore, discussion of the implementation of these estimators will be limited to Mplus. In Mplus, when ordered categorical dependent variables are analyzed, the default estimator is WLSMV. The syntax for using the WLSM and WLSMV is very similar to the Mplus WLS syntax reported in Appendix A. The categorical indicators are identified by including the heading "CATEGORICAL ARE:" when defining the "VARIABLES" command. Once the variables are defined to be categorical, CVM will be conducted. To request one of the robust estimation techniques, changes are made to the "ESTIMATOR IS:" heading under the "ANALYSIS" command. Insert "WLSM" to request the mean-adjusted WLS estimation procedure or "WLSMV" to use the mean- and variance-adjusted estimation technique. Appendix C outlines these procedures.

Bootstrapping

General Description

When a model is correctly specified and data are multivariate normal, the expected value of the χ^2 statistic equals the model's degrees of freedom. The degrees of freedom are used to identify the corresponding central χ^2 distribution necessary to evaluate the probability value associated with the obtained χ^2. The S-B adjustment described above also uses the theoretical χ^2 distribution in order to evaluate statistical significance even though data are not multivariate normal. Because data are not multivariate normal, the obtained ML-based χ^2 is adjusted when using the S-B correction in order to better approximate the expected χ^2 distribution under conditions of normality. Instead of using the theoretical sampling distribution and adjusting the obtained χ^2 for non-normality of the data, bootstrapping techniques can be conceptualized as using the obtained χ^2 and adjusting the sampling distribution used to compute the probability value. More specifically, bootstrapping can be used to construct an empirical distribution of model test statistics that incorporates the non-normality of the data and relieves researchers from relying on the theoretical χ^2 distribution and its underlying assumptions.

In general, bootstrapping is a resampling technique that treats the observed sample data as an estimate of the population (Efron & Tibshirani, 1993). A large number of cases are then drawn with replacement from the observed data (parent sample) in order to create B bootstrap samples of the same size (N). Sampling with replacement implies that a given case may appear more than once in the same bootstrap sample. Using the sample data, the statistic of interest is computed from each of the B bootstrap samples. The estimates computed from the B samples then form an empirical sampling distribution of the statistic of interest.

In SEM, there are two methods of conducting bootstrapping, the naive bootstrap and the Bollen-Stine bootstrap. They differ in the type of empirical distribution formed. We first discuss the naive bootstrap and its application to estimating standard errors and then discuss the Bollen-Stine bootstrap and its application to estimating the probability value associated with the obtained χ^2 value.

Assume we are estimating a structural model using non-normal data. In addition to testing model fit, we also wish to evaluate the statistical significance of the estimated parameters. Given non-normal data, the ML-based standard errors will be underestimated. In contrast, the bootstrap standard errors take into account the distribution of the data, producing more accurate standard errors. Using the naive approach, B samples of size N are drawn (with replacement), and the parameter estimate of interest (e.g., factor loading, error variance) is estimated using each of the B bootstrap samples. This distribution of bootstrap parameter estimates is then used to calculate the standard error for that parameter. Specifically, the standard deviation of the bootstrap estimates represents the bootstrap standard error.

While the procedure above may provide better estimates of standard errors, it is not appropriate for bootstrapping the empirical distribution of the χ^2 statistic in SEM. Specifically, the resulting sampling distribution is incorrect because it reflects not only non-normality and sampling variability, but also model misfit (i.e., the null hypothesis of perfect fit is false). As Bollen and Stine (1992) pointed out, the parent sample must first be transformed to reflect the covariance structure underlying the hypothesized model (i.e., bootstrap samples are drawn from a parent sample where the null hypothesis is true). Transforming the data so that the parent sample conforms to the specified model is necessary in order to generate χ^2 values that reflect only sampling variation and the impact of non-normality, and not model misfit.

For example, suppose we are estimating a model with 60 degrees of freedom using non-normal data. The sample data are first transformed to have the same covariance structure as the implied covariance matrix (using a matrix transformation given in Bollen & Stine, 1992), then B random samples of size N are taken from the transformed parent sample data matrix, and the χ^2 value for each sample is computed in order to form the empirical sampling distribution of χ^2 values. This empirical sampling distribution can then be used as the reference distribution to identify the probability of the ML-based χ^2. Suppose that the mean χ^2 value across the bootstrap samples equaled 72. This implies that the expected value of the empirical distribution is larger than that of the theoretical distribution (in this case 60). One would therefore expect differing probability values associated with the ML-based χ^2 when employing the empirical versus the theoretical distribution. The probability value associated with the obtained χ^2 using the empirical distribution is simply the proportion of bootstrap χ^2 values that exceed the ML-based χ^2 obtained from the original analysis (i.e., from fitting the proposed model to the untransformed sample data).

Bootstrapping with Continuous Non-Normal Data: Empirical Results

Chi-square. There have been very few studies that have examined the performance of the Bollen-Stine bootstrap for estimating the χ^2 probability value. Fouladi (1998) found that, similar to the S-B scaled χ^2, the Bollen-Stine bootstrap controlled Type I error rates better than NT methods when data were non-normal. In addition, when modeling non-normal data, the Bollen-Stine bootstrap generally provided more accurate probability values than the S-B scaled χ^2 in all conditions except when sample size was large. Fouladi did not necessarily advocate one method over the other but instead suggested that readers realize the liberal or conservative bias of each statistic and use this information to inform decisions pertaining to the plausibility of a model.

Similar to Fouladi (1998), Nevitt and Hancock (2001) found that the model rejection rates based on the Bollen-Stine bootstrap and the S-B χ^2

were more accurate than those from the ML-based χ^2 when estimating correctly specified models under conditions of moderate non-normality. Even under conditions of extreme non-normality and small sample sizes, the Type I error rate associated with the Bollen-Stine bootstrap was controlled, outperforming the S-B χ^2. It is important to note, however, that the bootstrap showed less power to identify misspecified models than the S-B χ^2. Again, this tradeoff between controlling Type I error rate and sensitivity to misspecification complicates the decision of choosing one of the two techniques over the other.

Standard Errors. Nevitt and Hancock (2001) also investigated the performance of the naive bootstrap standard errors. Under conditions of non-normality, the bootstrap standard errors displayed less bias than the ML-based standard errors, and to a lesser extent, the S-B robust standard errors. However, it must be noted that the bootstrap standard error estimates displayed notable variability, signifying a possible concern with the stability of these estimates. A related finding concerning the stability and bias of the naive bootstrap standard errors suggests that small samples ($N \leq 100$) should be avoided due to a dramatic increase in both the variability and bias of the bootstrap standard errors. In addition, increasing the number of bootstrap samples beyond 250 has seemingly no benefits in terms of decreasing the bias in standard errors or model rejection rates.

Bootstrap: Software Implementation

As noted by Fan (2003), there are very few applications of bootstrapping in substantive research, which may be due to the limited automated procedures in SEM software programs. Of the three software programs presented here (EQS 6.1, Mplus 3.01, and LISREL 8.54), EQS (version 6 or higher) is the only program that has an automated bootstrapping option that can produce Bollen-Stine χ^2 probability values, accompanied by naive bootstrap standard errors (see Appendix D for syntax). Mplus 3.01 has an automated bootstrapping option that can produce naive bootstrap standard errors and bootstrap confidence intervals for the parameter estimates (Muthén & Muthén, 2004). AMOS (Arbuckle & Wothke, 1999), another popular SEM software program, also has automated bootstrapping capabilities that can produce the Bollen-Stine χ^2 probability values and the naive bootstrap standard errors (Byrne, 2001).

SUGGESTIONS FOR DEALING WITH NON-NORMALITY AND ORDERED CATEGORICAL DATA

Recommendations

Because so much information is associated with issues surrounding non-normal and ordered categorical data, Table 9.5 summarizes our recommendations when analyzing data of this type. This table is not meant to trivialize the complex issues surrounding this topic; instead, it may be treated as a supplement to the information presented in this chapter. Also, as explained throughout the chapter, the type of estimation method or technique employed is closely tied to the degree of non-normality and/or the crudeness of the categorization. Therefore, researchers need to recognize the type of data they are modeling, in terms of both metric and distribution, before selecting a technique.

In brief, ML has been shown to be fairly robust if continuous data are only slightly non-normal; therefore, we recommended its use in this situation (e.g., Chou et al., 1991; Green et al, 1997). If data are continuous

Table 9.5. Recommendations for Dealing with Non-Normal and Ordered Categorical Data

Type of Data	*Suggestions*	*Caveats/Notes*
Continuous Data		
1. Approximately normally distributed	• Use ML estimation	• The assumptions of ML are met and estimates should be unbiased, efficient, and consistent.
2. Moderately non-normal (skew < 2, kurtosis < 7)	• Use ML estimation; fairly robust to these conditions • Use S-B scaling to correct χ^2 and standard errors for even slight non-normality	• Given the availability of S-B scaling methods in the software packages, one could always employ and report findings from both ML estimation and S-B scaling method.
3. Severely non-normal (skew > 2, kurtosis > 7)	• Use S-B scaling • Use bootstrapping	• S-B correction works well, currently much easier to implement than the bootstrap, and tends to be more sensitive to model misspecification than the bootstrap. • Fit indices that are not adjusted for the S-B correction should be adjusted by hand.

(Table continues)

and non-normally distributed, we recommend the use of either the S-B scaling method or bootstrapping. Given the availability of the S-B scaling method, the ease of its use, and the empirical studies showing promising results, we can easily understand why this method is becoming increasingly popular.

When modeling ordered categorical data, the research seems to indicate that if there are a large number of ordered categories the data could be treated as continuous in nature. If the variables have five categories or

Table 9.5. (Continued)

Type of Data	*Suggestions*	*Caveats/Notes*
Ordered Categorical Data		
1. Approximately normally distributed	• Use Mplus's WLSMV estimator • Use ML estimation if there are at least five categories • Use S-B scaling methods if there are at least four categories	• WLSMV will adjust the parameter estimates, standard errors, and fit indices for the categorical nature of the data. • Realize that if employing ML estimation that the parameter estimates will be attenuated. • Parameter estimates from S-B scaling equal ML-based estimates implying that they too will be attenuated.
2. Moderately non-normal (skew < 2, kurtosis < 7)	• Use Mplus's WLSMV estimator • Use ML estimation if there are at least five categories • Use S-B scaling methods if there are at least four categories	• WLSMV will adjust the parameter estimates, standard errors, and fit indices for the categorical nature of the data. • Realize that if employing ML estimation that the parameter estimates will be attenuated. • Parameter estimates from S-B scaling equal ML-based estimates, implying that they too will be attenuated.
3. Severely non-normal (skew > 2, kurtosis >7) or very few categories (e.g., 3)	• Use Mplus's WLSMV estimator • If Mplus is not available then employ S-B scaling method	• Fit indices recommended with WLSMV because they show promise with non-normal ordered categorical data: WRMR and RMSEA. • Realize that S-B correction doesn't correct parameters for attenuation.

more, the data are approximately normally distributed, and Mplus is not available, we recommend treating the data as continuous in nature and employing ML estimation. If the variables have five categories or more, the data are non-normally distributed, and Mplus is not available, we recommend S-B scaling methods or bootstrapping. On the other hand, if ordered categorical variables have fewer than five categories, we recommend employing CVM to address both the metric and distribution. We specifically recommend employing CVM using Mplus's robust estimator WLSMV, because unlike ADF/WLS, it avoids inverting a large asymptotic covariance matrix and has exhibited promising results.

Strategies Not Recommended

An obvious omission from the above recommendations table is the ADF/WLS estimator. Given the requirements of the ADF/WLS estimator (e.g., large sample size) and the lack of sensitivity to model misspecification (e.g., Olsson et al., 2000), we cannot recommend the use of this estimator as a method to analyze non-normal or ordered categorical data. As noted throughout the chapter, other techniques outperform this estimator and should be employed.

A second technique that we cannot recommend, though commonly used to construct "more normally distributed" data (e.g., Marsh, Craven, & Debus, 1991), involves parceling items together (e.g., sum or average a subset of items). For example, parceling items with opposite skew has been conducted in order for the resulting parcel to have a better approximation of a normal distribution. Following the same logic, the parceling of ordered categorical items has been conducted to achieve a more continuous normal distribution allowing for the use of NT methods. While it is true that the parcel may have properties that better approximate the assumptions underlying NT estimators, we cannot recommend the uncritical use of this technique as a strategy to deal with non-normal or categorical data because it results in ambiguous findings. As detailed at length in other sources (Bandalos, 2002, 2003; Bandalos & Finney, 2001), parceling can obscure the true relations among the variables leading to biased parameters estimates and fit indices.

DIRECTIONS FOR FUTURE RESEARCH AND CONCLUSIONS

The purpose of this chapter was to review techniques used to accommodate non-normal and categorical data and summarize previous research investigating their utility. Much of the previous research involving non-normal and/or categorical data was concerned with comparing the performance of different estimation techniques (e.g., ML, WLS) under various conditions such as model size, sample size, and the observed variable distribution characteristics. An appropriate question at this point is where do we go from here with respect to researching the effects of modeling categorical and non-normal data? We believe the most pressing questions concern the functioning of the robust estimators (WLSM and WLSMV) available in Mplus (Muthén & Muthén, 2004). Unlike ML and WLS, limited research has been conducted that evaluates the performance of these estimators. Additional studies exploring the functioning of WLSM and WLSMV under various conditions and in relation to other techniques are needed in order to better understand the utility of these estimators. Also, very recent advances in software allow categorical dependent variables to be analyzed using ML estimation techniques. Specifically, when using Mplus v3.0 to analyze categorical variables, a full-information ML estimator can be employed. This estimator uses information from the full multiway frequency table of all categorical variables, which is why it is referred to as a "full information" technique. This differs from WLS, which is a "limited-information technique" because it uses bivariate information, or two-way frequency tables between pairs of variables. This full-information estimator uses the two-parameter logistic model, common in item response theory, to describe the variation in the probability of the item response as a function of the factor(s) (L. Muthén, personal communication, November 14, 2003). The availability of the full-information ML estimation technique provides opportunities for new research in the area of analyzing categorical data (e.g., feasibility with large models, comparability of WLSMV-based versus ML-based parameter estimates and standard errors).

In closing, given the presence of non-normal and ordered categorical data in applied research, researchers need to not only recognize the properties of their data but to also utilize techniques that accommodate these properties. Simply using the default estimator from a computer package does not guarantee valid results. Understanding the issues surrounding various techniques, such as assumptions, robustness, and implementation in software programs, makes a researcher much more competent to handle issues that they may encounter.

ACKNOWLEDGMENTS

We thank Deborah Bandalos, Craig Enders, Gregory Hancock, David Kaplan, and especially Linda Muthén, for their helpful comments on this chapter.

APPENDIX A: ADF SYNTAX

The following syntax illustrates how to employ the ADF estimator with continuous and categorical data. The model being estimated is a two-factor model with six indicators per factor. Each observed variable serves as an indicator to only one factor, all error covariances are fixed at zero, and the factor correlation is freely estimated. The metric of the factor is set by constraining the factor variance to a value of 1.00.

Continuous Data

LISREL v8.54

```
! First run PRELIS to obtain covariance matrix and asymptotic covariance
   matrix
DA NI=12
LA
q1 q2 q3 q4 q5 q6 q7 q8 q9 q10 q11 q12
RA=example.dat
OR ALL
OU MA=CM SM=example.cov AC=example.acc BT XM
```

SIMPLIS command language employed using LISREL v8.54

```
!Second read matrices into SIMPLIS program
Title illustrating ADF estimation with continuous data
Observed Variables q1 q2 q3 q4 q5 q6 q7 q8 q9 q10 q11 q12
Covariance matrix from file example.cov
Asymptotic matrix from file example.acc
Sample size 1000
Latent variables: fact1 fact2
Relationships
q1 q2 q3 q4 q5 q6 = fact1
q7 q8 q9 q10 q11 q12 = fact2
Options: WLS
Path Diagram
End of Problem
```

EQS v6.1

```
/TITLE
 illustrating ADF estimation with continuous data
/SPECIFICATIONS
```

```
VARIABLES= 12; CASES= 1000; DATAFILE = 'example.ess';
MATRIX= raw; METHOD = agls;
EQUATIONS
V1 = *F1 + E1;
V2 = *F1 + E2;
V3 = *F1 + E3;
V4 = *F1 + E4;
V5 = *F1 + E5;
V6 = *F1 + E6;
V7 = *F2 + E7;
V8 = *F2 + E8;
V9 = *F2 + E9;
V10 = *F2 + E10;
V11 = *F2 + E11;
V12 = *F2 + E12;
/VARIANCES
F1 to F2 = 1;
E1 to E12 = *;
/COVARIANCES
F2, F1 = *;
/END
```

Mplus v3.01

```
TITLE: illustrating ADF estimation with continuous data
Data: FILE IS example.dat;
VARIABLE: NAMES ARE q1 – q12;
ANALYSIS: ESTIMATOR = WLS;
MODEL: f1 by q1* q2* q3* q4* q5* q6*;
f2 by q7* q8* q9* q10* q11* q12*;
f1 @ 1;
f2 @ 2;
```

Ordered Categorical Data

LISREL v8.54

```
!PRELIS run to obtain correct correlation matrix and asymptotic covari-
  ance matrix
Title ANALYZING ORDERED CATEGORICAL DATA
DA NI=12
LA
q1 q2 q3 q4 q5 q6 q7 q8 q9 q10 q11 q12
```

```
RA=cat.dat
OR ALL
OU MA=KM SM=poly.cm AC=catex.acc BT XM
```

SIMPLIS command language employed using LISREL v8.54

```
Title ANALYZING ORDERED CATEGORICAL DATA
Observed Variables q1 q2 q3 q4 q5 q6 q7 q8 q9 q10 q11 q12
Correlation matrix from File poly.cm
Asymptotic matrix from File catex.acc
Sample size 1000
Latent variables: fact1 fact2
Relationships
q1 q2 q3 q4 q5 q6 = fact1
q7 q8 q9 q10 q11 q12 = fact2
Options: WLS
Path diagram
End of problem
```

EQS v6.1

```
/TITLE
 illustrating CVM with categorical data
/SPECIFICATIONS
 VARIABLES= 12; CASES= 1000; DATAFILE = 'example.ess';
 MATRIX= RAW; METHOD = AGLS;
 CATEGORY=V1 V2 V3 V4 V5 V6 V7 V8 V9 V10 V11 V12;
/EQUATIONS
V1 = *F1 + E1;
V2 = *F1 + E2;
V3 = *F1 + E3;
V4 = *F1 + E4;
V5 = *F1 + E5;
V6 = *F1 + E6;
V7 = *F2 + E7;
V8 = *F2 + E8;
V9 = *F2 + E9;
V10 = *F2 + E10;
V11 = *F2 + E11;
V12 = *F2 + E12;
/VARIANCES
 F1 to F2 = 1;
 E1 to E12 = *;
/COVARIANCES
```

```
F2, F1 = *;
/END
```

Mplus v3.01[5]

```
TITLE: Mplus with ordered categorical data
DATA: FILE IS cat.dat;
VARIABLE: NAMES ARE q1-q12;
 CATEGORICAL ARE q1-q12;
ANALYSIS: ESTIMATOR = WLS;
MODEL: f1 BY q1* q2* q3* q4* q5* q6*;
 f2 BY q7* q8* q9* q10* q11* q12*;
 f1 @1;
 f2 @1;
```

APPENDIX B: S-B SCALING SYNTAX

The following syntax illustrates how to employ the S-B scaling methodology with continuous and ordered categorical data. The model being estimated is a two-factor model with six indicators per factor. Each observed variable serves as an indicator to only one factor, all error covariances are fixed at zero, and the factor correlation is freely estimated. The metric of the factor is set by constraining the factor variance to a value of 1.00.

Continuous and Ordered Categorical Data

SIMPLIS command language employed using LISREL v8.54

```
Title illustrating SB chisq and standard errors
Observed variables q1 q2 q3 q4 q5 q6 q7 q8 q9 q10 q11 q12
Covariance matrix from file example.cov
Asymptotic matrix from file example.acc
Sample size 1000
Latent variables: fact1 fact2
Relationships
q1 q2 q3 q4 q5 q6 = fact1
q7 q8 q9 q10 q11 q12 = fact2
Options: ML
Path diagram
End of problem
```

EQS v6.1

```
/TITLE
 illustrating SB chisq and standard errors
/SPECIFICATIONS
 VARIABLES= 12; CASES= 1000; DATAFILE = 'example.ess';
 MATRIX= raw; METHOD = ml, robust;
/EQUATIONS
V1 = *F1 + E1;
V2 = *F1 + E2;
V3 = *F1 + E3;
V4 = *F1 + E4;
V5 = *F1 + E5;
V6 = *F1 + E6;
V7 = *F2 + E7;
V8 = *F2 + E8;
V9 = *F2 + E9;
V10 = *F2 + E10;
V11 = *F2 + E11;
V12 = *F2 + E12;
/VARIANCES
 F1 to F2 = 1;
 E1 to E12 = *;
/COVARIANCES
 F2, F1 = *;
/END
```

Mplus v3.01

```
TITLE: illustrating SB chisq and standard errors
Data: FILE IS example.dat;
VARIABLE: NAMES ARE q1 – q12;
ANALYSIS: ESTIMATOR = MLM;
MODEL: f1 by q1* q2* q3* q4* q5* q6*;
       f2 by q7* q8* q9* q10* q11* q12*;
       f1 @ 1;
       f2 @ 1;
```

APPENDIX C: ROBUST WLS (WLSM, WLSMV) SYNTAX

The following syntax illustrates how to employ the robust estimation techniques using Mplus v2.11. The model being estimated is a two-factor model with six indicators per factor and the factor correlation is freely estimated. The metric of the factor is set by constraining the factor variance to a value of 1.00. It is noted that indicator error variance terms are not estimated in Mplus when indicators are identified as categorical. To employ either WLSM or WLSMV, simply type "WLSM" or "WLSMV" on the command line that specifies the estimation method. In the current example, WLSMV would be employed.

```
TITLE:      MPLUS with ordered categorical data – robust estimation
            procedures
DATA:       FILE IS cat.dat;
VARIABLE:   NAMES ARE q1-q12;
            CATEGORICAL ARE q1-q12;
ANALYSIS:   ESTIMATOR=WLSMV;
MODEL:      f1 by q1* q2* q3* q4* q5* q6*;
            f2 by q7* q8* q9* q10* q11* q12*;
            f1 @ 1;
            f2 @ 1;
```

APPENDIX D: BOOSTRAPPING SYNTAX

The following syntax illustrates how to employ the bootstrapping technique. The model being estimated is a two-factor model with six indicators per factor. Each observed variable serves as an indicator to only one factor, all error covariances are fixed at zero, and the factor correlation is freely estimated. The metric of the factor is set by constraining a path from a factor to an indicator equal to a value of 1.00 (see Hancock & Nevitt, 1999, for an explanation of why it is necessary).

EQS v6.1
Naive Bootstrap Standard Errors

```
/TITLE
naive bootstrap
/SPECIFICATIONS
VARIABLES= 12; CASES=1000; DATAFILE='example.ess';
MATRIX=RAW; METHOD=ML;
/EQUATIONS
V1 = 1F1 + E1;
V2 = *F1 + E2;
V3 = *F1 + E3;
V4 = *F1 + E4;
V5 = *F1 + E5;
```

```
V6 = *F1 + E6;
V7 = 1F2 + E7;
V8 = *F2 + E8;
V9 = *F2 + E9;
V10 = *F2 + E10;
V11= *F2 + E11;
V12 = *F2 + E12;
/VARIANCES
 F1 to F2 = *;
 E1 to E12 = *;
/COVARIANCES
 F1, F2 = *;
/technical              !itr increases number of iterations to increase the
itr = 500;                 bootstrap success rate
/SIMULATION             !The keyword bootstrap indicates the naïve boot-
bootstrap = 1000;          strap
replication = 250;      !1000 refers to the number of cases in the bootstrap
seed = 123456789;          samples
/OUTPUT                 !Number of replications equals 250
parameters;             !Default seed for the random number generator
/END                    !The output will contain the mean parameter esti-
                           mates and the standard
                        !deviations, which are the empirical standard errors
```

Bollen-Stine Bootstrap χ^2 *probability value*

```
/TITLE
  Bollen-Stine bootstrap chisq probability value
/SPECIFICATIONS
 VARIABLES= 12; CASES=1000; DATAFILE='example.ess';
 MATRIX=RAW; METHOD=ML;
/EQUATIONS
V1 = 1F1 + E1;
V2 = *F1 + E2;
V3 = *F1 + E3;
V4 = *F1 + E4;
V5 = *F1 + E5;
V6 = *F1 + E6;
V7 = 1F2 + E7;
V8 = *F2 + E8;
V9 = *F2 + E9;
V10 = *F2 + E10;
```

```
V11= *F2 + E11;
V12 = *F2 + E12;
/VARIANCES
F1 to F2 = *;
E1 to E12 = *;
/COVARIANCES
F1, F2 = *;
/technical
itr = 500;             !itr increases the number of iterations
/SIMULATION            !The keyword mbb indicates the model-based, or
mbb = 1000;               Bollen-Stine, bootstrap
replication = 250;     ! 1000 refers to the number of cases in the boot-
seed = 123456789;         strap samples
/OUTPUT                !Number of bootstrap samples (B) drawn equals
parameters;               250
/END                   !Default seed for the random number generator
                       !Output presents information concerning the
                          empirical distribution of
                       !the model-based chi-square values including the
                          value that represents
                       !the upper 5% of the distribution, which can be
                          used as the critical chi-!square value to assess sig-
                          nificance of the ML-based chi-square
```

Mplus v3.01

Naive Bootstrap Standard Errors and Confidence Intervals

```
TITLE:      MPLUS with naive bootstrap standard errors and CI
DATA:       FILE IS example.dat;
VARIABLE:   NAMES ARE q1-q12;
ANALYSIS:   BOOTSTRAP = 250;             !Number of bootstrap sam-
                                            ples (B) drawn = 250;
MODEL:      f1 BY q1 q2 q3 q4 q5 q6; !   !The size of the B samples=
                                            size of original sample;
            f2 BY q7 q8 q9 q10 q11 q12;  !Other sample sizes of B can-
                                            not be specified;
OUTPUT:     CINTERVAL;
```

NOTES

1. Technically, this gives the reweighted least squares fit function, which is asymptotically equivalent to ML's well-known fit function,

$F = \log|\Sigma(\hat{\theta})| + tr[\mathbf{S}\Sigma(\hat{\theta})^{-1}] - \log|\mathbf{S}| - p$, where p equals the number of observed variables.

2. The standard linear measurement model specifies that a person's score is a function of the relation (b) between the variable (y^*) and the factor (F) plus error (E): $y^* = b\text{F} + \text{E}$
3. This formulation is equivalent to the two-parameter normal ogive model (probit model) applied to dichotomous items in item response theory (Muthén & Asparouhov, 2002; Thissen & Orlando, 2001), $P(y \geq k|\text{F}) = \Phi[a\ (\text{F} - b_k)]$, where a is the item discrimination and b is the item difficulty.
4. The formula used to calculate these scaled standard errors is complex but can be found in Arminger and Schoenberg (1989) and Satorra and Bentler (1994).
5. A unique feature of Mplus concerns the indicator error variance terms. Whereas LISREL and EQS allow estimation of these parameters, Mplus does not estimate indicator error variance terms if indicators are identified as categorical. Muthén and Muthén (2001) state that this is related to the use of a correlation matrix as input for analyses. With correlation input, the diagonal elements (values of 1) do not enter into the computations. Values related to each item variance are considered by Mplus to be "residual correlations" rather than as item variance terms.

REFERENCES

Arbuckle, J., & Wothke, W. (1999). *AMOS 4.0 user's guide*. Chicago: Smallwaters Corporation.

Arminger, G., & Schoenberg, R. (1989). Pseudo maximum likelihood estimation and a test for misspecification in mean and covariance structure models. *Psychometrika, 54*, 409–425.

Austin, J. T., & Calderón, R. F. (1996). Theoretical and technical contributions to structural equation modeling: An updated annotated bibliography. *Structural Equation Modeling: A Multidisciplinary Journal, 3*, 105–175.

Babakus, E., Ferguson, C. E., & Jöreskog, K. G. (1987). The sensitivity of confirmatory maximum likelihood factor analysis to violations of measurement scale and distributional assumptions. *Journal of Marketing Research, 24*, 2228.

Bandalos, D. L. (2002). The effects of item parceling on goodness-of-fit and parameter estimate bias in structural equation modeling. *Structural Equation Modeling: A Multidisciplinary Journal, 9*, 78–102.

Bandalos, D. L. (2003, April). *Identifying model misspecification in SEM analyses: Does item parceling help or hinder?* Paper presented at the annual meeting of the American Educational Research Association, Chicago.

Bandalos, D. L., & Finney, S. J. (2001). Item parceling issues in structural equation modeling. In G. A. Marcoulides & R. E. Schumacker (Eds.), *New developments and techniques in structural equation modeling* (pp. 269–296). Mahwah, NJ: Erlbaum.

Bentler, P. M. (1990). Comparative fit indexes in structural equation models. *Psychological Bulletin, 107*, 238–246.

Bentler, P. M. (1995). *EQS: Structural equations program manual*. Encino, CA: Multivariate Software, Inc.

Bentler, P. M. (2004). *EQS for Window* (Version 6.1) [Computer software]. Encinco, CA: Multivariate Software, Inc.

Bentler, P. M., & Dudgeon, P. (1996). Covariance structure analysis: Statistical practice, theory, and directions. *Annual Review of Psychology, 47*, 563–592.

Bentler, P. M., & Wu, E. J. C. (2002). *EQS for Windows user's guide*. Encino, CA: Multivariate Software, Inc.

Bollen, K. A. (1989). *Structural equation modeling with latent variables*. New York: Wiley.

Bollen, K. A., & Stine, R. A. (1992). Bootstrapping goodness-of-fit measures in structural equation models. *Sociological Methods and Research, 21*, 205–229.

Browne, M. W. (1984). Asymptotic distribution-free methods in the analysis of covariance structures. *British Journal of Mathematical and Statistical Psychology, 37*, 62–83.

Byrne, B. M. (2001). *Structural equation modeling with AMOS: Basic concepts, applications, and programming*. Mahwah, NJ: Erlbaum.

Chou, C., & Bentler, P. M. (1995). Estimates and tests in structural equation modeling. In R. H. Hoyle (Ed.), *Structural equation modeling: Concepts, issues, and applications* (pp. 37–55). Thousand Oaks, CA: Sage.

Chou, C., Bentler, P. M., & Satorra, A. (1991). Scaled test statistics and robust standard errors for non-normal data in covariance structure analysis: A monte carlo study. *British Journal of Mathematical and Statistical Psychology, 44*, 347–357.

Curran, P. J., West, S. G., & Finch, J. F. (1996). The robustness of test statistics to non-normality and specification error in confirmatory factor analysis. *Psychological Methods, 1*, 16–29.

DiStefano, C. (2002). The impact of categorization with confirmatory factor analysis. *Structural Equation Modeling: A Multidisciplinary Journal, 9*, 327–346.

DiStefano, C. (2003, April). *Considering the number of categories and item saturation levels with structural equation modeling*. Paper presented at the annual conference of the American Educational Research Association, New Orleans, LA.

Dolan, C. V. (1994). Factor analysis of variables with 2, 3, 5, and 7 response categories: A comparison of categorical variable estimators using simulated data. *British Journal of Mathematical and Statistical Psychology, 47*, 309–326.

Efron, B., & Tibshirani, R. J. (1993). *An introduction to the bootstrap*. New York: Chapman & Hall.

Fan, X. (2003). Using commonly available software for bootstrapping in both substantive and measurement analyses. *Educational and Psychological Measurement, 63*, 24–50.

Finch, J. F., West, S. G., & MacKinnon, D. P. (1997). Effects of sample size and non-normality on the estimation of mediated effects in latent variable models. *Structural Equation Modeling: A Multidisciplinary Journal, 4*, 87–107.

Fouladi, R. T. (1998, April). *Covariance structure analysis techniques under conditions of multivariate normality and non-normality—Modified and bootstrap based test statistics*. Paper presented at the annual meeting of the American Educational Research Association, San Diego, CA.

Gierl, M. J., & Mulvenon, S. (1995, April). *Evaluating the application of fit indices to structural equation models in educational research: A review of the literature from 1990 through 1994.* Paper presented at the annual meeting of the American Educational Research Association, San Francisco.

Green, S. B., Akey, T. M., Fleming, K. K., Hershberger, S. L., & Marquis, J. G. (1997). Effect of the number of scale points on chi-square fit indices in confirmatory factor analysis. *Structural Equation Modeling: A Multidisciplinary Journal, 4*, 108–120.

Hancock, G. R., & Nevitt, J. (1999). Bootstrapping and the identification of exogenous latent variables within structural equation models. *Structural Equation Modeling: A Multidisciplinary Journal, 6*, 394–399.

Hoogland, J. J., & Boomsma, A. (1998). Robustness studies in covariance structure modeling: An overview and a meta-analysis. *Sociological Methods & Research, 26*, 329–367.

Hu, L., & Bentler, P. M. (1998). Fit indices in covariance structure modeling: Sensitivity to underparameterized model misspecification. *Psychological Methods, 3*, 424–453.

Hu, L., & Bentler, P. M. (1999). Cutoff criteria for fit indexes in covariance structure analysis: Conventional criteria versus new alternatives. *Structural Equation Modeling: A Multidisciplinary Journal, 6*, 1–55.

Hu, L., Bentler, P. M., & Kano, Y. (1992). Can test statistics in covariance structure analysis be trusted? *Psychological Bulletin, 112*, 351–362.

Hutchinson, S. R., & Olmos, A. (1998). Behavior of descriptive fit indexes in confirmatory factor analysis using ordered categorical data. *Structural Equation Modeling: A Multidisciplinary Journal, 5*, 344–364.

Jöreskog, K., & Sörbom, D. (1996). *PRELIS 2: User's reference guide*. Chicago: Scientific Software International.

Jöreskog, K., & Sörbom, D. (2004). *LISREL 8.54 for Windows* [Computer software]. Lincolnwood, IL: Scientific Software International, Inc.

Kaplan, D. (2000). *Structural equation modeling: Foundations and extensions*. Thousand Oaks, CA: Sage.

MacCallum, R. C., & Austin, J. T. (2000). Applications of structural equation modeling in psychological research. *Annual Review of Psychology, 51*, 201–226.

Marsh, H. W., Craven, R. G., & Debus, R. (1991). Self-concepts of young children 5 to 8 years of age: Measurement and multidimensional structure. *Journal of Educational Psychology, 83*, 377–392.

Micceri, T. (1989). The unicorn, the normal curve, and other improbable creatures. *Psychological Bulletin, 105*, 156–166.

Muthén, B. O. (1984). A general structural equation model with dichotomous, ordered categorical, and continuous latent variable indicators. *Psychometrika, 49*, 115–132.

Muthén, B. O. (1993). Goodness of fit with categorical and other non-normal variables. In K. A. Bollen & J. S. Long (Eds.), *Testing structural equation models* (pp. 205–243). Newbury Park, CA: Sage.

Muthén, B. O., & Asparouhov, T. (2002). *Latent variable analysis with categorical outcomes: Multiple-group and growth modeling in Mplus.* Retrieved March 15, 2004, from http://www.statmodel.com/mplus/examples/webnotes/CatMGLong.pdf

Muthén, B. O., du Toit, S., & Spisic, D. (1997). *Robust inference using weighted least squares and quadratic estimating equations in latent variable modeling with categorical and continuous outcomes.* Unpublished manuscript.

Muthén, B. O., & Kaplan, D. (1985). A comparison of some methodologies for the factor analysis of non-normal Likert variables. *British Journal of Mathematical and Statistical Psychology, 38*, 171–189.

Muthén, B. O., & Kaplan, D. (1992). A comparison of some methodologies for the factor analysis of non-normal Likert variables: A note on the size of the model. *British Journal of Mathematical and Statistical Psychology, 45*, 19–30.

Muthén, L. (1999, August 10). Mplus and non-normal data. Message posted to SEMNET discussion list, archived at http://bama.ua.edu/cgi-bin/wa?A2=ind9908&L=semnet&D=0&T=0&P=9802

Muthén, L. (2003, April 29). Mplus and other questions. Message posted to SEMNET discussion list, archived at http://bama.ua.edu/cgi-bin/wa?A2=ind0304&L=semnet&D=0&I=1&P=52337.

Muthén, L. K., & Muthén, B. O. (2001). *Mplus user's guide* (2nd ed). Los Angeles: Authors.

Muthén, L. K., & Muthén, B. O. (2004). *Mplus user's guide* (3rd ed). Los Angeles: Authors.

Nevitt, J., & Hancock, G. R. (2000). Improving the root mean square error of approximation for non-normal conditions in structural equation modeling. *Journal of Experimental Education, 68*, 251–268.

Nevitt, J., & Hancock, G. R. (2001). Performance of bootstrapping approaches to model test statistics and parameter standard error estimation in structural equation modeling. *Structural Equation Modeling: A Multidisciplinary Journal, 8*, 353–377.

Olsson, U. (1979). Maximum likelihood estimation of the polychoric correlation coefficient. *Psychometrika, 44*, 443–460.

Olsson, U., Foss, T., Troye, S. V., & Howell, R. D. (2000). The performance of ML, GLS, and WLS estimation in structural equation modeling under conditions of misspecification and non-normality. *Structural Equation Modeling: A Multidisciplinary Journal, 7*, 557–595.

Olsson, U. H., Troye, S. V., & Howell, R. D. (1999). Theoretical fit and empirical fit: The performance of maximum likelihood versus generalized least squares estimation in structural equation models. *Multivariate Behavioral Research, 34*, 31–58.

Potthast, M. J. (1993). Confirmatory factor analysis of ordered categorical variables with large models. *British Journal of Mathematical and Statistical Psychology, 46*, 273–286.

Rosenberg, M. (1989). *Society and the adolescent self-image* (Rev. ed.). Middletown, CT: Wesleyan University Press.

Satorra, A., & Bentler, P. M. (1994). Corrections to test statistics and standard errors in covariance structure analysis. In A. von Eye & C. C. Clogg (Eds.), *Latent variables analysis: Applications for developmental research* (pp. 399–419). Thousand Oaks, CA: Sage.

Satorra, A., & Bentler, P. M. (2001). A scaled difference chi-square test statistic for moment structure analysis. *Psychometrika, 66*, 507–514.

Thissen, D., & Orlando, M. (2001). Item response theory for items scored in two categories. In D. Thissen & H. Wainer (Eds.), *Test scoring* (pp. 73–140). Mahwah, NJ: Erlbaum.

Tremblay, P. F., & Gardner, R. C. (1996). On the growth of structural equation modeling in psychological journals. *Structural Equation Modeling: A Multidisciplinary Journal, 3*, 93–104.

West, S. G., Finch, J. F., & Curran, P. J. (1995). Structural equation models with non-normal variables: Problems and remedies. In R. H. Hoyle (Ed.) *Structural equation modeling: Concepts, issues, and applications* (pp. 56–75). Thousand Oaks, CA: Sage.

Yu, C., & Muthén, B. (2002, April). *Evaluation of model fit indices for latent variable models with categorical and continuous outcomes*. Paper presented at the annual meeting of the American Educational Research Association, New Orleans, LA.

CHAPTER 10

ANALYZING STRUCTURAL EQUATION MODELS WITH MISSING DATA

Craig K. Enders

A wealth of options exists for analyzing structural equation models (SEM) with missing data, including the expectation maximization (EM) algorithm, full information maximum likelihood (FIML) estimation, and multiple imputation (MI). Although these so-called "modern" missing data methods (Collins, Schafer, & Kam, 2001) have been known in the statistical literature for some time (e.g., Dempster, Laird, & Rubin, 1977; Finkbeiner, 1979; Rubin, 1987), they have only recently become widely available in software packages. As late as the mid-1990s there were relatively few sophisticated options for performing an SEM analysis with missing data. Muthén, Kaplan, and Hollis (1987) had earlier proposed a fairly complex multiple group procedure for missing data, and AMOS (Arbuckle, 2003) was the only commercial SEM software package that implemented FIML estimation. These modern missing data methods have subsequently been implemented in additional software packages, and the types of models to which these methods can be applied have grown rapidly. The missing data literature experienced concurrent growth during the last decade, and many familiar SEM procedures for

Structural Equation Modeling: A Second Course, 315–344

complete data have been extended to the missing data context (e.g., rescaled test statistics, robust standard errors, and bootstrapping).

The primary goals of this chapter are to summarize recent methodological advances, and to provide the reader with a thorough overview of the analytic options for conducting SEM analyses with missing data. The chapter begins with a brief overview of Rubin's (1976) theoretical framework, as this provides the basis for understanding how missing data procedures affect SEM parameter estimates. Next, I describe several widely available missing data techniques. Although brief descriptions of traditional methods (e.g., listwise deletion) are provided, more emphasis is given to the modern techniques: EM, FIML, and MI. These methods are currently considered to be the "state of the art" (Schafer & Graham, 2002), and are justified by a wealth of theoretical and empirical work. In discussing each of the modern methods, I also demonstrate the use of an "inclusive" strategy (Collins et al., 2001) that incorporates auxiliary variables into the analysis. An auxiliary variable can be thought of as any variable that would not appear in the substantive model if one were analyzing a complete dataset. The use of such variables can mitigate estimation bias and increase the likelihood that missing data assumptions required by the modern approaches are met. Finally, an example analysis is provided for each of the three modern techniques.

The concepts presented in this chapter are illustrated using a confirmatory factor analysis (CFA) of the Eating Attitudes Test (EAT; Garner, Olmsted, Bohr, & Garfinkel, 1982), a widely used self-report instrument used to assess eating disorder risk. For simplicity, a subset of 10 manifest indicators was extracted from a model proposed by Doninger, Enders, and Burnett (2005). These indicators measured two latent factors, Drive for Thinness and Food Preoccupation. A single background variable, body mass index (BMI), is also used in the subsequent examples. The original data subset had no missing values, so missingness was imposed on five items (EAT1, EAT10, EAT12, EAT24, and EAT18) such that cases in the upper quartile of the BMI were assigned a 50% probability of deletion; as discussed below, this type of selective missingness is termed missing at random. This process resulted in the deletion of 11% to 17% of the responses on these five items. The raw data (N = 200) used in the subsequent analyses can be obtained from the author. A graphical depiction of the two-factor CFA model is shown in Figure 10.1.

Before proceeding, a brief clarification about the terminology used in this chapter is warranted. The term FIML is frequently used to describe maximum likelihood estimation in the missing data context (*direct maximum likelihood* is another common term found in the literature). To be precise, complete-data maximum likelihood is also a full information estimator, so it would be appropriate to use the term FIML in this con-

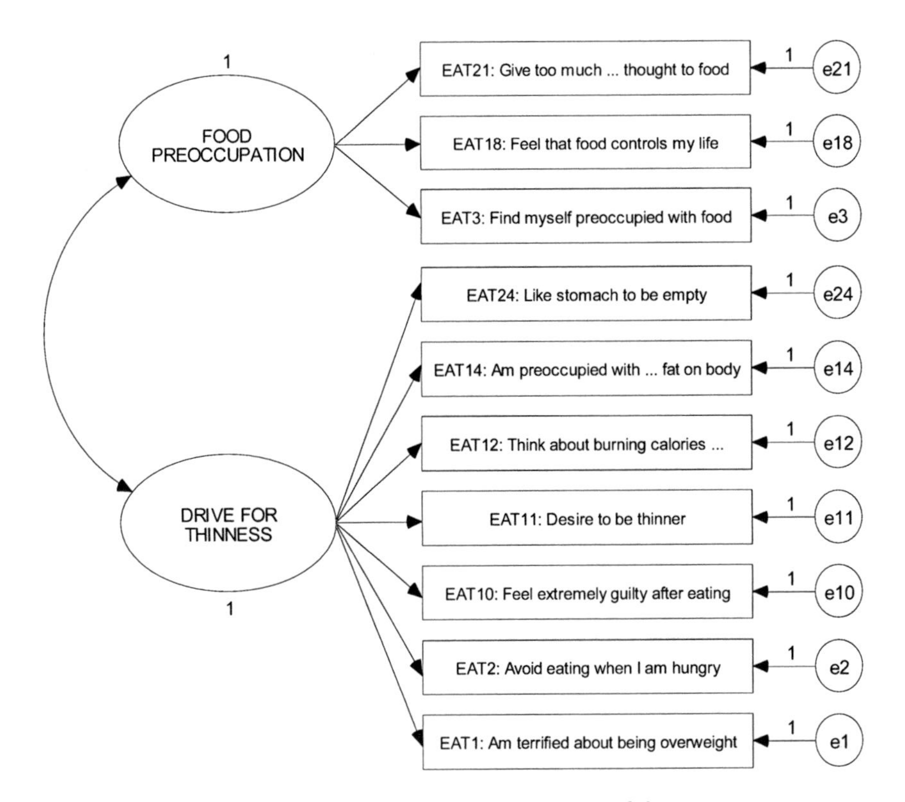

Figure 10.1. Two-factor confirmatory factor analysis model.

text as well. However, to be consistent with existing SEM literature, I use the term FIML to denote maximum likelihood estimation for missing data.

THEORETICAL BACKGROUND

In order to appreciate how the choice of missing data procedure might impact SEM parameter estimates, it is necessary to understand Rubin's (1976) taxonomy of missing data mechanisms. These mechanisms offer different explanations for how missingness is probabilistically related to the values of variables in the dataset, and have important implications for the performance of the techniques outlined in the next section. As we will see, missing data techniques differ in their assumptions about the missing data mechanism, and may provide biased results when these assumptions are not met.

According to Rubin, data are *missing completely at random* (MCAR) when missingness on a variable *X* is unrelated to the values of other observed variables as well as to the values of *X* itself. In this case the observed data are essentially a random sample of the hypothetically complete dataset. MCAR is the condition generally required by traditional missing data techniques such as listwise and pairwise deletion. It is important to note that MCAR is the only one of Rubin's (1976) mechanisms that can be formally tested—the other mechanisms would require knowledge of the missing values. A SAS program for conducting Little's (1988) test of MCAR can be downloaded at http://www.asu.edu/clas/psych/people/faculty/enders.htm or obtained from the author, and Little's test is commercially available in the SPSS Missing Value Analysis (MVA) module as well.

The *missing at random* (MAR) condition is less restrictive in the sense that missingness on a variable *X* is related to one or more other observed variables in the model, but is unrelated to the values of *X* itself. To illustrate, suppose a positive relationship existed between BMI scores and missingness, such that higher rates of nonresponse on EAT items measuring Food Preoccupation are associated with higher BMI scores. This situation would be described as MAR if no residual relationship existed between missingness and Food Preoccupation after controlling for BMI. In this case, the missing data mechanism is said to be *ignorable*, as unbiased parameter estimates can be obtained from the modern missing data methods (EM, FIML, and MI).

A subtle point is that MAR is defined with respect to the variables in the statistical model. This implies that a CFA of the EAT would need to include BMI scores if MAR is to hold, which appears to be an awkward situation, as BMI is a background variable that does not appear in the latent variable model shown in Figure 10.1. I return to this issue in a subsequent section, and demonstrate how to incorporate auxiliary variables such as BMI into the SEM.

Finally, data are said to be *missing not at random* (MNAR) when missingness on a variable *X* is dependent on the values of *X* itself. Returning to the CFA example, an MNAR mechanism would result if the propensity to respond to the food preoccupation items was systematically related to one's underlying level of preoccupation (e.g., highly preoccupied individuals tend to skip these items). The modern missing data methods discussed below assume MAR data, and will likely produce biased results when data are MNAR. MNAR missing data methods have recently been developed, but these approaches are quite sensitive to model misspecification, and may produce substantial parameter estimate bias. For now, the modern missing data techniques discussed in this chapter represent the "practical state of the art" (Schafer & Graham, 2002, p. 173).

TRADITIONAL MISSING DATA TECHNIQUES

Having discussed Rubin's (1976) theoretical work, I now outline some traditional missing data techniques. Space limitations preclude an exhaustive review of these methods, and more comprehensive overviews can be found elsewhere in the literature (e.g., Schafer & Graham, 2002).

Listwise Deletion

Listwise deletion (LD) removes all cases with missing values from the dataset, and subsequent analyses are performed using the remaining complete cases. The simplicity of performing analyses using a complete dataset is appealing, but LD can potentially result in a dramatic reduction in sample size, and thus power. Perhaps more problematic, LD typically requires MCAR missingness (the most stringent assumption) in order to produce unbiased parameter estimates. One notable exception to this requirement occurs in multiple regression models where missingness on the predictor variables does not depend on the values of the outcome variable. In this case, LD produces unbiased estimates of regression parameters under MAR (Allison, 2002).

Pairwise Deletion

Pairwise deletion (PD) attempts to use as much of the observed data as possible, and does so by removing cases on an analysis-by-analysis basis. In the context of SEM, each element of the input covariance matrix is computed using cases with complete data on a given variable (for variances) or variable pair (for covariances). Like LD, PD requires an MCAR mechanism. Widely documented problems with PD include the specification of a single value of N and the possibility of nonpositive definite matrices (i.e., different subsamples of cases are used to compute covariance matrix elements, resulting in impossible combinations of variances and covariances).

Mean Imputation

Different variations of mean imputation have been discussed in the literature, the most common of which replaces missing values with the arithmetic mean of the available cases (i.e., arithmetic mean imputation, or AMI). Under MCAR, AMI will provide unbiased estimates of the mean, but variance and covariance terms will be negatively biased (any missing observation adds a value of zero to the numerator of these formulas). Mean imputation has consistently produced biased parameter estimates in simulation studies (e.g., Wothke, 2000), and should be avoided.

Regression Imputation

Conditional mean, or regression, imputation imputes missing values with the predicted values from a linear regression. Because imputed values fall directly on a regression surface (thus lacking the variation present in the hypothetically complete data), a random residual term can be added to each imputed value to reduce the bias in variance and covariance terms (i.e., stochastic regression imputation). This method produces consistent, but inefficient, parameter estimates, and can be fairly complicated with arbitrary patterns of missing data, as different regression equations must be constructed for each unique pattern.

Hot Deck Imputation

Hot deck (HD) imputation replaces missing values with the observed data from another case, randomly selected from a group of individuals who are similar with respect to a set of covariates. Brown (1994) studied HD imputation in the context of SEM, and reported biased parameter estimates under an MCAR mechanism. A variation of hot deck imputation, similar response pattern imputation (SRPI), is available in the PRELIS software package (Jöreskog & Sörbom, 1999). Like HD, SRPI replaces missing values with the observed data from another case having similar responses on a set of user-defined matching variables. However, a minimization criterion is used to identify a *single* individual that most closely resembles the case with missing data. Little is known about the conditions under which SRPI might provide optimal performance; it appears that the method functions well under MCAR, but not MAR (Enders & Bandalos, 2001).

I now turn to the so-called "modern" missing data techniques: EM, FIML, and MI. Recent empirical studies have consistently demonstrated the superiority of these methods, both with respect to parameter estimate bias and efficiency (e.g., Arbuckle, 1996; Enders, 2001; Enders & Bandalos, 2001; Graham & Schafer, 1999; Muthén et al., 1987). These methods are also theoretically attractive because parameter estimates are unbiased and efficient under MAR. As noted previously, traditional methods generally require the more stringent MCAR assumption.

EXPECTATION MAXIMIZATION

The expectation maximization (EM) algorithm produces a maximum likelihood estimate of the covariance matrix and of the mean vector from a saturated model. The EM covariance matrix can, in turn, be used as input data for SEM or other multivariate analyses. EM is a general approach that has been adapted to direct estimation problems (e.g., multilevel models, mixture models), but the algorithm described here is an indirect approach in the sense that missing values must be preprocessed prior to the SEM analysis.

EM relies on a two-step iterative procedure whereby missing values (or functions of the missing values) are imputed, and the covariance matrix and mean vector are estimated. The EM analysis begins with an initial estimate of the covariance matrix, which can be obtained using any number of traditional methods (Little & Rubin, 2002, p. 225). The goal of the *expectation*, or E, step is to obtain estimates of the sufficient statistics (i.e., the variable sums and sums of products, ΣX_{ij} and $\Sigma X_{ij}X_{ik}$, respectively, where i denotes cases and j and k denote variables) required to compute the covariance matrix in the subsequent *maximization*, or M, step. To accomplish this, a series of regression equations are constructed from the current estimate of the covariance matrix, and the contribution of each missing X_{ij} to the sufficient statistics is the predicted value from a regression equation. A residual term is also added to the variable sums and sums of products in order to adjust for the fact that imputed values fall directly on a regression surface. At the M step, a new estimate of the mean vector and of the covariance matrix are computed using the sufficient statistics from the previous E step. There is nothing special about M step computations, as standard complete data formulae are used. This updated covariance matrix is passed along to the next E step, and the two-step algorithm is repeated until the difference between covariance matrices in adjacent M steps becomes trivially small.

To illustrate, consider a simple situation with three observed variables: X_1, X_2, and X_3. Furthermore, suppose that some cases have complete data on all variables, while other individuals are missing X_1. Yet a third group of cases are missing both X_1 and X_2. The EM analysis begins with an initial estimate of the covariance matrix, perhaps obtained via listwise deletion. The purpose of the E step is to estimate the sufficient statistics. The cases with no missing values simply contribute their observed values to the variable sums and sums of products. For cases missing X_1, the predicted values from the regression of X_1 on X_2 and X_3 are used in the sufficient statistics. To restore residual variation to these estimated quantities (conditional means falling directly on a regression surface), each of these cases also contributes an adjustment term to $(X_{i1})^2$ which is the residual variance from the regression of X_1 on X_2 and X_3. Thus, the missing value is replaced with a residual-adjusted conditional mean. In a similar fashion, the regression of X_1 on X_3 and of X_2 on X_3 are used to generate predicted values for the cases with missing X_1 and X_2 scores. In this case, the residual covariance between X_1 and X_2, controlling for X_3, is added to the sums of products computations involving X_{i1} and X_{i2}. Once the sufficient statistics (i.e., ΣX_{ij} and $\Sigma X_{ij}X_{ik}$) are estimated, an updated covariance matrix is computed in the M step using standard formulae. This new covariance matrix is fed into the next E step, where new estimates of the missing values are generated, and the two-step process is repeated.

The use of an EM input matrix should produce SEM parameter estimates that are nearly identical to those of FIML. However, standard errors and model fit statistics will likely be incorrect because no single value of N is applicable to the entire EM covariance matrix. Enders and Peugh (2004) studied this issue using MCAR data and obtained accurate standard errors using the harmonic mean of the number of complete observations per variable (i.e., the per-variance Ns). However, the model fit (i.e., χ^2) statistic was quite sensitive to the choice of N, and the optimal value was highly dependent on the missing data pattern. Their results suggested that caution is warranted when assessing fit using an EM input matrix. Finally, it is important to note that EM assumes both multivariate normality and an MAR pattern of missing data.

Incorporating Auxiliary Variables

The incorporation of auxiliary variables can play an important role when estimating a model with missing data. The definition of MAR stipulates that missingness is related to the values of observed variables, but unrelated to the underlying values that are missing. In most cases we do not know which variables are related to the missingness mechanism, but we do know that biased parameter estimates may result if parameter estimates are not conditioned on these variables. As such, the inclusion of auxiliary variables may both (a) increase the likelihood that MAR will hold, at least partially, and (b) serve to mitigate estimation bias and improve efficiency (i.e., power) when MAR does not hold. The results of Collins and colleagues (2001) suggested that researchers can be liberal about adding auxiliary variables without incurring negative consequences.

The incorporation of auxiliary variables is quite simple with EM, and is perhaps the only reason one would choose this approach over FIML estimation. To incorporate auxiliary variables, the EM covariance matrix is estimated using a superset of the variables appearing in the substantive model, and the relevant covariance matrix elements are subsequently extracted for the SEM analysis. Because the pertinent covariance matrix elements are conditioned on the auxiliary variables, the substantive model need not include the additional variables.

Example Analysis

EM is available in both SAS (PROC MI) and SPSS (MVA procedure), but in the latter case is sold as an additional module. As such, Schafer's (1999) freeware program NORM was used to obtain an EM estimate of the covariance matrix, and this matrix was subsequently used as input data for the CFA. NORM is discussed briefly in a subsequent section, and a detailed illustration of its use can be found in Schafer and Olsen (1998). The EM covariance matrix obtained from NORM was subsequently used as input for the CFA estimation using Mplus (Muthén & Muthén, 2004).

As mentioned previously, one of the difficult issues regarding the use of an EM input matrix is the specification of the sample size. Following recommendations from Enders and Peugh (2004), the harmonic mean of the number of complete observations per variable (i.e., the harmonic mean of the per variance *N*s) was used ($N = 186$), as this value produced honest confidence intervals in their simulations. Note that the bootstrap could also be used to obtain standard errors when using an EM input matrix (see, e.g., Graham, Hofer, Donaldson, MacKinnon, & Schafer, 1997). Enders and Peugh reported that the model fit statistic was quite sensitive to the choice of *N* (a dramatic increase in model rejection rates was generally observed for other choices of *N*), so caution is warranted when assessing model fit. Finally, the initial EM analysis included BMI scores, so it was not necessary to include BMI in the subsequent CFA model.

Selected parameter estimates and standard errors from the EM analysis are given in Table 10.1. The fit indices from the EM analysis were as follows: $\chi^2(34) = 69.64$, $p < .001$, CFI = .964, TLI = .953, RMSEA = .075, and SRMR = .043. Consistent with Enders and Peugh (2004), the magnitude of the χ^2 statistic was somewhat larger than that from FIML ($\chi^2 = 65.99$, $p = .001$), which is discussed below. In situations where miss-

Table 10.1. Selected Parameter Estimates and Standard Errors from Heuristic Analysis

Estimate	*Complete Data*	*FIML Estimation* *AMOS*	*EQS*	*LISREL*	*Mplus*	*MI*	*EM*
		Drive for Thinness Factor Loadings					
EAT1	1.173	1.154	1.154	1.150	1.154	1.163	1.153
	(.105)	(.110)	(.110)	(.110)	(.110)	(.114)	(.109)
EAT2	.538	.542	.542	.540	.542	.540	.541
	(.058)	(.058)	(.058)	(.058)	(.058)	(.058)	(.060)
EAT10	.725	.749	.749	.750	.749	.753	.743
	(.060)	(.064)	(.064)	(.063)	(.063)	(.066)	(.062)
		Food Preoccupation Factor Loadings					
EAT3	.913	.901	.901	.900	.901	.901	.899
	(.080)	(.081)	(.081)	(.080)	(.080)	(.081)	(.082)
EAT18	.896	.908	.908	.910	.908	.915	.910
	(.067)	(.071)	(.071)	(.071)	(.071)	(.074)	(.069)
		Factor Correlation					
r_{F1F2}	.760	.760	.760	.760	.760	.759	.759
	(.041)	(.042)	(.042)	(.040)	(.042)	(.043)	(.042)

Note. Standard errors are given in parentheses. Missing data were imposed on EAT1, EAT10, and EAT18. MI = multiple imputation; EM = expectation maximization. Parameter estimates were generated using AMOS 5.0, EQS 6.1, LISREL 8.54, and Mplus 3.01.

ing values were differentially dispersed across manifest indicators, Enders and Peugh recommended the use of the minimum N per variable when assessing model fit using the χ^2 statistic. Following this suggestion, a second CFA analysis was performed specifying $N = 174$. The resulting EM fit indices were as follows: $\chi^2(34) = 64.89$, $p = .001$, CFI = .967, TLI = .956, RMSEA = .072, and SRMR = .043. Note that the EM χ^2 statistic from this analysis was quite similar to that obtained with full information maximum likelihood (FIML).

FULL INFORMATION MAXIMUM LIKELIHOOD

Structural equation models have traditionally been estimated using covariance matrices. Most readers are probably familiar with the usual maximum likelihood fitting function, which reflects the discrepancy between the observed and reproduced covariance matrices. With missing data, the log-likelihood is computed from the observed data points, necessitating the use of raw data. In this situation, a log-likelihood is computed for each case, and these individual likelihoods are summed to produce the log-likelihood for the entire sample, log L. Assuming a multivariate normal distribution in the population, the log-likelihood for case i is

$$\log L_i = K_i - \frac{1}{2}\log\left|\boldsymbol{\Sigma}_i\right| - \frac{1}{2}\left(\mathbf{x}_i - \boldsymbol{\mu}_i\right)' \boldsymbol{\Sigma}_i^{-1}\left(\mathbf{x}_i - \boldsymbol{\mu}_i\right) \tag{1}$$

where $\mathbf{x}_i$ is the vector of raw data, $\boldsymbol{\mu}_i$ is the vector of population means, $\boldsymbol{\Sigma}_i$ is the population covariance matrix, and K_i is a constant that depends on the number of observed values for case i. In the missing data context the size and content of the arrays in Equation 1 depend on the number of observed values for case i.

To illustrate, consider a simple situation with three observed variables: X_1, X_2, and X_3. For cases missing X_2, these arrays contain elements corresponding to X_1 and X_3 as follows,

$$\boldsymbol{\mu}' = \begin{bmatrix}\mu_1 & \mu_3\end{bmatrix} \text{ and } \boldsymbol{\Sigma} = \begin{bmatrix}\sigma_{11} & \\ \sigma_{31} & \sigma_{33}\end{bmatrix},$$

and the contribution to the sample log-likelihood for these cases is.

$$\log L_i = K_i - \frac{1}{2}\log\begin{vmatrix}\sigma_{11} & \sigma_{13} \\ \sigma_{31} & \sigma_{33}\end{vmatrix} - \frac{1}{2}\left(\begin{bmatrix}X_1 \\ X_3\end{bmatrix} - \begin{bmatrix}\mu_1 \\ \mu_3\end{bmatrix}\right)\begin{bmatrix}\sigma_{11} & \sigma_{13} \\ \sigma_{31} & \sigma_{33}\end{bmatrix}^{-1}\left(\begin{bmatrix}X_1 \\ X_3\end{bmatrix} - \begin{bmatrix}\mu_1 \\ \mu_3\end{bmatrix}\right).$$

In a similar vein, the arrays for cases missing X_1 will contain elements corresponding to X_2 and X_3 as follows:

$$\boldsymbol{\mu}' = \begin{bmatrix}\mu_2 & \mu_3\end{bmatrix} \text{ and } \boldsymbol{\Sigma} = \begin{bmatrix}\sigma_{22} & \\ \sigma_{32} & \sigma_{33}\end{bmatrix}.$$

Estimates of the saturated and latent variable model are obtained by maximizing the sample log-likelihood (i.e., log L, the sum of the log L_i), but in the latter case $\boldsymbol{\mu}_i$ and $\boldsymbol{\Sigma}_i$ are functions of the hypothesized model's parameters. A χ^2 test of model fit is obtained by taking the difference between -2 log L values from the saturated and latent variable models, with degrees of freedom equal to the difference in the number of estimated parameters.

Some final points should be made regarding FIML. First, missing values are not imputed. Rather, parameter estimates and standard errors are estimated directly from the observed data by applying iterative computational algorithms to the sample log-likelihood. Like complete-data maximum likelihood, Equation 1 is derived from the multivariate normal density, so the usual normality concerns are applicable. Also, when given the option, FIML standard errors should be estimated using the observed, rather than expected, information matrix, as the latter is only appropriate under MCAR, and may produce biased standard errors under MAR (Kenward & Molenberghs, 1998). Finally, FIML parameter estimates are unbiased and efficient under MAR (this also includes the more stringent MCAR condition). Interested readers can consult Arbuckle (1996) for additional details on FIML estimation.

A Note on Non-normality

A considerable body of research has documented the impact of non-normality on SEM estimation with complete data, and three points have consistently emerged: (1) parameter estimates are generally unbiased, (2) standard error estimates are negatively biased, and (3) the model fit (i.e., χ^2) statistic yields excessive rejection rates. Enders (2001) examined the impact of non-normality on FIML estimation and, not surprisingly, replicated these three findings under MCAR and MAR mechanisms.

With complete data, it is known that robust standard errors and rescaled test statistics perform well when normality assumptions are violated (see Finney and DiStefano, Chapter 9, this volume). These rescaling methods have been extended to the missing data context by Yuan and Bentler (2000), and are currently available in Mplus (version 2.1 or higher) and EQS (version 6 or higher; Bentler & Wu, 2002). While ML estimation assumes MAR, robust standard errors for missing data require the more stringent MCAR condition. However, limited simulation results suggest these standard errors are relatively accurate under MAR (Enders, 2001).

Bootstrapping is another corrective procedure that has received attention in the SEM literature. As described in Chapter 9 in this volume, two forms of the bootstrap are applicable to SEM, the naïve bootstrap and the Bollen-Stine bootstrap (the former is appropriate for estimating standard errors, while the latter can be used to obtain standard errors or an adjusted *p*-value for the model fit statistic). Enders (2002) extended the Bollen-Stine bootstrap to the missing data context, and simulation results suggest this procedure effectively controls the Type I error inflation associated with non-normal data (Enders, 2001, 2002), although the procedure is conservative under small *N*s. Finally, Enders (2001) reported that standard error estimates obtained from the naive bootstrap were relatively accurate under MAR. The most recent release of Mplus (version 3) provides bootstrap estimates of standard errors for missing data, and custom

SAS programs for performing both forms of the bootstrap can be downloaded at http://www.asu.edu/clas/psych/people/faculty/enders.htm, or obtained from the author.

Incorporating Auxiliary Variables

Graham (2003) proposed two methods for including auxiliary variables into a FIML analysis: the "extra DV model" and the "saturated correlates" model. Because Graham's simulation results seem to favor the saturated correlates model, I limit the discussion to this approach. Graham outlined three formal rules for incorporating auxiliary variables into an SEM analysis: an auxiliary variable must be modeled so that it is permitted to be (1) correlated with all other auxiliary variables, (2) correlated with any exogenous manifest (i.e., observed) variable, and (3) correlated with the residual of any manifest indicator (i.e., an indicator of a latent variable) or endogenous manifest variable. Note that Graham's rules are general, and can be applied to a variety of models (e.g., SEMs, path models, multiple regression models, etc.).

Returning to the CFA of the Eating Attitudes Test, it is quite straightforward to incorporate BMI as an auxiliary variable. A graphical depiction of the saturated correlates model for this example is shown in Figure 10.2. Following Graham's rules, BMI scores are correlated with the residual terms for each of the eight manifest indicator variables (i.e., rule "c" above). While this is an oversimplified example, additional auxiliary variables could be readily be added by invoking rules "a" and "c" above (but not "b," as there are no other exogenous manifest variables within the model).

Example Analysis

The use of FIML is largely transparent to the user, and requires only minor modifications to the basic syntax of commercial SEM software packages. One goal of this chapter is to stress the importance of incorporating auxiliary variables into missing data analyses, so all FIML analyses illustrated here incorporate BMI as an auxiliary variable using Graham's (2003) saturated correlates approach. A graphical depiction of this model is shown in Figure 10.2, and the corresponding syntax for Mplus, EQS, and LISREL (Jöreskog, Sörbom, Du Toit, & Du Toit, 2001) is given in the Appendices. Using the AMOS graphical interface, one would simply draw the diagram shown in Figure 10.2, select "Analysis Properties" from the "View/Set" pull-down menu, then select "Estimate means and intercepts"

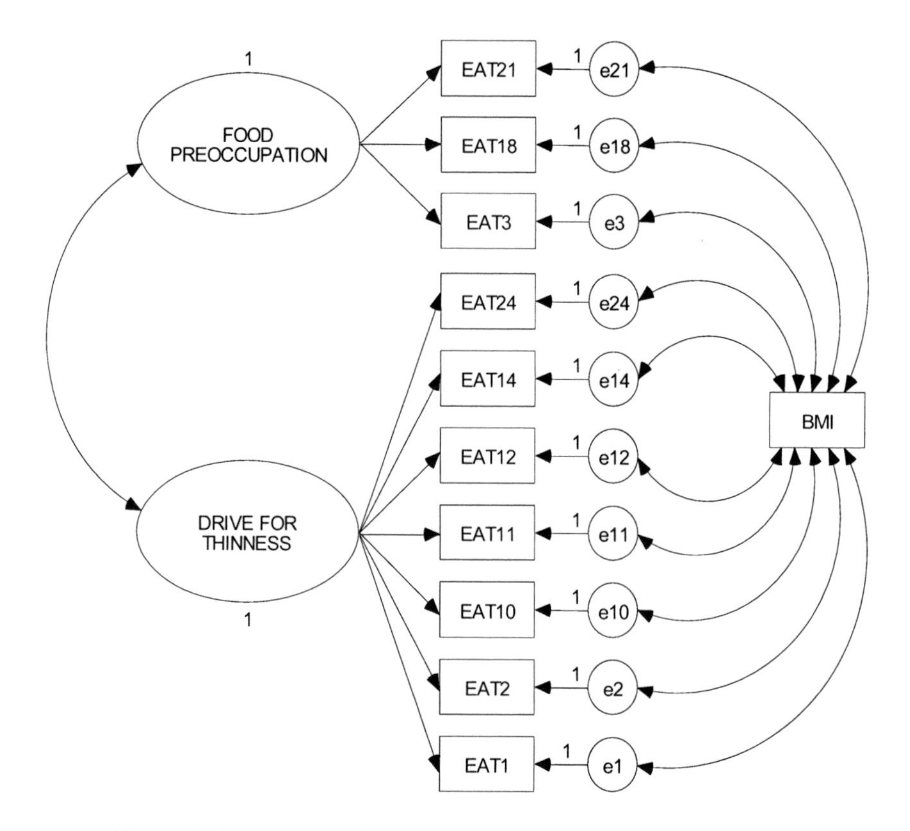

Figure 10.2. Graham's (2003) "saturated correlates" model.

from the "Estimation" tab and "Observed information matrix" from the "Output" tab.

Results from the FIML analyses are shown in Table 10.1. As would be expected, FIML parameter estimates were nearly identical across software packages. LISREL currently estimates standard errors using the expected, rather than observed, information matrix, but standard errors were trivially different in this example. Although not shown in the table, fit indices generally suggested adequate model fit, CFI = .968, TLI = .948, RMSEA = .069, and SRMR = .037. The log-likelihood values for the saturated and latent variable models were -2954.428 and -2987.422, respectively, so the test of model fit was obtained as $\chi^2(34) = 65.99$, $p = .001$. The degrees of freedom value for this test is simply the difference between the number of estimated parameters for these two models; the saturated model had 77 estimated parameters (66 unique covariance matrix elements and 11 means), and the CFA model had 43 estimated parameters

(10 loadings, 10 intercepts, 10 residual variances, 10 error covariances, 1 factor correlation, 1 mean, and 1 variance).

A brief discussion of estimation options is warranted, if only to underscore the need to examine software options thoroughly before choosing an analysis platform. At the time of this writing, all commercial SEM packages offer FIML estimation, but differences do exist among packages. For example, Mplus (version 2.1 and higher) and EQS (version 6.1) offer rescaled test statistics and robust standard errors for missing data, while LISREL (version 8.54) and AMOS (versions 4 and 5) do not. Also, Mplus, EQS, and AMOS can compute standard errors using either the expected or observed information matrix (as noted previously, the latter is preferred), while LISREL does not currently offer such an option.

SEM packages currently differ in the number of fit indices reported with missing data, but this is likely to change rapidly as new software is released. While there is no reason to expect fit indices to differ across packages, one important exception is worth noting. In AMOS version 4, comparative fit indices are computed using a null model where the means are constrained to zero. In many SEM applications, the specification of a mean structure in the missing data likelihood is a technical, rather than substantive, necessity, and will produce a gross misspecification of the null model when manifest indicators have nonzero means. The end result of this misspecification is an inflation of the comparative fit indices—models with missing data will appear to fit very well! AMOS 5 no longer incorporates this constraint in the null model, so comparative fit indices will be consistent with those obtained from other software packages. Users who wish to "correct" this problem can (a) center the manifest indicators at their respective means, (b) manually estimate an alternate null model where means are not constrained to zero and compute the comparative fit indices by hand, or (c) download a fit index calculator at http://www.amos-development.com/SemFitCalculator.

MULTIPLE IMPUTATION

Multiple imputation (MI) creates *m* copies of the sample dataset, each of which is imputed with different estimates of the missing values. The SEM (or any other analysis) of interest is subsequently fit to the *m* complete datasets, and parameter estimates and standard errors are obtained from each. A final set of parameter estimates and standard errors is subsequently obtained by combining the *m* sets of estimates using arithmetic rules outlined by Rubin (1987). Different MI algorithms have been proposed in the literature, but Schafer's (1997) data augmentation procedure is the approach outlined here.

Like EM, data augmentation involves a two-step (*imputation* and *posterior*) iterative procedure. The imputation (I) step entails random draws of the missing values (conditional on the current covariance matrix), while the posterior (P) step involves random draws of the parameters (conditional on the imputed values). The process begins with a preliminary estimate of the covariance matrix, typically obtained via the EM algorithm. In the first I step, a series of regression equations is computed from the EM covariance matrix, and missing values are filled in with predicted scores from these equations. Because the predicted values fall directly on a regression surface, residual variation is restored to the imputed data points by adding a random draw from the normal distribution of residuals for a given variable. The covariance matrix and mean vector are subsequently computed from the imputed data using standard formulae. Second, in the P step, Bayesian methods are used to construct a new estimate of the population covariance matrix by randomly sampling elements from a posterior distribution based on the means and covariances computed at the previous I step. Sampling from the posterior distribution requires the specification of a prior distribution for the means and covariances. According to Schafer (1997, p. 154), a noninformative (i.e., Jeffreys) prior distribution that contains little or no information about the parameters can be used for this purpose (e.g., a normal distribution might be used as a prior distribution for the means). The new covariance matrix elements sampled from the P step are carried forward to the next imputation step, where a new set of regression equations is constructed, and the cycle is repeated.

An important way that MI differs from EM and FIML is in the definition of convergence. EM and FIML parameter estimates converge to a single set of values, so convergence is attained when parameter estimates no longer change from one iteration to the next. In contrast, MI data augmentation converges when the *distribution* of parameter estimates (means and covariances) is stable across P steps. This definition of convergence is quite different, as the covariance matrix elements (and thus the regression equations used to impute missing values) continue to vary randomly across iterations, even after convergence is attained.

Conducting a preliminary EM analysis is often a good way to informally gauge the convergence behavior of the MI data augmentation process, because data augmentation generally converges more quickly than EM (Schafer, 1997). As a conservative guideline, it is frequently suggested that the number of data augmentation iterations should be at least twice as large as the number of cycles required for EM to converge (i.e., the "2 ×

EM iterations" guideline). An initial EM analysis is also useful because it provides start values for the first data imputation step.

MI convergence can be assessed more formally by examining the serial dependence of the means and covariances drawn from the posterior distribution at each P step. The degree of serial dependence is assessed by saving the means and covariances at each step, and using these values to construct autocorrelation function (ACF) and time series plots. For any given parameter, the lag-k autocorrelation is the Pearson correlation between values separated by k iterations of the data augmentation procedure. For example, if 100 iterations of data augmentation were performed, the lag-3 autocorrelation would be obtained by correlating parameter values from iterations 1–97 with those from iterations 4–100 (i.e., the correlation between parameter values separated by 3 iterations). Convergence is assessed by graphing autocorrelation values, and identifying the lag k at which the autocorrelation drops to within sampling error of zero. Time series plots display the value of a given parameter at each iteration (the iterations are displayed on the horizontal axis, and the corresponding parameter estimate is plotted on the vertical axis). Convergence is attained for a given parameter if the plotted values form a horizontal band that doesn't vertically drift or wander. Example plots were generated from the example analysis presented later in this section.

After identifying the number of data augmentation steps required for convergence, the final step is to create the m imputed datasets. This can be accomplished by running separate data augmentation chains of k iterations for each of the m imputed datasets (i.e., parallel), or running a single chain where an imputed dataset is created after every kth iteration (i.e., sequential). When using the sequential approach, it is important that the m datasets be statistically independent of one another (i.e., represent random draws), and this is accomplished by allowing a sufficient number of data augmentation cycles to lapse between the creation of each imputed dataset.

Some final points are important regarding MI. First, like EM and FIML, MI relies on the multivariate normal model. However, there is evidence to suggest that MI performs well even when data are non-normal (e.g., Graham & Schafer, 1999), and variables can be transformed prior to imputation to enhance MI performance further. Schafer (1997) argued that the normal model can also be used with nominal and ordinal variables. MI software packages (e.g., NORM, SAS PROC MI) offer the user a number of options that are useful in this regard, including transformations, dummy coding, rounding (for the imputation of discrete values), and the ability to specify minimum and maximum values for imputed variables, to name a few. Finally, MI will produce unbiased parameter estimates under an MAR mechanism.

Incorporating Auxiliary Variables

MI imputes missing values from an unstructured covariance structure (i.e., values are imputed using the implied covariance matrix from a saturated model), so auxiliary variables can readily be incorporated into the imputation process. Schafer (1997, p. 143) provided a discussion of this issue, and suggested that the imputation model should include variables that are related to the variable being imputed, and potentially related to missingness on that variable. Also, the model should be at least as general as the model used to address one's substantive hypotheses (e.g., if the analysis model includes an interaction, the imputation model should as well). In the CFA example, the MI procedure would incorporate 11 variables (10 manifest indicators plus BMI scores). However, the subsequent CFA model need not include BMI, as the imputed data points are already conditioned on this variable.

Example Analysis

Schafer's (1999) NORM program was used to create $m = 10$ multiply imputed datasets. To illustrate the inclusion of auxiliary variables, the imputation model included the set of 10 manifest indicators as well as BMI scores. Because the CFA modeled the linear relations among the variables, no additional terms (e.g., interactions, nonlinear terms) were included in the imputation model. Consistent with Schafer's (1997, p. 148) recommendations, imputed values were rounded to the nearest integer, as this was consistent with the original metric of the manifest indicators (imputed BMI scores would not be rounded, had they been missing).

Space limitations preclude a detailed overview of NORM, but a brief description of the package is warranted at this point. NORM is a standalone freeware package for Windows that offers a straightforward point-and-click interface that requires no syntax. NORM offers the user a variety of options, including the ability to perform EM analyses, apply transformations, dummy code categorical variables, and specify formats for imputed values, to name a few. Text data are easily imported using space-delimited free format, provided that a missing value code is used (e.g., -9). Interested readers are encouraged to consult Schafer and Olsen (1998) for a more detailed illustration of the NORM package. The same data augmentation algorithm implemented in NORM is also available in SAS PROC MI (SAS version 8.2 and higher).

An EM analysis was used as an initial gauge of MI convergence behavior. EM converged in 20 iterations, which suggests that data augmentation might converge in fewer iterations (Schafer, 1997). To assess convergence more formally, 200 iterations of data augmentation were performed (this number was chosen somewhat arbitrarily). The elements of $\boldsymbol{\mu}$ and $\boldsymbol{\Sigma}$ were saved after every iteration, and these values were used to generate time series and ACF plots for every parameter (NORM saves parameter estimates and creates these plots upon request). The ideal time series plot forms a horizontal band that does not vertically wander or drift, and this was the case for all simulated parameters. Consistent with expectations, the ACF plots suggested that convergence was attained (i.e., the autocorrelations reduced to within sampling error of zero) after only a few iterations. To illustrate, Figure 10.3 displays the time series and ACF plots for the worst linear function (WLF) of the parameters, a summary value based on the elements in $\boldsymbol{\mu}$ and $\boldsymbol{\Sigma}$ that converged most slowly (Schafer, 1997, pp. 129–131). Both plots in Figure 10.3 indicate that convergence was attained.

Finally, $m = 10$ imputed datasets were created using a sequential data augmentation chain. Although it appeared that MI converged rather quickly, I abandoned the "2 × EM iterations" guideline in favor of a more conservative approach whereby an imputed dataset was created after every 100th cycle of the data augmentation chain, beginning with the 200th iteration (i.e., a "burn-in" period of 200 iterations was used). Given the convergence diagnostics, it appeared that the distribution of parameters would likely stabilize after 200 iterations, and allowing 100 iterations to lapse between each imputed dataset would ensure that the imputations were statistically independent (i.e., mimicked independent random draws). Up to this point, there is nothing in the MI procedure that is unique to the SEM context. In fact, the resulting datasets could be used for any number of analyses (e.g., compute internal consistency reliability estimates, scale score means, SEM analyses, etc.).

Continuing with the heuristic analysis, the two-factor CFA model shown in Figure 10.1 was fit to each of the $m = 10$ imputed datasets using Mplus. Unlike the FIML analyses, it was not necessary to include BMI in the CFA model, as the imputed values were already conditioned on BMI scores. This series of analyses resulted in 10 sets of parameter estimates and standard errors that were subsequently combined into a single set of estimates using Rubin's (1987) rules.

The MI parameter estimate is simply the arithmetic average of that parameter across the m analyses. That is,

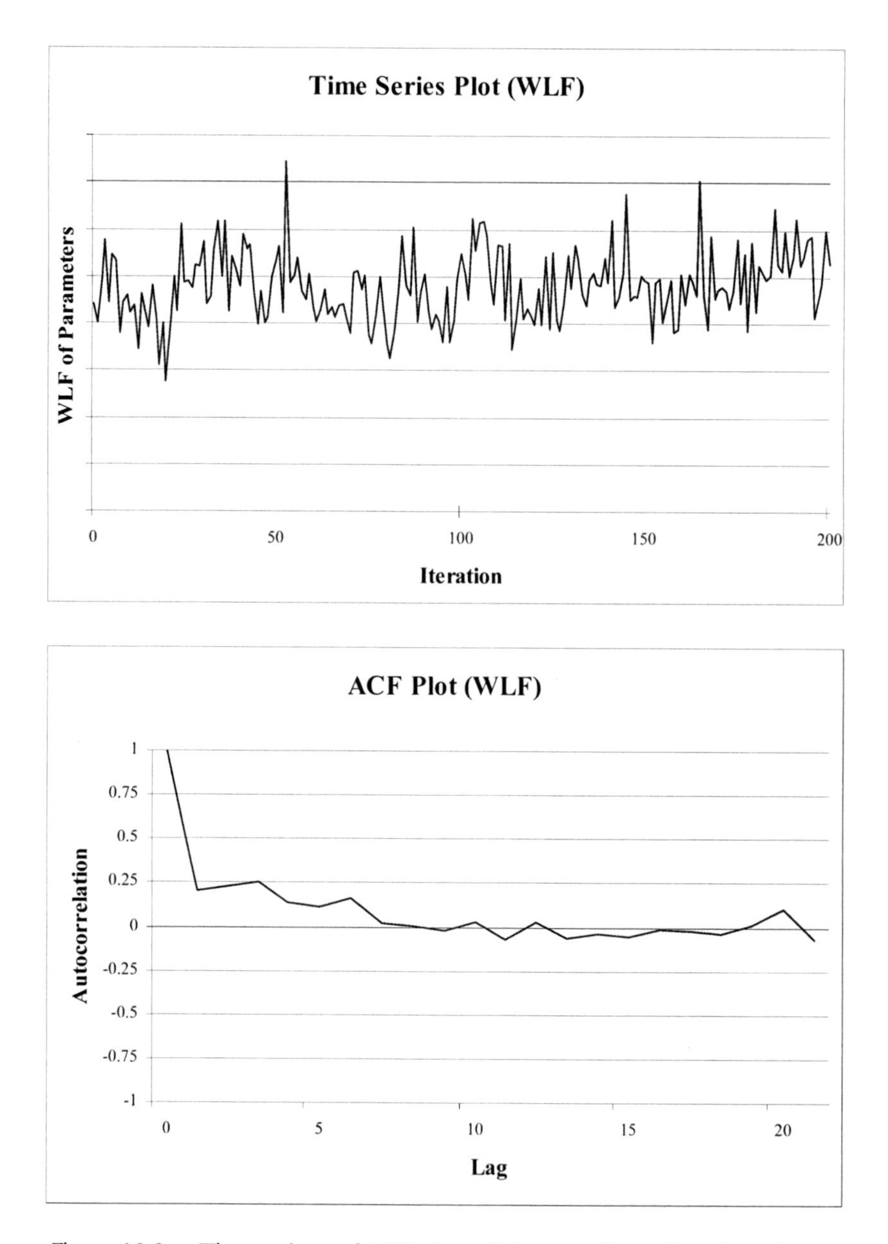

Figure 10.3. Time series and ACF plots of the worst linear function of parameters.

$$\overline{Q} = \frac{1}{m}\sum_{i=1}^{m}\hat{Q}_i, \quad (2)$$

where m is the number of imputations and $\hat{Q}_i$ is the parameter estimate from the ith imputed dataset. To illustrate the application of Rubin's rules, the set of parameter estimates and standard errors for the EAT1 loading are given in Table 10.2. The MI estimate of the EAT1 loading was obtained by averaging the 10 loading values in Table 10.2, which resulted in an estimate of 1.163. Other model parameters could be combined in a similar fashion.

Combining a set of standard errors is only slightly more complicated than combining point estimates. The within-imputation variance for a given parameter is defined as the average of its m variance estimates (i.e., squared standard errors), or

$$\overline{U} = \frac{1}{m}\sum_{i=1}^{m}\hat{U}_i, \quad (3)$$

where $\hat{U}_i$ is the variance estimate from the ith imputed dataset, and m is the number of imputations. From Table 10.2, the arithmetic average of the $m = 10$ variance estimates is .0113. The between-imputation variance is the variance of the parameter estimate itself across the m imputations, or

Table 10.2. Multiple Imputation Estimates for the EAT1 Factor Loading

m	b_{V1F1}	*S.E.*	*Var.*
1	1.190	.105	.0111
2	1.214	.108	.0116
3	1.090	.105	.0110
4	1.182	.109	.0119
5	1.121	.105	.0109
6	1.202	.105	.0111
7	1.189	.105	.0110
8	1.153	.109	.0120
9	1.165	.105	.0110
10	1.128	.107	.0113

$$B = \frac{1}{m}\sum_{i=1}^{m}\left(\hat{Q}_i - \overline{Q}\right)^2 . \tag{4}$$

From Table 10.2, the between-imputation variance of the regression weights was .0016. Finally, the total variance associated with a given parameter estimate is a function of the within- and between-variance estimates,

$$T = \overline{U} + \left(1 + \frac{1}{m}\right)B . \tag{5}$$

Using the within- and between-imputation variance components calculated previously, $T = .013$. Thus, the MI standard error estimate for this parameter is simply $\sqrt{T}$, or .114.

A custom program could easily be written to perform these computations, or parameter estimates and standard errors could be combined using facilities in NORM or SAS (i.e., PROC MIANALYZE). The most recent release of Mplus (version 3) also includes a facility for analyzing and combining parameter estimates from multiply imputed datasets, the syntax for which is given in Appendix D. Two features of this program are worth noting, both of which are specified in the DATA command. First, an analysis of multiply imputed datasets is requested using the TYPE = IMPUTATION option. Second, the file named in the FILE option is not a data file, but is a text file containing the names of the imputed datasets to be analyzed. The resulting Mplus output file contains the combined parameter estimates and standard errors described above.

MI parameter estimates and standard errors are given in Table 10.1. Collins and colleagues (2001) suggested that FIML and MI will tend to produce similar results when using the same input data, even when the MI model is more general (i.e., has more parameters); in this case, the MI model is more general because the imputations were created under a saturated covariance structure. While FIML and MI produced nearly identical results in this case, this may not always be true. Collins and colleagues also suggested that MI standard errors will tend to be somewhat larger than those of FIML, and this was generally true for the indicators with missing data, although the difference was slight. More research is needed to delineate situations where these two approaches differ.

The assessment of model fit when using MI has not received attention in the SEM literature, and the development of rules for combining model fit indices would certainly be an important contribution. It is fairly straightforward to combine likelihood ratio χ^2 values (or Wald tests) using an approximate method proposed by Li, Meng, Raghunathan, and Rubin (1991). A more precise method that utilizes the covariance matrix of parameter estimates is also available, but is not as computationally attractive. The test statistic outlined by Li and colleagues uses an F reference distribution, and is computed as a function of the number of imputations, the m model fit statistics, and the model degrees of freedom. Returning to the example analysis, the $m = 10$ model fit statistics were combined to produce $F(34, 229.08) = 2.00, p = .002$. Li and colleagues cautioned that this procedure is an approximation to the true probability, which could range between one half and twice as large as the obtained p. Given the approximate nature of this test and the fact that the model fit statistic has been downplayed in favor of myriad fit indices, the practical utility of combining multiply imputed χ^2 statistics into a single inference seems questionable at this point. A SAS program for combining likelihood ratio χ^2 statistics can be downloaded at http://www.asu.edu/clas/psych/people/faculty/enders.htm, or obtained from the author.

SUMMARY AND CONCLUSIONS

The primary goals of this chapter were to summarize recent methodological advances in missing data analyses, and to provide the reader with a thorough overview of the analytic options for conducting SEM analyses using "modern" missing data approaches. To this end, three approaches, EM, FIML, and MI, were described in some detail, and a heuristic analysis was conducted to demonstrate their use. Given that the heuristic analyses produced relatively trivial differences among the three approaches, the reader might be left wondering, "Which approach should I use?" To answer this question better, I conclude with a brief summary of the pros and cons associated with each approach.

The sufficient statistics required by many multivariate procedures are contained within the covariance matrix and mean vector, so EM provides a straightforward approach for obtaining an input matrix for a variety of analyses. The ease with which auxiliary variables can be added to the EM analysis is also beneficial. Perhaps the biggest drawback associated with using EM is the specification of a single N. Somewhat surprisingly, Enders and Peugh (2004) found that this does not seem to affect standard error estimates as much as it does the model χ^2 statistic. Although they did not examine the behavior of model fit indices, it seems safe to conclude that certain fit statistics would be similarly affected. Given the problems associated with assessing model fit, there appears to be no compelling advantage to EM if one is willing to incorporate auxiliary variables into an FIML analysis using Graham's (2003) saturated correlates approach.

FIML has the advantage of being widely available in all commercial SEM packages. The method is straightforward to use, and is virtually transparent to the user. Aside from requiring raw data, perhaps the most noticeable consequence of using FIML is the estimation of parameters that might not normally appear in a latent variable model (i.e., indicator intercepts and manifest variable means). However, these parameters can simply be ignored if they are not of substantive interest. The incorporation of auxiliary variables is a potential drawback for FIML estimation. Variables that satisfy the MAR assumption may not appear in the substantive model, yet unbiased estimation requires the inclusion of these variables. However, with some extra work, Graham's (2003) saturated correlates approach can be used to incorporate auxiliary variables into the estimation process without affecting the parameters of interest. As a final caveat, it should be noted that FIML estimation may not be available for certain classes of models. However, the number and types of models that can be estimated using this approach has grown substantially since Arbuckle's (1996) chapter, and will likely continue to grow (e.g., a FIML estimator is not available for logistic regression models in Mplus version 2, but is available in version 3).

One feature that makes MI very attractive is its flexibility. Because the imputation process is not model-specific (i.e., imputation is based on a saturated covariance structure), the same set of imputed datasets can be used for any number of analyses. As such, the use of MI would allow one to estimate models that are not yet amenable to FIML estimation. The ease of including auxiliary variables is certainly another benefit of MI, and it appears that liberal use of such variables is not problematic. As noted previously, the imputed data points are conditioned on the auxiliary variables, so these variables need not be included in the model used to address one's substantive hypotheses. Despite the considerable benefits, there are downsides to using MI for SEM analyses. From the space devoted to the procedure in this chapter, one might (correctly) conclude that MI is relatively labor intensive. SEM analyses frequently entail the estimation of many parameters, and combining *m* sets of estimates and standard errors can be time consuming. However, the most recent release of Mplus (version 3) includes a facility for analyzing and combining parameter estimates from multiply imputed datasets, so this phase of the analysis can be completed quite quickly. Finally, although likelihood ratio χ^2 statistics can be combined into a single inference, many SEM users may be troubled by the fact that no rules exist for combining fit indices from multiply imputed datasets.

In sum, the use of ML estimation (either FIML or an indirect approach using an EM input matrix) and MI will, in many cases, produce similar results. Given identical assumptions (i.e., multivariate normality and MAR), the main considerations appear to be the (a) availability of FIML estimation for the latent variable model of interest, (b) ease of incorporating auxiliary variables, (c) assessment of model fit, and (d) ease of use. A wealth of computer simulation studies suggest that ML and MI are superior to the traditional missing data approaches outlined earlier (e.g., Arbuckle, 1996; Enders, 2001; Enders & Bandalos, 2001; Graham & Schafer, 1999; Muthén et al., 1987), both with respect to bias and efficiency, so choosing among the "modern" techniques is probably more a matter of personal preference and convenience.

APPENDIX A: MPLUS SYNTAX FOR THE EXAMPLE ANALYSIS

```
TITLE:

Eating Attitudes Test Saturated Correlates CFA Model With Missing Data;

DATA:

file is c:\eatexample.txt;

VARIABLE:

names are eat1 eat2 eat3 eat10 eat11 eat12 eat14 eat18 eat21 eat24 bmi;
usevariables are eat1 – bmi;
missing are all (-9);

ANALYSIS:

type = h1 missing;
estimator = ml;
information = observed;

MODEL:

driveft by eat1* eat2* eat10* eat11* eat12* eat14* eat24*;
foodpre by eat3* eat18* eat21*;
driveft@1 foodpre@1;
driveft with foodpre*;
bmi with eat1 - eat24*;
```

```
OUTPUT:

sampstat standardized tech5;
```

APPENDIX B: SIMPLIS SYNTAX FOR THE EXAMPLE ANALYSIS

```
Observed Variables: eat1 eat2 eat3 eat10 eat11 eat12 eat14 eat18 eat21
eat24 bmi
Missing Value Code -9
Sample Size: 200
Raw Data from File c:\eatexample.txt
Latent Variables: driveft foodpre

Relationships:
  eat1  = const + driveft
  eat2  = const + driveft
  eat10 = const + driveft
  eat11 = const + driveft
  eat12 = const + driveft
  eat14 = const + driveft
  eat24 = const + driveft
  eat3  = const + foodpre
  eat18 = const + foodpre
  eat21 = const + foodpre
  bmi   = const

Let the errors of bmi and eat1 - eat24 correlate
Set the error variances of eat1 - eat24 free
Set the variance of bmi free
Set the variance of driveft to 1
Set the variance of foodpre to 1

Options: sc
Path Diagram
End of Problem
```

APPENDIX C: EQS SYNTAX FOR THE EXAMPLE ANALYSIS

```
/TITLE
 Eating Attitudes Test Saturated Correlates CFA Model With Missing Data
/SPECIFICATIONS
 DATA= 'C:\Eat Example Data.ESS';
 VARIABLES=11; CASES=200;
 METHOD=ML; ANALYSIS=MOMENT; MATRIX=RAW;
 MISSING=ML; SE=OBSERVED;
/LABELS
 V1=EAT1; V2=EAT2; V3=EAT3; V4=EAT10; V5=EAT11;
 V6=EAT12; V7=EAT14; V8=EAT18; V9=EAT21; V10=EAT24;
 V11=BMI;
/EQUATIONS
 V1 = *V999 + *F1 + E1;
 V2 = *V999 + *F1 + E2;
 V3 = *V999 + *F2 + E3;
 V4 = *V999 + *F1 + E4;
 V5 = *V999 + *F1 + E5;
 V6 = *V999 + *F1 + E6;
 V7 = *V999 + *F1 + E7;
 V8 = *V999 + *F2 + E8;
 V9 = *V999 + *F2 + E9;
 V10 = *V999 + *F1 + E10;
 V11 = *V999 + E11;
 F1 = D1;
 F2 = D2;
/VARIANCES
 V999 = 1;
 E1 = *;
 E2 = *;
 E3 = *;
 E4 = *;
 E5 = *;
 E6 = *;
 E7 = *;
 E8 = *;
 E9 = *;
 E10 = *;
 E11 = *;
 D1 = 1;
 D2 = 1;
/COVARIANCES
```

```
 E11,E1 = *;
 E11,E2 = *;
 E11,E3 = *;
 E11,E4 = *;
 E11,E5 = *;
 E11,E6 = *;
 E11,E7 = *;
 E11,E8 = *;
 E11,E9 = *;
 E11,E10 = *;
 D2,D1 = *;
/PRINT
 FIT=ALL;
 TABLE=EQUATION;
/END
```

APPENDIX D: MPLUS SYNTAX FOR ANALYZING AND COMBINING MULTIPLY IMPUTED DATASETS

```
TITLE:

Analyze and combine estimates from 10 imputed files;

DATA:

file is c:\imputedfiles.txt;
type = imputation;

VARIABLE:

names are eat1 eat2 eat3 eat10 eat11 eat12 eat14 eat18 eat21 eat24 bmi;
usevariables are eat1 - eat24;

ANALYSIS:

estimator = ml;

MODEL:

driveft by eat1* eat2* eat10* eat11* eat12* eat14* eat24*;
foodpre by eat3* eat18* eat21*;
driveft@1 foodpre@1;
driveft with foodpre*;
```

REFERENCES

Allison, P. D. (2002). *Missing data.* Thousand Oaks, CA: Sage.

Arbuckle, J. L. (1996). Full information estimation in the presence of incomplete data. In G. A. Marcoulides & R.E. Schumacker (Eds.), *Advanced structural equation modeling* (pp. 243–277). Mahwah, NJ: Erlbaum.

Arbuckle, J. L. (2003). *Amos 5.0 update to the Amos user's guide* [Computer software and manual]. Chicago: SmallWaters.

Bentler, P. M., & Wu, E. J. C. (2002). *EQS 6 for Windows user's guide* [Computer software and manual]. Encino, CA: Multivariate Software.

Brown, R. L. (1994). Efficacy of the indirect approach for estimating structural equation models with missing data: A comparison of five methods. *Structural Equation Modeling: AnInterdisciplinary Journal, 4,* 287–316.

Collins, L. M., Schafer, J. L., & Kam, C.-H. (2001). A comparison of inclusive and restrictive strategies in modern missing data procedures. *Psychological Methods, 6,* 330–351.

Dempster, A. P., Laird, N. M., & Rubin, D. B. (1977). Maximum likelihood from incomplete data via the EM algorithm. *Journal of the Royal Statistical Society, Ser. B, 39,* 1–38.

Doninger, G. L., Enders, C. K., & Burnett, K. F. (2005). Validity evidence for Eating Attitude Test scores in a sample of female college athletes. *Measurement in Physical Education and Exercise Science, 9,* 35-50.

Enders, C. K. (2001). The impact of nonnormality on full information maximum likelihood estimation for structural equation models with missing data. *Psychological Methods, 6,* 352–370.

Enders, C. K. (2002). Applying the Bollen-Stine bootstrap for goodness-of-fit measures to structural equation models with missing data. *Multivariate Behavioral Research,* 37, 359–377.

Enders, C. K., & Bandalos, D. L. (2001). The relative performance of full information maximum likelihood estimation for missing data in structural equation models. *Structural Equation Modeling: A Multidisciplinary Journal, 8,* 430–457.

Enders, C. K., & Peugh, J. L. (2004). Using an EM covariance matrix to estimate structural equation models with missing data: Choosing an adjusted sample size to improve the accuracy of inferences. *Structural Equation Modeling: A Multidisciplinary Journal, 11,* 1–19.

Finkbeiner, C. (1979). Estimation for the multiple factor model when data are missing. *Psychometrika, 44,* 409–420.

Garner, D. M., Olmsted, M. P., Bohr, Y., & Garfinkel, P. E. (1982). The Eating Attitudes Test: Psychometric features and clinical correlates. *Psychological Medicine, 12,* 871–878.

Graham, J. W. (2003). Adding missing-data relevant variables to FIML-based structural equation models. *Structural Equation Modeling: A Multidisciplinary Journal,* 10, 80–100.

Graham, J. W., Hofer, S. M., Donaldson, S. I., Mackinnon, D. P., & Schafer, J. L. (1997). Analysis with missing data in prevention research. In K. J. Bryant & M. Windel (Eds.), *The science of prevention: Methodological advances from alcohol and substance abuse research* (pp. 325–366). Washington, DC: American Psychological Association.

Graham, J. W., & Schafer, J. L. (1999). On the performance of multiple imputation for multivariate data with small sample size. In Rick H. Hoyle (Ed), *Statistical strategies for small sample research* (pp. 1–29). Thousand Oaks, CA: Sage.

Jöreskog, K. G., & Sörbom, D. (1999). *PRELIS 2: User's reference guide* [Computer software and manual]. Chicago: Scientific Software.

Jöreskog, K. G., Sörbom, D., Du Toit, S., & Du Toit, M. (2001). *LISREL 8: New statistical Features* [Computer software and manual]. Chicago: Scientific Software.

Kenward, M. G., & Molenberghs, G. (1998). Likelihood based frequentist inference when data are missing at random. *Statistical Science, 13*, 236–247.

Li, K. H., Meng, X. L., Raghunathan, T. E., & Rubin, D. B. (1991). Significance levels from repeated p-values with multiply-imputed data. *Statistica Sinica, 1*, 65–92.

Little, R. J. A. (1988). A test of missing completely at random for multivariate data with missing values. *Journal of the American Statistical Association, 83*, 1198–1202.

Little, R. J. A., & Rubin, D. B. (2002). *Statistical analysis with missing data* (2nd ed.). Hoboken, NJ: Wiley.

Muthén, B., Kaplan, D., & Hollis, M. (1987). On structural equation modeling with data that are not missing completely at random. *Psychometrika, 52*, 431–462.

Muthén, L. K., & Muthén, B. O. (2004). *Mplus user's guide* [Computer software and manual]. Los Angeles: Authors.

Rubin, D. B. (1976). Inference and missing data. *Biometrika, 63*, 581–592.

Rubin, D. B. (1987). *Multiple imputation for nonresponse in surveys*. New York: Wiley.

Schafer, J. L. (1999). *NORM: Multiple imputation of incomplete multivariate data under anormal model* [Computer software]. University Park: Department of Statistics, The Pennsylvania State University.

Schafer, J. L. (1997). *Analysis of incomplete multivariate data*. New York: Chapman & Hall.

Schafer, J. L., & Graham, J. W. (2002). Missing data: Our view of the state of the art. *Psychological Methods, 7*, 147–177.

Schafer, J. L., & Olsen, M. K. (1998). Multiple imputation for multivariate missing data problems: A data analyst's perspective. *Multivariate Behavioral Research, 33*, 545–571.

Wothke, W. (2000). Longitudinal and multi-group modeling with missing data. In T. D. Little, K. U. Schnabel, & J. Baumert (Eds.), *Modeling longitudinal and multiple group data: Practical issues, applied approaches and specific examples* (pp. 219–240). Mahwah, NJ: Erlbaum.

Yuan, K.-H., & Bentler, P. M. (2000). Three likelihood-based methods for mean and covariancestructure analysis with nonnormal missing data. In M. Becker & M. Sobel (Eds.), *Sociological methodology 2000* (pp. 165–200). Malden, MA: Blackwell.

CHAPTER 11

USING MULTILEVEL STRUCTURAL EQUATION MODELING TECHNIQUES WITH COMPLEX SAMPLE DATA

Laura M. Stapleton

As researchers become aware of the great wealth of large-scale datasets available at the national and international level (within the United States, for example, data are collected by the National Center for Education Statistics, the National Science Foundation, and the Center for Disease Control), questions of how to analyze such data will be frequent. The choice of analysis depends first and foremost on the research question being asked. Once decided, the researcher typically needs to utilize special techniques in order to estimate model parameters and standard errors appropriately. In this chapter I introduce the types of complex sample data that exist and illustrate the structural equation modeling (SEM) analysis tools that are available to researchers using such data. The focus is on analyzing multilevel models, but researchers should be aware that multilevel models answer specific research questions. Multilevel models allow the researcher to examine relations among variables within a nested structure (such as students within schools) as well as relations at a group level (in this case,

Structural Equation Modeling: A Second Course, 345–383

school). For other research questions, a single-level modeling approach may be appropriate and strategies for such modeling is briefly discussed.

The first section outlines the characteristics of complex samples and typical sampling methods are described. Consequences of naively analyzing complex sample data using traditional statistical methods that assume simple random sampling are then discussed. The next section introduces three different methods to model data from complex samples and the choice of method depends on the research question to be answered. Examples of research questions for each method are provided and example programs for the multilevel analyses are discussed in detail. The last section of the chapter points the reader toward outside resources for both applied examples of multilevel SEM analyses and methodological extensions of multilevel structural equation modeling.

REVIEW OF COMPLEX SAMPLING DESIGNS

National surveys and testing programs generally do not collect data based on simple random samples (SRSs). Longford (1995) and Skinner, Holt, and Smith (1989) provided reviews of the special characteristics of such complex sample designs, including clustering, stratification, unequal selection probability, and nonresponse and post-stratification adjustment. In general, each of these elements of complex sampling designs requires attention during the analysis stage, but it is difficult to assess the overall impact of the sampling design on the statistics of analysis when there are many elements that require accommodation. Presented below is a treatment of selected complex sampling characteristics and their individual effects on model parameter estimation at the analysis stage.

Cluster Sampling

An assumption in the use of traditional analysis techniques, such as single-level SEM, is that observations are independently sampled. However, most large data sets are collected using a cluster sampling technique—it is more efficient to collect data from 20 students in one school, for example, than to travel to 20 different schools to obtain data from one student at each. Technically, cluster sampling refers to the random selection of clusters and then subsequent inclusion of all individuals within the cluster into the sample. When only a portion of the elements in the cluster are included in the sample, the sampling strategy is referred to as *multistage* sampling. However, the phrase *cluster sampling* has taken on a connotation of both types of strategies and is used here. In cluster sampling, the first-stage unit (or the first level of selection) is termed the *primary sampling unit* (PSU). Two-stage cluster selection examples include random selection of schools and then the selection of students from each school (here, schools are the PSUs). Another example is random selection of classrooms in a school and then the sampling of all students in the selected classrooms. As a result of this cluster sampling, data usually have some degree of dependence among observations (e.g., students who are in the same classroom tend to be more like each other than like students in another classroom). However, formulas for calculating standard errors that are incorporated into most statistical packages are based on an SRS design where independent observations are assumed (Lee, Forthofer, & Lorimor, 1989). Because these formulas assume that the correlation of the residuals is zero, a researcher may underestimate the sampling variance, resulting in Type I error rates of unacceptably high levels. Dramatic examples in an analysis of variance context are available in Kish and Frankel (1974) and Scariano and Davenport (1987). Likewise, in an SEM context, bias due to violation of the assumption of independent observations has been discussed in the estimation of standard errors of parameters (Hox, 2002; Kaplan & Elliot, 1997a) and χ^2 statistics (Muthén & Satorra, 1995). An analysis of clustered data ignoring the dependencies in the data, therefore, may lead to the misidentification of statistically significant path coefficients where only random covariation exists and may lead to inappropriate rejection of hypothesized models.

Stratification

Stratification in the sampling design refers to the division of all elements in the population into mutually exclusive categories, where sampling is performed within each category. There are two reasons why researchers might choose a sample using stratification. First, a researcher might state, "I want to make sure I get a representative sample" and subsequently sample to ensure, for example, that the selected elements include some boys and some girls. Although the researcher might not realize it, the sample drawn was stratified by gender. This desire for representativeness is an informal expression of the second reason why stratified samples are taken. Stratification decreases the chances of getting a "bad" sample (the sample is guaranteed to be representative on the variable or variables chosen for stratification). Therefore, the statistical estimates from this sample will tend to have more precise standard errors than if the sample were collected based on a completely random process. This advantage assumes, however, that the stratification variable is correlated with the response variable (Kalton, 1983). No standard software for SEM currently has the ability to take advantage of this strata information and estimate the more precise standard errors; however, the strata can be modeled as fixed effects, either by incorporating dummy variables to represent strata or modeling the strata using a multisample approach (although this would be difficult as there are usually dozens of strata).

Any time sampling elements are stratified, a choice of whether to use *proportionate* or *disproportionate* sampling must be made. In proportionate sampling, the same selection probability is used in each strata. In disproportionate sampling, the researcher can choose to use different selection probabilities in each stratum. If disproportionate sampling is used, however, the use of sample weights is required to obtain unbiased parameter estimates in the analysis. When proportionate sampling is used, weights are not necessary. For example, suppose there are 1000 students in a population (500 girls and 500 boys) and a sample of 100 is taken, making sure that 50 of the students are girls and 50 of the students are boys. A sampling rate of .1 (50/500) was used in both the girls' group and the boys' group and, because this same sampling probability (π) was used within both strata, the use of sampling weights to adjust for differential selection probabilities is not required. In this case, the sampling technique is referred to as *proportionate stratified sampling*. However, if 100 boys were sampled and only 50 girls were sampled, then $\pi = .2$ (i.e., 100/500) for boys and $\pi = .1$ (i.e., 50/500) for girls. This process would be described as *disproportionate stratified sampling*. Strategies for analyzing these data are described in the next section.

Unequal Selection Probabilities

Elements in the population are associated with unequal selection probabilities when those elements are sampled at different rates across strata. (Also, elements can be selected with differential probabilities when using cluster sampling if clusters are of unequal sizes; however, most sampling techniques use *probability proportionate to size* sampling to avoid this situation.) Analyses ignoring these unequal inclusion probabilities can lead to biased parameter estimates when the probability of selection is correlated with the response variable. Kaplan and Ferguson (1999), Korn and Graubard (1995), and Lee and colleagues (1989) demonstrated, using SEM and regression techniques, the parameter estimate bias that can result when unequal probability of selection is ignored in traditional, single-level analyses. Often, the easiest way to incorporate unequal selection probabilities is to use design weights for observations. These weights can be built to reflect the original unequal sample inclusion probabilities, and help to compensate for differential response rates. As an example, assuming no nonresponse, if a particular observation's probability of selection into the sample is 1/10 ($\pi_i = .1$), the raw weight is the reciprocal of that probability: $w_i = 1/\pi_i = 10$ (each person in the sample represents 10 people in the population). The estimation of sampling variance when using weights is not invariant, however, with respect to constant multiplicative factors of weights, and so a choice of normalization for the weights w_i is essential (Longford, 1995). The use of a normalized weight such that the sum of the weights equals the sample size was shown to result in negatively biased standard errors in single-level SEM (Kaplan & Ferguson, 1999) and Stapleton (2002) proposed the use of an effective sample size weight, adjusting for nonoptimal weighting, which was demonstrated to result in unbiased estimates of standard errors.

An example dataset might help to clarify the sampling terms introduced above. The Early Childhood Longitudinal Study (ECLS), in the base year, used a three-stage stratified sampling design (U.S. Department of Education, 2001). First, all schools in the target population were divided into sampling "groups" based on geographic area, size, and income (usually counties or sets of counties) and these groups were placed in 62 mutually exclusive strata. These strata were defined by region of the nation, population size, race/ethnicity concentration, and per capita income. Twenty-four of these strata were "certainty" strata and contained just one geographic group (because there was only one group in the stratum, the group would automatically be selected into the sample, for example, New York City). In the remaining 38 strata, just two geographic groups were selected in each stratum. Once these 100 geographic groups were selected (as the PSUs), all the schools within each of the selected geographic groups were sorted by school characteristics and were selected using systematic sampling (a form of stratification called *implicit*). Selection of schools was therefore the second stage in the sampling design. Finally, in the third stage, approximately 24 children were sampled from each selected school. Students were placed into two different explicit strata: Asian/Pacific Islander students and all other students. Asian/Pacific Islander students were sampled at a rate about three times the sampling rate for the other students. Thus, there are two levels of clustering: schools within geographic groups and children within schools. There are also several different types of stratification: areas by geographic region, size, and income; schools by school characteristics; and children by race/ethnicity. In addition, the children were sampled at disproportional sampling rates: Asian/Pacific Islander students were sampled at a rate higher than the other children, so they will be "overrepresented" in the sample. Given these issues, a statistical technique that assumes simple random sampling should not be used. Each of the characteristics of complex samples treated above—clustering, stratification, and disproportionate selection probabilities—can affect standard error and parameter estimates. The choice of analysis technique will define how each is accommodated at the analysis stage.

National data collection agencies are not the only organizations collecting data using complex sampling methods. Within a university, for example, a thesis student might want to obtain a sample of undergraduates and therefore a sample is chosen of 50 freshmen, 50 sophomores, 50 juniors, and 50 seniors. This is a stratified sample and that fact would need to be addressed to calculate accurate standard errors for parameter estimates. If the number of students at each class level (freshman, sophomore, etc.) in the population is not equal, then the sample was chosen using disproportionate selection rates and sampling weights would need to be utilized to obtain accurate estimates of overall population characteristics—so, even what might appear to be a "simple" sampling process might make the analysis not so simple after all. In the next section, strategies to accommodate these aspects of complex sampling designs are discussed.

ANALYSIS OPTIONS

Three different analysis options are discussed in this section, the choice of which hinges on the research question to be addressed: aggregate modeling, pooled within-group covariance matrix modeling, and multilevel modeling. These options are shown using the simple bivariate case in the graphs in Figure 11.1 where in each graph a variable representing hours of TV watching is shown on the horizontal axis and a hypothetical latent factor, achievement, is shown on the vertical axis.

Aggregate Modeling

Suppose we have data from students who are clustered in schools (Figure 11.1a) and one is interested in overall population characteristics such as whether the number of hours spent watching TV has an impact on a latent construct of achievement. In this case, the interest is not in disentangling the effects at the school level from the effects at the individual level (it is possible that within schools there is no relation between TV watching and achievement but across schools there is a relation). The interest, therefore, is in examining the overall relation between TV watching and achievement, regardless of the level of the relation, and for this research question a typical single-level SEM model would be appropriate. Muthén and Satorra (1995) term this type of approach with clustered data an *aggregate* analysis (shown in Figure 11.1b): a single regression line is estimated, but appropriate standard errors must be estimated. Given this example, the sampling design would need to be addressed: the cluster

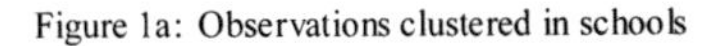

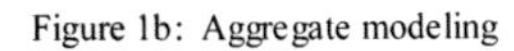

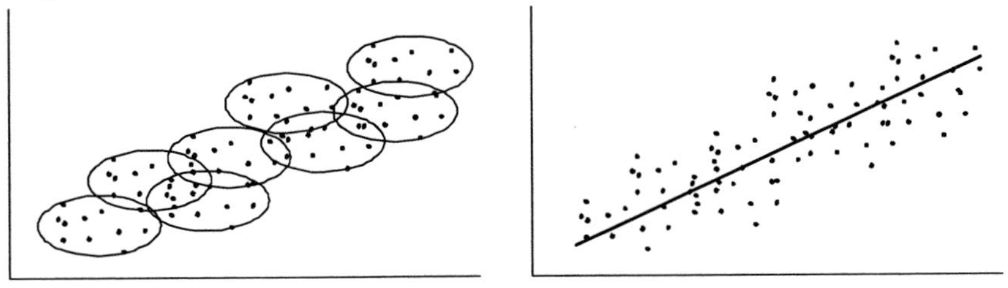

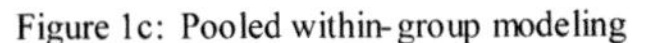

Figure 1d: Multilevel modeling

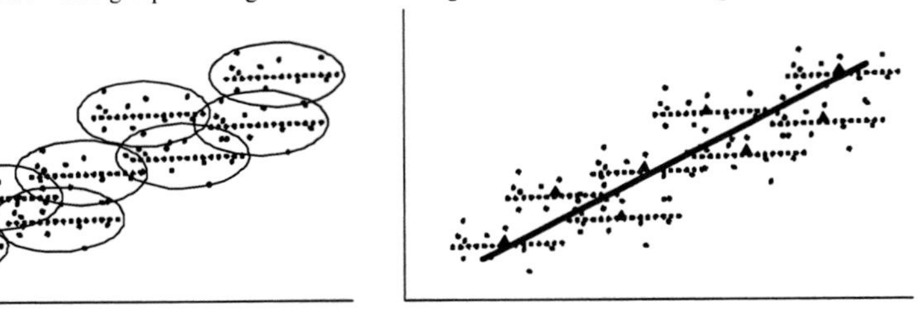

Figure 11.1. Graphical representation of modeling options.

sampling of schools in geographic regions and students in schools, the disproportionate sampling of the students, and the stratification. An analysis that adjusts estimates of standard errors given a sampling design but does not explicitly model the sampling design is referred to as a *design-based* analysis (Kalton, 1977). Most of the existing software programs to undertake design-based analyses produce either tables of data (comparison of means and proportions) or regression analyses. Such survey software (i.e., SUDAAN, WESVAR, STATA, SAS) uses either repeated replications or the Taylor Series approximation to obtain accurate estimates of parameters and standard errors (for descriptions of these techniques, see Lee et al., 1989). Currently, there are just two viable approaches to running design-based aggregate analyses for SEM researchers: use design-effect adjusted weights to create the covariance matrix or use a special analysis available in the Mplus software.

The first approach is advocated in National Center for Education Statistics training sessions (K. Rust, personal communication, June 18, 2002) and involves creating a design-effect adjusted weight by dividing the normalized sampling weight for each sample element by the average design effect for the variables in the analysis. The design effect can be seen as an adjustment factor. It is the ratio of the estimate of the sampling variance of a parameter taking into account the sampling design over the estimate of the sampling variance under simple random sampling. The average design effect for selected groups of variables in a dataset is generally available in user's guides and technical reports for national large-scale surveys. A secondary researcher using this approach must determine the most appropriate average design effect or composite of average design effects for the study. Rust noted, however, that this approach does not completely resolve the issues in sampling variance estimation; the standard errors will likely be somewhat overestimated.

A second approach for aggregate modeling is offered in the Mplus software, which has an estimation alternative for those using complex sample data, *TYPE=COMPLEX* (Muthén & Muthén, 1998). This estimation includes a Taylor series–like function to provide a normal theory covariance matrix for analysis. This matrix is created by obtaining a weighted covariance matrix that combines the variances and covariances of the PSUs. Note that for this technique, the *CLUSTER* command refers to the PSU and not the ultimate cluster in the data (if sampling involves three stages or more). For example, for the ECLS dataset, the *CLUSTER* variable would be the geographic group and not the school.

Pooled Within-Group Covariance Matrix Modeling

If one's research interest is in testing a model that explains the relations *within* a school between hours of TV watching and a latent construct of achievement and ignores between-school relations, then data could be modeled as suggested by Hox (2002), using a pooled within-school covariance matrix. This type of analysis is shown in Figure 11.1c; note that a common slope is estimated in each of the school groups (and in this example, there appears to be no relation between TV watching and the latent achievement construct within groups). Calculation of the pooled within-group covariance matrix would take place outside of the SEM software and the matrix would be analyzed as if it were a simple covariance matrix but interpretation would be focused on relative relations within schools. The pooled within-group covariance matrix can be calculated as:

$$\mathbf{S}_{\text{PW}} = (N-G)^{-1} \sum_{g=1}^{G} \sum_{i=1}^{n_g} (\mathbf{y}_{gi} - \bar{\mathbf{y}}_g)(\mathbf{y}_{gi} - \bar{\mathbf{y}}_g)'$$

where G is the total number of groups, N is the total sample size, $\mathbf{y}_{gi}$ is the vector of responses for person i within cluster g, and where $\bar{\mathbf{y}}_g$ is the vector of cluster means. When modeling the covariance matrix, the analyst should indicate to the software that the appropriate number of cases is $(N-G)$, not N. Note that for the ECLS example, weights (with an appropriate scaling) would still need to be used in the calculation of the matrix because children were selected at disproportionate rates within schools. Also, since there is stratification in the sampling design, estimates of standard errors likely will be somewhat positively biased.

Multilevel Modeling

A third choice in analyzing data with a nested structure is to use multilevel SEM. This type of analysis allows the researcher to examine within-school relations (e.g., If a student watches more TV than other students in his/her school, how does this relate to his/her relative achievement?) and between-school relations (e.g., What is the relation between the average hours of TV watched at a school and the average achievement in the school?). The first question is similar to the question answered by the previously discussed analysis technique of modeling the pooled within-group covariance matrix. However, multilevel modeling provides two advantages: (1) the researcher obtains an estimate of the amount of variance that exists in the measured variables between schools and can model that variance, and (2) at the within-school level the analysis will be more powerful than an analysis using only the within-school covariance matrix because the between-school covariance matrix provides additional information about the within-school relations. Figure 11.1d shows conceptually the modeling that is done. In this analysis, within-group relations are estimated and relations among the school means (indicated by the triangles) are also examined; in this example, there is no relation between X and Y within groups but there is a positive relation between the average X and average Y across the groups. The next section develops the logic of multilevel SEM and presents example analyses.

THE MULTILEVEL STRUCTURAL EQUATION MODELING PROCESS

For the moment, let us concentrate on cluster sampling, setting aside other complex sampling issues, and assume that we have sample data from a two-stage random sampling process: first schools were randomly selected and then students within those schools were randomly selected with equal probability. For this chapter, a dataset representing a population of over 700,000 observations was created that reflects a hierarchy (students within 1,000 schools) and the relation between hours of TV watching (a measured variable) and achievement (a latent variable indicated by three test scores) was varied both within and between schools. To supply example data, a two-stage sample was drawn from this population. The sample consists of 100 schools selected at random with subsequent random selection of 2% of the students within each selected school. Note that π_g, the probability of a school being selected, was .1 and $\pi_{i|g}$, the probability of a student being selected given the school's selection, was .02, resulting in an overall probability of selection for each student ($\pi_g \pi_{i|g}$) of .002 and the inverse of this probability is 500 (the raw weight for each student). Because neither clusters nor students were selected with differential selection probabilities, all weights are equal and there is no reason to include them in the analysis. This dataset will be used for all analyses and is available from the author. The dataset contains eight variables:

SCHLID	The ID number for the school
STUDID	The ID number for the student
TV	Number of hours of television typically watched during a weekend
TEST1	Score on a standardized exam (e.g., letters)
TEST2	Score on a standardized exam (e.g., shapes)
TEST3	Score on a standardized exam (e.g., sounds)
GENDER	Sex of the student (0 = male, 1 = female)
MOS	Measure of size for the PSU (school size)

The number of students per school in the sample ranged from 6 to 27 and descriptive statistics for the student-level data are shown in Table 11.1.

Suppose our interest is to test the model shown in Figure 11.2. TV represents "hours of TV per weekend" (which we assume is measured without error and thus is modeled as a manifest variable). TEST1, TEST2, and TEST3 are all test scores on various knowledge tests and the latent construct of "achievement" (ACHV) is assumed to be indicated by these three observed variables.

A first question that might arise with these clustered data before modeling is, "How much of the variance in the four manifest variables—TV watching and the three test scores—can be attributed to school differences as opposed to individual differences?" The intraclass correlation (ICC),

Table 11.1. Selected Descriptive Statistics for the Example Dataset

	Mean	*Standard Deviation*	*Minimum*	*Maximum*
TV	6.09	1.12	2	10
TEST1	49.86	1.10	46.13	53.95
TEST2	49.94	1.13	45.46	53.33
TEST3	49.99	1.09	46.16	53.45
GENDER	0.50	0.50	0	1

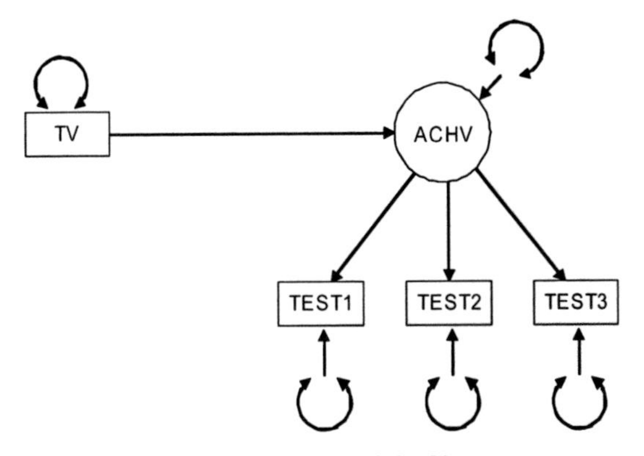

Figure 11.2. Theoretical model of interest.

$$\rho = \frac{\sigma_B^2}{\sigma_B^2 + \sigma_W^2},$$

is defined for a population as a ratio of the univariate between-school variance over the total variance and can be estimated from the components of an ANOVA for a given sample:

$$\hat{\rho} = \frac{MS_B - MS_W}{MS_B + (n.-1)MS_W}$$

where $n.$ is the sample size per group if balanced. If clusters do not contain equal numbers of students, then $n.$ is calculated such that

$$n. = \frac{N^2 - \sum_{g=1}^{G} n_g{}^2}{N(G-1)}$$

where G is the number of clusters, N is the total sample size, and n_g is the number of elements within cluster g (Kenny & Judd, 1986).

In this example, because there are varying numbers of students per school, an estimate for $n.$ is needed. The total number of students in the sample is 1,415 and the number of schools is 100. The sum of the squared cluster sizes is 22,197, thus we obtain an $n.$ of 14.13, quite close to the average cluster size of 14.15. Components from an ANOVA, using SCHLID as the grouping variable, and the resulting estimates of the ICC for each of the four variables are shown in Table 11.2.

Table 11.2. ANOVA Components and Estimates of ICCs for the Example Dataset

	MS_W	MS_B	ICC
TV	1.06	3.86	.157
TEST1	1.02	3.69	.156
TEST2	0.99	5.13	.228
TEST3	0.96	4.19	.192

The ICCs contained in Table 11.2 are all moderately sized (Kreft & de Leeuw, 1989), indicating that around 20% of the variance in each of the variables can be attributed to school or contextual effects and 80% to individual effects. Previous research indicates that with geographically determined clusters, the ICC tends to be relatively low on demographic variables (such as age and gender) and higher for socioeconomic variables and attitudes (Kalton, 1977). In educational studies involving intact classrooms, ICC values have been found to be rather high, such as between .3 and .4 for mathematics achievement (Muthén, 1997).

The focus of multilevel SEM has been to parse out the within-school and between-school variance for each variable and then model within-school relations on the within-school variance components while simultaneously modeling between-school relations on the between-school variance components. A requirement in this type of model is that the within-school covariance matrices are assumed to be equal across all schools (i.e., $\mathbf{\Sigma}_{W1} = \mathbf{\Sigma}_{W2} = \ldots = \mathbf{\Sigma}_{Wg}$), allowing for a single pooled within-school matrix ($\mathbf{\Sigma}_W$) to be used in estimation. This restriction can be thought of as random effects modeling; an assumption is made that the relations between variables is the same across all clusters. The statistical theory behind this assumption is available in Goldstein and McDonald (1988) and McDonald and Goldstein (1989).

As an example of this multilevel variance parsing, imagine it is hypothesized that TV is comprised of two parts. There is variance due to differences in schools: perhaps some schools are in neighborhoods where children are not closely observed and are free to watch television as much as they desire and other schools are in neighborhoods where children tend to be scheduled with extracurricular activities and do not watch much television on average. Also, there is variance due to differences in individuals: students in the various schools are different from their own classmates due to factors such as intelligence, personality, and parenting styles. So, the variance in the TV variable, σ^2_{TV}, is comprised of some function of σ^2_B (between-school variance) and some function of σ^2_W (within-school variance), as shown in Figure 11.3, with the between- and within-

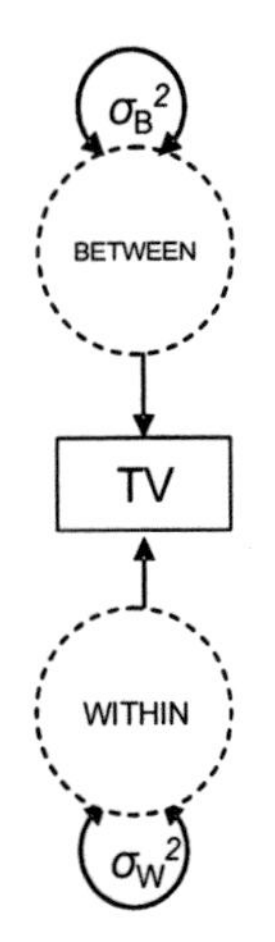

Figure 11.3. Parsing variance of observed variables into between and within components.

processes denoted as factors (shown with dotted lines to differentiate them from traditional "construct" factors).

Seemingly, this model is underidentified with only one variable (at least two variances will need to be estimated). However, with the use of both the sample within- and between-school covariance matrices, the model becomes identified. Muthén (1994) and Muthén and Satorra (1995) discussed the theoretical background of the estimation and proposed to use multiple group modeling to model sample within- and between-group matrices simultaneously. Notice that this is not a true multiple group situation—there are not two separate groups. The estimates of the between- and pooled within-school covariance matrices are calculated such that:

$$\mathbf{S}_{\mathrm{B}} = (G-1)^{-1} \sum_{g=1}^{G} n_g (\bar{\mathbf{y}}_g - \bar{\mathbf{y}})(\bar{\mathbf{y}}_g - \bar{\mathbf{y}})'$$

$$\mathbf{S}_{\mathrm{PW}} = (N-G)^{-1} \sum_{g=1}^{G} \sum_{i=1}^{n_g} (\mathbf{y}_{gi} - \bar{\mathbf{y}}_g)(\mathbf{y}_{gi} - \bar{\mathbf{y}}_g)'$$

where $\bar{\mathbf{y}}$ is the vector of grand mean responses. The information contained in $\mathbf{S}_{\mathrm{PW}}$ includes unbiased estimates of within-school variance and covariance ($E[\mathbf{S}_{\mathrm{PW}}] = \mathbf{\Sigma}_{\mathrm{W}}$) but the information in $\mathbf{S}_{\mathrm{B}}$ contains both within- and between-school variance and covariance information ($E[\mathbf{S}_{\mathrm{B}}] = \mathbf{\Sigma}_{\mathrm{W}}$ +

$n.\Sigma_B$). Muthén (1994) showed that multilevel SEM parameters can be estimated in the balanced case (equal number of students in each school) by minimizing the following fit function:

$$F_{ML} = G\{\log|\hat{\Sigma}_W + n.\hat{\Sigma}_B| + tr[\hat{\Sigma}_W + n.\hat{\Sigma}_B]^{-1} - \log|\mathbf{S}_B| - p\} + (N - G)\{\log|\hat{\Sigma}_W| + tr[\hat{\Sigma}_W^{-1}] - \log|\mathbf{S}_{PW}| - p\}$$

where $n.$ is the common group size, where $\hat{\Sigma}_W$ and $\hat{\Sigma}_B$ represent the model-implied within- and between-school covariance matrices, and where Σ_W is assumed equivalent across clusters in the population. It also has been suggested that this fit function can be used in the more typical unbalanced case. This estimator has been termed MUML by some (Muthén & Satorra, 1995); however, it has also been referred to as limited information maximum likelihood and pseudo-balanced maximum likelihood (Hox & Maas, 2001). This multilevel SEM estimation can be undertaken, in either the balanced or unbalanced case, using conventional multiple-group structural equation software such as EQS, AMOS, and LISREL, and will be demonstrated using EQS in this chapter. In order to estimate the fit function, the two components of $\mathbf{S}_B$ and $\mathbf{S}_{PW}$ must be modeled simultaneously with specific constraints and the G, N-G, and $n.$ values provided in the syntax.

Manual Multiple Group Setup

While most SEM software applications have multilevel functions that set up the multiple group modeling for the researcher, it is instructive to walk through an example manually setting up the multiple group syntax. I will use EQS version 5.7b (Bentler, 1995) to examine a series of models; version 6.1 of EQS will still require the user to undertake some of this process manually (if using the MUML approach) but other software programs automate this process. After the EQS analyses, the simpler multilevel analysis setup using the Mplus and LISREL software is described.

Hox (2002) proposed five steps for undertaking multilevel SEM analyses and we will use his steps as a guide. After examining each of these five steps, a flowchart will be provided to clarify the goals of and actions undertaken for each of these steps. Before starting, the $\mathbf{S}_B$, $\mathbf{S}_{PW}$, and $n.$ values for our data must be calculated. These matrices can be computed using outside software such as SAS and SPSS. Note, however, that many programs will calculate the $\mathbf{S}_B$ using a denominator that is not equivalent to G-1 and this matrix will need to be adjusted. The values for $\mathbf{S}_B$, $\mathbf{S}_{PW}$, and $n.$ for our sample data were calculated in SAS and are used in the EQS syntax in the following steps.

Step 1. First, Hox suggested that the theoretical model be tested on the pooled within-group covariance matrix ($\mathbf{S}_{PW}$) only. Recall that the theoretical model included hours of TV having an effect on achievement, as indicated by three test scores. This model is shown in Figure 11.4 and the necessary equations are in the EQS syntax in Figure 11.5. Two aspects of the figure have been drawn unconventionally, however, to prepare the model for future steps using multiple group modeling. First, factors are created to represent each variable (e.g., V1 is just a function of one unit of F1), thus having no residuals attached to the variables. The labels in the EQS syntax indicate that F1 to F4 are "within" processes of each of the variables. Second, the model has been flipped so that the latent construct of interest appears at the bottom of the figure with its indicator variables on top. We will be adding to the diagram on top of the indicator variables in a later step. The syntax should appear to be just a simple analysis of our theoretical model except for one important distinction: note that the number of cases is set to *N-G*.

The results of this model indicate acceptable model fit ($\chi^2_{(2)}$ = 1.827, CFI = 1.000, SRMR = .007). If our model fit had been worse, at this point we might have considered making theoretically meaningful modifications.

Step 2. After settling on a within-school model using the pooled within-school covariance matrix, Hox suggested that the pooled within- and between-school covariance matrices should be modeled simultaneously, imposing the within-school model (Figure 11.4) on both matrices with

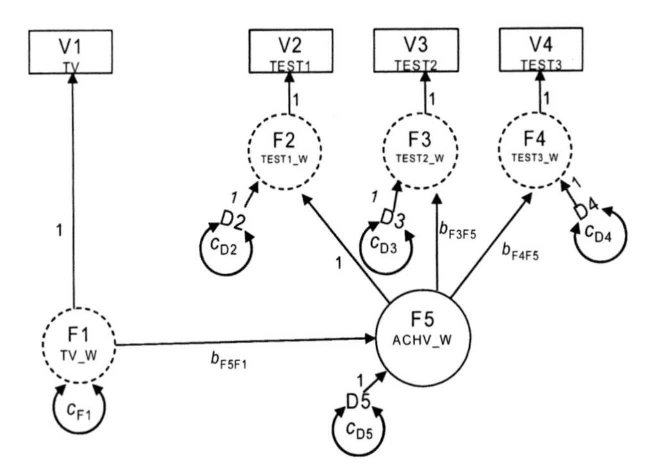

Figure 11.4. Theoretical model for the pooled within-school matrix.

```
/TITLE
 ANALYSIS ON POOLED WITHIN-SCHOOL MATRIX
/SPECIFICATIONS
 CAS=1315; VAR=4; MATRIX=COV; METHOD=ML;
/LABELS
 V1=TV; V2=TEST1; V3=TEST2; V4=TEST3;
 F1=TV_W; F2=TEST1_W; F3=TEST2_W; F4=TEST3_W;
 F5=ACHV_W;
/EQUATIONS
 V1=F1;
 V2=F2;
 V3=F3;
 V4=F4;
 F2=1F5+D2;
 F3=*F5+D3;
 F4=*F5+D4;
 F5=*F1+D5;
/VARIANCES
 F1=*;
 D2 TO D5=*;
/MATRIX
1.057
-0.308    1.015
-0.322    0.492    0.994
-0.325    0.472    0.459    0.963
/PRINT
 FIT=ALL;
/END
```

Figure 11.5. EQS syntax for Hox's first step in multilevel SEM.

constraints across all estimates from the within- and between-sources of data. This model is referred to as the multilevel *null* model. Recall that the expected value of $\mathbf{S}_{\mathrm{PW}}$ is $\boldsymbol{\Sigma}_{\mathrm{W}}$ and thus, the way we are modeling the within sources of variance should fit the data adequately. However, recall that the expected value of $\mathbf{S}_{\mathrm{B}}$ is $(\boldsymbol{\Sigma}_{\mathrm{W}} + n.\boldsymbol{\Sigma}_{\mathrm{B}})$ and we are currently modeling $\mathbf{S}_{\mathrm{B}}$ as being comprised of only within-school variation ($\boldsymbol{\Sigma}_{\mathrm{W}}$) and no between source of variation (thus the term *null* model). Given the moderate level of our intraclass correlations, it is expected that this model will fit poorly—and we would conclude that the sample between-school covariance matrix ($\mathbf{S}_{\mathrm{B}}$) contains more than just the within-group relations. This simultaneous modeling is demonstrated in the EQS syntax shown in Figure 11.6. Note that the between- and within- covariance matrices are treated as if they are two separate samples but this is just a "trick" to model them simultaneously.

There are several things worth noting in this new syntax. First, two "groups" are now shown—one group for the pooled within-school matrix and one for the between-school matrix. Second, the modeling of the pooled within-school matrix remains exactly as it was in the previous EQS syntax in Step 1. Next, the between-school matrix is modeled

```
/TITLE
 NULL MODEL
 GROUP1 POOLED WITHIN SCHOOL MATRIX
/SPECIFICATIONS
 CAS=1315; VAR=4; MATRIX=COV; METHOD=ML; GROUPS=2;
/LABELS
 V1=TV; V2=TEST1; V3=TEST2; V4=TEST3;
 F1=TV_W; F2=TEST1_W; F3=TEST2_W; F4=TEST3_W;
 F5=ACHV_W;
/EQUATIONS
 V1=F1;
 V2=F2;
 V3=F3;
 V4=F4;
 F2=1F5+D2;
 F3=*F5+D3;
 F4=*F5+D4;
 F5=*F1+D5;
/VARIANCES
 F1=*;
 D2 TO D5=*;
/MATRIX
1.057
-0.308    1.015
-0.322    0.492    0.994
-0.325    0.472    0.459    0.963
/END
/TITLE
 NULL MODEL
 GROUP2 BETWEEN SCHOOL MATRIX
/SPECIFICATIONS
 CAS=100; VAR=4; MATRIX=COV; METHOD=ML;
/LABELS
 V1=TV; V2=TEST1; V3=TEST2; V4=TEST3;
 F1=TV_W; F2=TEST1_W; F3=TEST2_W; F4=TEST3_W;
 F5=ACHV_W;
/EQUATIONS
 V1=F1;
 V2=F2;
 V3=F3;
 V4=F4;
 F2=1F5+D2;
 F3=*F5+D3;
 F4=*F5+D4;
 F5=*F1+D5;
/VARIANCES
 F1=*;
 D2 TO D5=*;
/MATRIX
3.860
-1.526    3.694
-1.304    2.151    5.128
-1.424    1.541    1.813    4.191
/CONSTRAINTS
 (1,F1,F1)=(2,F1,F1);
 (1,D2,D2)=(2,D2,D2);
 (1,D3,D3)=(2,D3,D3);
 (1,D4,D4)=(2,D4,D4);
 (1,D5,D5)=(2,D5,D5);
 (1,F3,F5)=(2,F3,F5);
 (1,F4,F5)=(2,F4,F5);
 (1,F5,F1)=(2,F5,F1);
/PRINT
 FIT=ALL;
/END
```

Figure 11.6. EQS syntax for Hox's second step in multilevel SEM—the *null* model.

with *CASES=G*, or 100 in this case. Also note that the */EQUATIONS* and */VARIANCES* section of the model for the between-school matrix is specified exactly as it is for the within-school matrix (as it must be) and every parameter that is estimated in both models is constrained to be solved to the same value in the */CONSTRAINTS* section. Although in typical multiple group modeling the process of setting and releasing constraints would be used to test for invariance, in multilevel modeling these constraints are necessary to model the two matrices simultaneously.

As expected, this *null* model fit quite poorly with a $\chi^2_{(12)}$ value of 608.439, CFI of .460, and SRMR of .368. The poor fit in the multilevel *null* model (after finding good fit for the $\mathbf{S}_{PW}$ matrix alone) is an indicator that there is, indeed, between-school variance to be explained. By modeling the between-school covariance matrix as if the only variance that contributed to it was within-school variance, the fit indices alert us to a disconfirmation of this model and provide an indication that there is between-school variance to be modeled. If this *null* model had adequate fit, the researcher could tentatively conclude that there appears to be no statistically significant group-level variance; however, it may be informative to undertake Step 3 as described below before proceeding to interpret the within-school path estimates.

Step 3. If adequate fit is not obtained at Step 2, the next step in Hox's proposed process is to model the hypothesized between-school variance. In this *independence* model, the variances of all of the variables in the between-school covariance matrix are partitioned into between-school and within-school variance. In addition, the within-school model is imposed on the within-school process and the between-school processes are modeled not to covary at all. This lack of covariance among the between-process variables is where the term *independence* is derived. Figure 11.7 shows the model that is applied to the between-school covariance matrix and includes both the hypothesized within-school model and the between-school model that will be tested in this *independence* model.

First, recall that the expected value of $\mathbf{S}_B$ is $(\Sigma_W + n.\Sigma_B)$. This expectation states that variance is comprised of one unit of within-school variance and $n.$ units of between-school variance. Thus, Figure 11.7 shows each manifest variable as a scaled function of these two sources of variance. Note that the lower portion of the model (below the observed variables) remains the same as shown in Figure 11.3—this is the within-school portion of the between-group variability. The between-school portion of the model (above the observed variables) provides an interesting addition. First, four new "factors" have been added; each manifest variable now is being modeled as a function of both within-school processes ("factors" F1

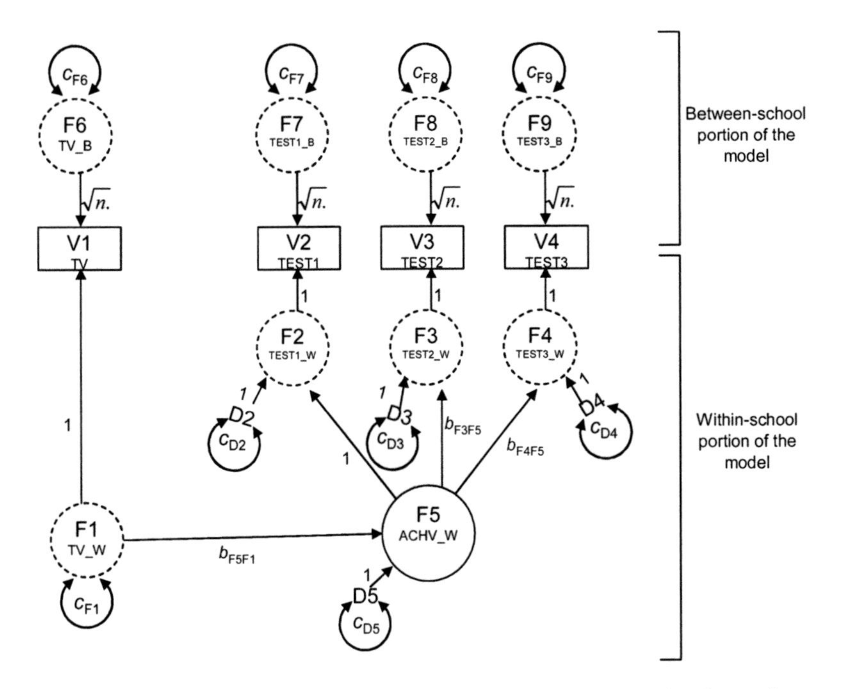

Figure 11.7. *Independence* model imposed on the between-school covariance matrix.

to F4) and between-school processes ("factors" F6 to F9). The paths from the between-school process to the observed variables must be set to the square root of *n.* (the "average" group size) for scaling purposes (Muthén, 1994). As was calculated earlier, this value is the square root of 14.13, or 3.760. The syntax required to run this model is similar to the syntax for the *null* model but with some additions to the modeling of the between-school covariance matrix, so only the latter half of the syntax is provided in Figure 11.8.

Note that there are two types of additions to the syntax, both in the Group 2 modeling section of the between-school covariance matrix. First, within the measurement equations each of the manifest indicator variables is now defined as having components of both within- and between-school processes: one unit of within-school process and 3.760 units of between-school process (the 3.760 is the value of the square root of *n.* and must be included). Second, the variances F6 to F9 (our "between-school process" factors) are being estimated. Hox suggested not to model the covariance of these factors in this step to determine, via badness of fit, if there is between-school covariance to be modeled. No new information has been provided, so we should expect our degrees of freedom to

```
.
. <<first half not shown because it is the same as for the null model>>
.
/TITLE
 INDEPENDENCE MODEL
 GROUP2 BETWEEN SCHOOL MATRIX
/SPECIFICATIONS
 CAS=100; VAR=4; MATRIX=COV; METHOD=ML;
/LABELS
 V1=TV; V2=TEST1; V3=TEST2; V4=TEST3;
 F1=TV_W; F2=TEST1_W; F3=TEST2_W; F4=TEST3_W;
 F5=ACHV_W;
 F6=TV_B; F7=TEST1_B; F8=TEST2_B; F9=TEST3_B;
/EQUATIONS
 V1=F1 + 3.760F6;
 V2=F2 + 3.760F7;
 V3=F3 + 3.760F8;
 V4=F4 + 3.760F9;
 F2=1F5+D2;
 F3=*F5+D3;
 F4=*F5+D4;
 F5=*F1+D5;
/VARIANCES
 F1=*;
 D2 TO D5=*;
 F6 TO F9=*;
/MATRIX
3.860
-1.526    3.694
-1.304    2.151    5.128
-1.424    1.541    1.813    4.191
/CONSTRAINTS
 (1,F1,F1)=(2,F1,F1);
 (1,D2,D2)=(2,D2,D2);
 (1,D3,D3)=(2,D3,D3);
 (1,D4,D4)=(2,D4,D4);
 (1,D5,D5)=(2,D5,D5);
 (1,F3,F5)=(2,F3,F5);
 (1,F4,F5)=(2,F4,F5);
 (1,F5,F1)=(2,F5,F1);
/PRINT
 FIT=ALL;
/END
```

Figure 11.8. EQS syntax for Hox's third step in multilevel SEM—the *independence* model.

decrease by four because we are estimating these four new variances. Note that there are no changes in the */CONSTRAINTS* section because the new "factor" variances that are estimated were added only to the between-school modeling process (in Group 2) and therefore cannot be constrained across the two groups.

The fit of this model is improved over the *null* model ($\Delta\chi^2_{(4)} = 571.768$) and this improvement in fit supports our hypothesis that there is between-school variance to be captured. (If there were no significant difference in fit between the *independence* and *null* models, the researcher could return to Step 2 and interpret the resulting estimates.) The overall fit of this model is not unilaterally outstanding due to a lack of exact fit and a high SRMR value ($\chi^2_{(8)} = 36.671$, CFI = .974, SRMR = .175) and suggests that there may be relations to be modeled among the school averages of the variables. If the overall fit were acceptable at this point, a researcher could interpret this model, discussing the amount of variance estimated at the between-group level and the strength of the estimated within-group paths.

Step 4. Assuming that the fit of the *independence* model is not acceptable, the next model that Hox suggested in the process is termed the *maximal* model; it can be thought of as a saturated model at the between-school processing level; all between-school processing "factors" are allowed to covary. The fit from this saturated model provides a baseline for assessing the model fit of a theoretical model at the between-school level (if one will be hypothesized in Step 5). This syntax is the same as the syntax for the *independence* model but with the inclusion of the following statement in the between-school model:

```
/COVARIANCES
 F6 TO F9=*;
```

Six covariances are now being estimated and we will return to a model with two degrees of freedom. The between-school process is completely saturated and within-school processes are modeled. The fit is very similar to the fit obtained when modeling the pooled within-school matrix alone (this will usually be the case, although it will not be exact): $\chi^2_{(2)} = 1.827$, CFI = 1.000, and SRMR = .005. If the researcher is interested in modeling the relations between these school variables, Hox suggested proceeding to model the hypothesized relations in Step 5.

Step 5. With the fit of the *maximal* model as a baseline for our between-school processing, we then model the between-school process elements of our variables as theory dictates. Suppose that we believe there is an underlying general achievement factor at the school level that can be indicated by the average scores on the three tests. In addition, we hypothesize that the average hours of TV watched by children in a school has an impact on the average achievement of the students in the school. This hypothesis suggests that we believe that the same pro-

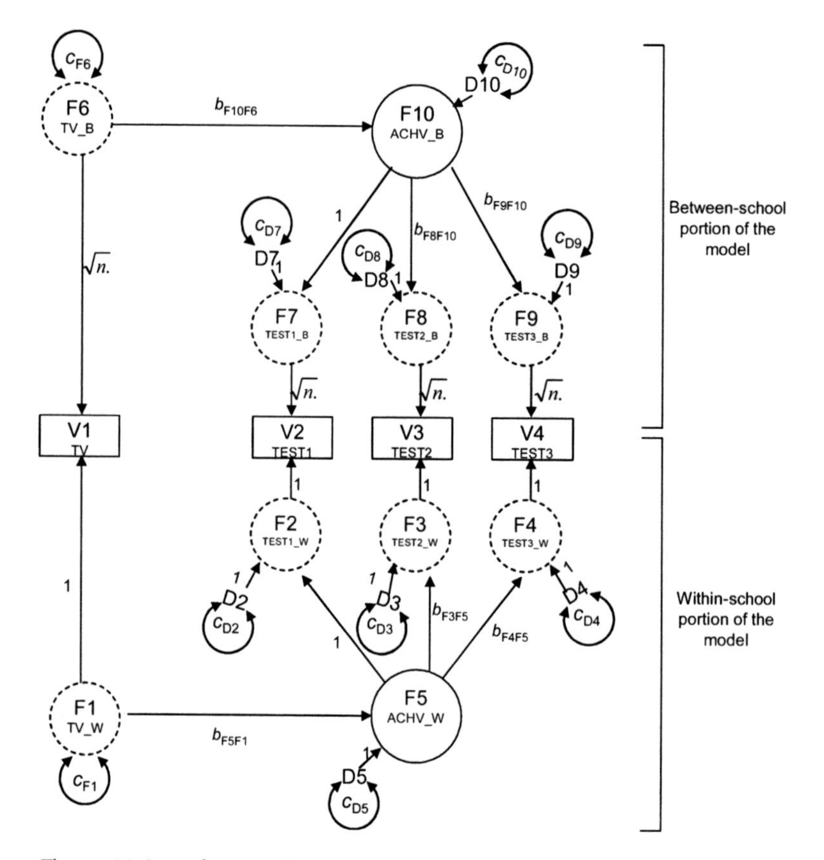

Figure 11.9. *Theoretical* model imposed on the between-school covariance matrix.

cess that occurs within schools (TV watching affects achievement) occurs between schools and we will set up our model accordingly, as shown in Figure 11.9.

The only changes necessary in the syntax are the removal of the */COVARIANCES* statement that was added for the *maximal* model, the replacement of *F7 to F9* with *D7 to D9* in the */VARIANCES* statement in the between-school section of the model, and the addition of the following equations

```
/EQUATIONS
 F7=1F10+D7;
 F8=*F10+D8;
 F9=*F10+D9;
 F10=*F6+D10;
```

The fit of this model is excellent ($\chi^2_{(4)}$ = 3.642, CFI=1.000, SRMR=.019) and we can proceed to interpret the estimates, which will be discussed in the next section. Before moving to interpretation, however, a summary flowchart of Hox's five steps of multilevel model testing is shown in Figure 11.10. At each step, adequate fit could be assessed by

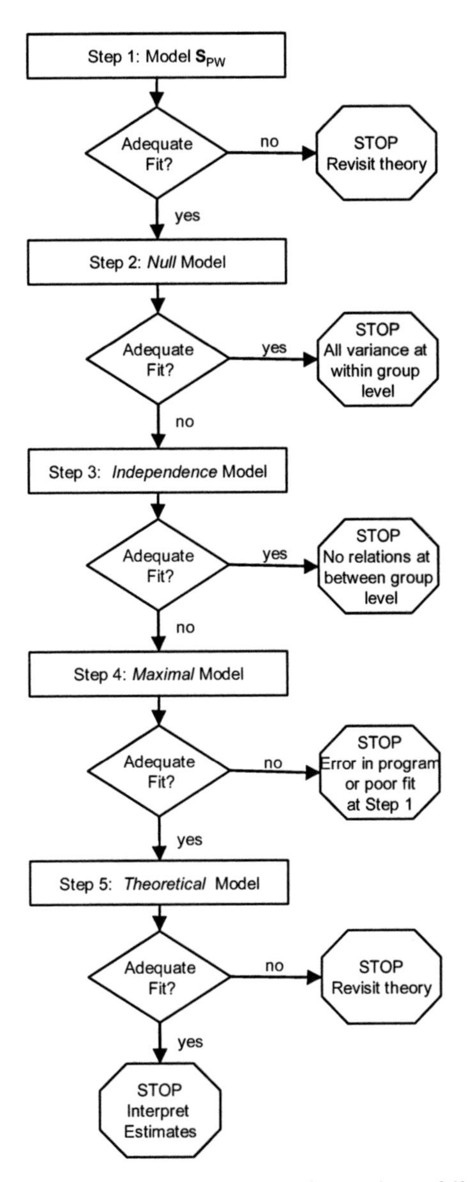

Figure 11.10. Flowchart of Hox's proposed steps in multilevel modeling.

using global criteria for fit indices as well as undertaking χ^2 difference testing against the model from the previous step.

Model Interpretation. Although the EQS output provides estimates and tests of those estimates for both the pooled within- and between-school "groups," due to the constraints the information that is contained in the second group contains all of the estimates from both parts of the model. The following section explains selected parts of what EQS calls the "Group 2" portion of the output and interprets the information contained therein.

First, from the estimates shown in Figure 11.11, we can see that the within-school loadings from ACHV_W (F5) to TEST2_W (F3) and TEST3_W (F4) are significantly different from zero (with *z*-statistics greater than 18). To obtain a statistical test of the loading from ACHV_W (F5) to TEST1_W (F2), we would need to rerun the model with this loading free and another indicator chosen as the referent with a loading fixed

```
TEST1_W =F2  =    1.000 F5     + 1.000 D2


TEST2_W =F3  =     .989*F5     + 1.000 D3
                   .055
                 18.102

TEST3_W =F4  =     .958*F5     + 1.000 D4
                   .053
                 18.022

ACHV_W  =F5  =   -.307*F1      + 1.000 D5
                   .022
                -13.652

TEST1_B =F7  =    1.000 F10    + 1.000 D7


TEST2_B =F8  =    1.049*F10    + 1.000 D8
                   .303
                  3.467

TEST3_B =F9  =     .798*F10    + 1.000 D9
                   .246
                  3.249

ACHV_B  =F10 =   -.412*F6      + 1.000 D10
                   .119
                 -3.449
```

Figure 11.11. EQS output of unstandardized estimates for Hox's fifth step in multilevel SEM—the *theoretical* model.

to 1.0. Next, we can see that *within* schools, the path from TV_W (F1) to ACHV_W (F4) is estimated to be -.307 and significantly different from zero (z = -13.652). This finding suggests that, given this model, for each additional hour of TV that a student watches (relative to other students in the school), that student's latent achievement is expected to decrease by about .3 units (where achievement is scaled in the metric of Test 1). Turning to the between-school "factors" in the model, we can see that the loadings from ACHV_B (F10) to TEST2_B (F8) and TEST3_B (F9) are significantly different from zero (z = 3.467 and 3.249, respectively). Remember that we have less information at the between-school level—data from just 100 schools—so tests of these estimates generally will be less powerful, resulting in smaller test statistics. Next we can see that between schools, the path from TV_B (F6) to ACHV_B (F10) is estimated to be -.412 and significantly different from zero (z = -3.449). This finding suggests that for each additional hour of TV watched on average by students at one school, the school's average achievement is expected to decrease by about .4 units.

Turning to the standardized solution, shown in Figure 11.12, the information of interest is provided after the first four lines (these first lines just contain standardized information regarding the parsing of variance into between- and within-school processes). Using the within-school standardized loadings, we could assess the construct reliability of our achievement factor within schools. Additionally, we could use the between-school standardized loading estimates to obtain reliability information for a school-level construct of achievement. Looking at the structural relations, given this model, a one standard deviation increase in TV watching (relative to the other students within a school) is expected to result in a .450 standard deviation drop in achievement. As modeled, we have explained about 20%

```
STANDARDIZED SOLUTION:                                          R-SQUARED

TV        =V1  =    .524 F1     + .852 F6
TEST1     =V2  =    .523 F2     + .852 F7
TEST2     =V3  =    .441 F3     + .898 F8
TEST3     =V4  =    .479 F4     + .878 F9
TEST1_W   =F2  =    .696 F5     + .718 D2                             .484
TEST2_W   =F3  =    .695*F5     + .719 D3                             .484
TEST3_W   =F4  =    .685*F5     + .729 D4                             .469
ACHV_W    =F5  =  -.450*F1      + .893 D5                             .202
TEST1_B   =F7  =    .751 F10    + .660 D7                             .564
TEST2_B   =F8  =    .637*F10    + .771 D8                             .405
TEST3_B   =F9  =    .547*F10    + .837 D9                             .299
ACHV_B    =F10 =  -.559*F6      + .829 D10                            .312
```

Figure 11.12. EQS output of standardized estimates for Hox's fifth step in multilevel SEM—the *theoretical* model.

of the within-school variance in achievement. Similarly, given this model, a one standard deviation increase in *average* TV watching for a school (relative to the averages for other schools) is expected to result in a .559 standard deviation drop in average achievement. As modeled, we have explained over 30% of the between-school variance in average achievement. So, not only is there a negative relation between hours of TV watched and achievement for students within schools, but a negative relation between a school's average hours of TV watched and the school's average achievement, above and beyond the relation at the within-school level.

According to Muthén (1994), this pseudo-balanced maximum likelihood estimator is fairly robust in the unbalanced case, but the possibility remains that it can provide incorrect standard errors and χ^2 tests of fit. In fact, this issue would lead us to be cautious in the practice of examining differences in χ^2 values between steps in the Hox-proposed process. A version of the pseudo-balanced estimator used above but with robust estimates of standard errors and χ^2 test statistics is available in the Mplus software beginning with version 2.1. Let us turn now to running our same example analysis using other software programs.

Automatic Multilevel Modeling Setup

When using Mplus version 2.14 or LISREL version 8.54, there is no need to calculate between- and within-group covariance matrices outside the software and use multiple group modeling to "trick" the software to run multilevel models. The Mplus analysis option, *TYPE IS TWOLEVEL*, and the SIMPLIS syntax, *$CLUSTER*, undertake the variance partitioning and the multiple group modeling for the user behind the scenes. This automatic processing also means that Hox's proposed steps of *null*, *independence*, *maximal*, and *full* modeling are not straightforward to implement. In this section only our final theoretical model will be shown using the Mplus and LISREL software.

Mplus. The syntax for Mplus appears in Figure 11.13 and most of it should be straightforward. Note that although the dataset contained eight variables, the user should indicate which variables will be used in the analysis. Also, the variable that indicates the "level 2" grouping, or school in our case, must be identified with the *CLUSTER IS* option in the *VARIABLE:* section. In the *ANALYSIS:* section, the *TYPE IS TWOLEVEL* option indicates that this will be a multilevel analysis. The *ESTIMATOR IS MUML* option requests that Mplus produce the pseudo-balanced estimates. The estimates that we get for this run are very similar to the estimates we obtained with our more manual process using EQS version 5.7b and these estimates appear in the first and second columns of Table 11.3

```
TITLE: THEORETICAL (FULL) MODEL

DATA:
 FILE IS "C:\TEMP\TS_SRS.dat";
 NOBSERVATIONS = 1415;

VARIABLE:
 NAMES ARE SCHLID STUDID TV TEST1 TEST2 TEST3 GENDER MOS;
 USEVARIABLES ARE SCHLID TV TEST1 TEST2 TEST3;
 CLUSTER IS SCHLID;

ANALYSIS:
 TYPE IS TWOLEVEL;
 ESTIMATOR IS MUML;
 ITERATIONS = 1000;
 CONVERGENCE = 0.00005;

MODEL:
 %WITHIN%
 F1 BY TEST1 TEST2 TEST3;
 F1 ON TV;

 %BETWEEN%
 F2 BY TEST1 TEST2 TEST3;
 F2 ON TV;

OUTPUT: STANDARDIZED SAMPSTAT;
```

Figure 11.13. Mplus syntax for Hox's fifth step in multilevel SEM—the *theoretical* model.

for comparison. With the *TYPE IS TWOLEVEL* option, the researcher models the relations that are hypothesized to exist within schools in the *%WITHIN%* section of the *MODEL:* statement and the relations that are hypothesized to exist among schools in the *%BETWEEN%* section of the *MODEL:* statement. Although the same observed variables are referenced at each level of the model in the Mplus syntax (such as TEST1), any factors modeled must have unique references across the levels (i.e., F1 and F2).

The output obtained from Mplus is fairly similar to the EQS output but with additional information regarding the multilevel nature of the data. First the data are summarized, as shown in Figure 11.14, including the number of clusters and the school IDs of the clusters at each size level, the *quasi-average cluster size* (this is our $n.$ value), and estimates of the ICCs (but for the endogenous variables only). These ICC values match our manually calculated estimates exactly except for TEST1 (.156 as compared with .158). To obtain ICC values for the exogenous variables, data provided as a result of specifying the option *SAMPSTAT* in the *OUTPUT:*

Table 11.3. A Comparison of Parameter Estimates (and Standard Error Estimates) Across Software for Multilevel SEM

	EQS 5.7b	*Mplus 2.14 (MUML)*	*Mplus 2.14 (MUMLM)*	*Mplus 2.14 (ML)*	*LISREL 8.54*
Between-group estimates					
b_{F8F10}	1.049	1.048	1.048	1.011	1.011
	(.303)	(.301)	(.282)	(.294)	(.294)
b_{F910}	0.798	0.798	0.798	0.800	0.800
	(.246)	(.244)	(.261)	(.246)	(.247)
b_{F10F6}	-.412	-.412	-.412	-.437	-.437
	(.119)	(.119)	(.138)	(.126)	(.126)
c_{D7}	.083	.083	.083	.082	.082
	(.031)	(.031)	(.031)	(.032)	(.031)
c_{D8}	.173	.174	.174	.169	.169
	(.042)	(.042)	(.043)	(.041)	(.041)
c_{D9}	.160	.160	.160	.158	.158
	(.034)	(.034)	(.030)	(.034)	(.034)
c_{D10}	.074	.074	.074	.073	.073
	(.032)	(.032)	(.025)	(.033)	(.033)
c_{F6}	.198	.198	.198	.187	.186
	(.039)	(.039)	(.048)	—	(.038)
Within-group estimates					
b_{F3F5}	.989	.989	.989	.988	.988
	(.055)	(.055)	(.064)	(.055)	(.055)
b_{F4F5}	.958	.958	.958	.957	.957
	(.053)	(.053)	(.052)	(.053)	(.053)
b_{F5F1}	-.307	-.307	-.307	-.307	-.307
	(.022)	(.022)	(.021)	(.022)	(.022)
c_{D2}	.523	.523	.523	.523	.523
	(.031)	(.031)	(.025)	(.031)	(.031)
c_{D3}	.513	.514	.514	.514	.514
	(.030)	(.030)	(.032)	(.030)	(.030)
c_{D4}	.512	.512	.512	.512	.512
	(.029)	(.029)	(.026)	(.029)	(.029)
c_{D5}	.392	.392	.392	.393	.393
	(.034)	(.034)	(.035)	(.034)	(.034)
c_{F1}	1.057	1.057	1.057	1.055	1.055
	(.041)	(.041)	(.039)	—	(.041)
χ^2 value	3.642	3.663	3.487	3.430	3.430

statement can be used. This option provides the estimates of the unrestricted covariance matrix for both within-school and between-school relations. The within-school matrix is an estimate of $\mathbf{\Sigma}_W$ while the between-school matrix is an estimate of $\mathbf{\Sigma}_B$ and therefore the elements on the diagonal of the matrices can be used to calculate ICCs for variables of interest using the equation: $\hat{\rho} = \frac{\hat{\sigma}_B^2}{\hat{\sigma}_B^2 + \hat{\sigma}_W^2}$.

```
SUMMARY OF DATA

   Number of clusters           100

     Size (s)    Cluster ID with Size s

        6           303      464
        7             9       33      660      885
        8           613      386      666      763      801       80      971
        9           241      886      966      402
       10           877      112      713      959      168      432
       11           707      416        7      922      953      799       18      855
       12            40      718      895      908      182      202      236      683
                    138
       13           208      393      604      731      606      697
       14           472      150      309      612      714      333      645      732
                    657      349      255
       15           244      931      950      411      474      591      602
       16           330      460      297       13      221      493      616      983
       17           147      779      207      282      164      977      608
       18            64      116      982      543
       19           531      296      495
       20           284      816
       21           933      675      519
       22           637      174      599      342
       23           332      310
       24           220
       26           500
       27           939

   Quasi-average cluster size   14.134

   Estimated Intraclass Correlations for the Y Variables

               Intraclass                Intraclass                Intraclass
   Variable  Correlation    Variable  Correlation    Variable  Correlation

   TEST1         0.158      TEST2         0.228      TEST3         0.192
```

Figure 11.14. Mplus data summary output for Hox's fifth step in multilevel SEM—the *theoretical* model.

After this preliminary information and the measures of model fit are presented, the estimates are provided in five columns (as shown in Figure 11.15): the unstandardized parameter estimate, the standard error estimate, the associated z-statistic, and two types of standardized parameter estimates. For our purposes, the standardized estimates in the last column (*StdYX*) are appropriate. The values from this run match the manually set-up analysis using EQS down to the third decimal point (see Table 11.3). Muthén (1994) indicated that the standard errors and χ^2 values may not be correct using MUML estimation with unbalanced data and starting with version 2.1 in Mplus the estimator option of MUMLM is available to estimate robust standard errors and mean-adjusted robust χ^2 values. This

```
MODEL RESULTS

                         Estimates     S.E.  Est./S.E.      Std      StdYX

Within Level

 F1         BY
    TEST1                    1.000    0.000      0.000    0.701      0.696
    TEST2                    0.989    0.055     18.106    0.693      0.695
    TEST3                    0.958    0.053     18.024    0.672      0.684

 F1         ON
    TV                      -0.307    0.022    -13.659   -0.437     -0.450

 Variances
    TV                       1.057    0.041     25.642    1.057      1.000

 Residual Variances
    TEST1                    0.523    0.031     17.049    0.523      0.516
    TEST2                    0.514    0.030     17.078    0.514      0.516
    TEST3                    0.512    0.029     17.596    0.512      0.532
    F1                       0.392    0.034     11.633    0.798      0.798

Between Level

 F2         BY
    TEST1                    1.000    0.000      0.000    0.328      0.752
    TEST2                    1.048    0.301      3.484    0.344      0.636
    TEST3                    0.798    0.244      3.266    0.262      0.547

 F2         ON
    TV                      -0.412    0.119     -3.465   -1.256     -0.558

 Variances
    TV                       0.198    0.039      5.116    0.198      1.000

 Residual Variances
    TEST1                    0.083    0.031      2.673    0.083      0.435
    TEST2                    0.174    0.042      4.133    0.174      0.595
    TEST3                    0.160    0.034      4.735    0.160      0.701
    F2                       0.074    0.032      2.295    0.688      0.688
```

Figure 11.15. Mplus estimates output for Hox's fifth step in multilevel SEM—the *theoretical* model.

estimation can be run by including the option *ESTIMATOR IS MUMLM* or by having no *ESTIMATOR IS* option as it is the default estimation method for *TYPE IS TWOLEVEL*. The software developers note, however, that this scaled χ^2 value should not be used with traditional χ^2 difference testing (Muthén & Muthén, 1998). Instead, an alternate procedure has been developed and should be used to compare these scaled χ^2 values (Satorra & Bentler, 2001). Parameter estimates for our sample data using the *ESTIMATOR IS MUMLM* option are shown in Table 11.3 and, as expected, differ from the *MUML* estimates only for the standard errors and the χ^2 value.

A final type of estimation available in Mplus for multilevel random effects models is running full-information maximum likelihood (ML) estimation for unbalanced data. This estimation has been implemented in the most recent versions of both Mplus and LISREL software and minimizes the following fit function:

$$F_{\mathrm{ML}} = \sum_{g=1}^{G}\{\log|\hat{\Sigma}_{\mathrm{B}} + n_g^{-1}\hat{\Sigma}_{\mathrm{W}}| + tr[(\hat{\Sigma}_{\mathrm{B}} + n_g^{-1}\hat{\Sigma}_{\mathrm{W}})^{-1}(\bar{\mathbf{y}}_g - \hat{\mu})(\bar{\mathbf{y}}_g - \hat{\mu})']\} +$$

$$(N-G)\log|\hat{\Sigma}_{\mathrm{W}}| + (N-G)tr[\hat{\Sigma}_{\mathrm{W}}^{-1}\mathbf{S}_{\mathrm{PW}}]$$

Note that there are now terms for each cluster (g) within this function and when cluster sizes are balanced it simplifies to the MUML version. Although studies have suggested that when using unbalanced data, MUML (or MUMLM) estimates are fairly robust to variations in sample size (Hox & Maas, 2001; Muthén, 1994), when using Mplus there appears to be no disadvantage to using the *ESTIMATOR IS ML* option. This example with 100 clusters was solved in very little computational time. It is possible, however, that data with many more clusters, small clusters, or a more complex model may present difficulties in ML estimation. Parameter and standard error estimates with ML estimation were slightly different than our previous runs and are shown in Table 11.3. When using the *ESTIMATOR IS ML* option, the model is estimated conditional on X, so exogenous variables are not a part of the model and therefore the output does not provide estimates of the variance of the exogenous factor, TV (F1 and F6). The values shown in Table 11.3 are the sample variances obtained from the unrestricted covariance matrices provided from the *OUTPUT: SAMPSTAT;* option.

LISREL. When using SIMPLIS syntax in LISREL, the only estimation option for multilevel analyses is full-information maximum likelihood estimation as discussed above (du Toit & du Toit, 2001). The syntax for running our example model is shown in Figure 11.16. The analyst needs to specify the relations that exist at the between- and within-school levels (in this case, by using the *GROUP* statements to separate the two sections of the model). When SIMPLIS encounters the *$CLUSTER* statement (as seen on the third line of the syntax), a multilevel model is run. Note that LISREL (unlike Mplus and EQS) requires that variables that are presumed to be measured without error still need to be modeled as latent variables. Thus, in the syntax above, the factor FTV is indicated by the manifest variable TV (note that the loading is set to 1) and the residual

```
GROUP BETWEEN SCHOOLS
  RAW DATA FROM FILE TS_SRS.PSF
  $CLUSTER SCHLID
  LATENT VARIABLES
    FTV ACHIEVE
  RELATIONSHIPS
    TV=1*FTV
    TEST1=1*ACHIEVE
    TEST2 TEST3 = ACHIEVE
    ACHIEVE = FTV

  SET THE ERROR VARIANCE OF TV TO 0
  SET THE VARIANCE OF FTV FREE
  SET THE ERROR VARIANCE OF ACHIEVE FREE
  SET THE ERROR VARIANCE OF TEST1 FREE
  SET THE ERROR VARIANCE OF TEST2 FREE
  SET THE ERROR VARIANCE OF TEST3 FREE

GROUP WITHIN SCHOOLS
  RAW DATA FROM FILE TS_SRS.PSF
  RELATIONSHIPS
    TV=1*FTV
    TEST1=1*ACHIEVE
    TEST2 TEST3 = ACHIEVE
    ACHIEVE = FTV

  SET THE ERROR VARIANCE OF TV TO 0
  SET THE VARIANCE OF FTV FREE
  SET THE ERROR VARIANCE OF ACHIEVE FREE
  SET THE ERROR VARIANCE OF TEST1 FREE
  SET THE ERROR VARIANCE OF TEST2 FREE
  SET THE ERROR VARIANCE OF TEST3 FREE

  LISREL OUTPUT: ND=3 SC
```

Figure 11.16. SIMPLIS syntax for Hox's fifth step in multilevel SEM—the *theoretical* model.

variance for TV is set to zero. The parameter estimates from LISREL are shown in Table 11.3 and match the estimates from Mplus ML estimation nearly exactly. Some explanation of the LISREL output is warranted. LISREL first calculates the unrestricted covariance matrix, as shown in Figure 11.17, for both within-school and between-school relations (as was provided in Mplus) and therefore the elements on the diagonal of the matrices can be used to calculate ICCs for variables of interest as shown earlier. The parameter estimates and their standard errors are shown in the output in a very similar fashion as in EQS output and will not be shown here.

It is hoped that the sections above, while limited, will help the researcher who is just getting started with multilevel structural equation modeling. All three software packages provide multilevel modeling capa-

```
GROUP BETWEEN SCHOOLS

        Covariance Matrix

               TEST1      TEST2      TEST3         TV
             --------   --------   --------   --------
   TEST1      0.190
   TEST2      0.114      0.281
   TEST3      0.077      0.095      0.226
      TV     -0.087     -0.068     -0.076      0.187

 GROUP WITHIN SCHOOLS

        Covariance Matrix

               TEST1      TEST2      TEST3         TV
             --------   --------   --------   --------
   TEST1      1.016
   TEST2      0.492      0.994
   TEST3      0.473      0.459      0.964
      TV     -0.308     -0.322     -0.324      1.055
```

Figure 11.17. Selected LISREL data summary output for Hox's fifth step in multilevel SEM—the *theoretical* model.

bilities that are fairly robust and users should use the software with which they are most comfortable. Version 6.1 of EQS is reported to have another multilevel estimation function proposed for use with unbalanced cluster sizes (Bentler & Liang, 2002). The approaches shown above can be generalized to other software programs not specifically addressed here (e.g., AMOS; Arbuckle, 2003).

ADDITIONAL ISSUES IN MULTILEVEL STRUCTURAL EQUATION MODELING

Given this brief introduction to multilevel models in an SEM framework, the reader likely has questions regarding specific issues in multilevel modeling. In addition, the dataset used in the examples contained clustering only; usually we would need to accommodate other aspects of complex samples. Therefore, the following topics are briefly addressed in this section: incorporating sampling weights (from disproportionate stratified sampling), modeling with school-level variables, modeling with missing data, and modeling random slopes. For more detailed explanations of analyses and extensions, suggested readings are provided at the end of the chapter.

First, sampling weights can be accommodated with each of the methods of multilevel modeling addressed above. When undertaking the multiple-group manual setup, sampling weights should be incorporated in the calculation of $\mathbf{S}_{PW}$ and $\mathbf{S}_B$ outside of the SEM software. Stapleton (2002) outlined the procedures for doing this, demonstrated the effect of the choice of normalization of the weights, and suggested that the weights be normalized such that the sum of the weights equals the sum of the squares of the weights (this method is different from the typical normalization process, which yields a sum of weights that is equal to the sample size). When using Mplus, the option *WEIGHT IS [variable name]* can be used in the *VARIABLE:* statement. Note that no matter what scale has been applied to the weight (i.e., raw, normalized, effective sample size), Mplus automatically scales the weight so that the sum of the weights is equivalent to the sample size. This scaling may result in underestimated standard errors (Stapleton, 2002) so caution is urged. To incorporate sampling weights in the LISREL software, the user will need to use the LISREL programming language and not the more user-friendly SIMPLIS syntax. In the LISREL syntax, the user can specify *WEIGHT=* [variable name] in the DA section.

The inclusion of school-level variables also can be accomplished with all of the approaches to multilevel SEM that have been addressed in this chapter. For example, a researcher may want to theorize that school size has an effect on average school achievement. For the manual setup of multiple group modeling, the analyst can include the school-level variable (e.g., school size–MOS in the sample dataset) in the calculation of $\mathbf{S}_{PW}$ and $\mathbf{S}_B$ ($\mathbf{S}_{PW}$ should contain zeroes for the elements of the matrix associated with school size). The school-level variable is then not modeled on the within-school side of the model, but is modeled in the between-school side of the model. When modeling with a school-level variable using the Mplus *TYPE IS TWOLEVEL* option, the analyst must declare in the syntax that school size is a school-level variable by including the option *BETWEEN IS MOS* in the *VARIABLE:* statement. Once identified as a school-level variable, the school-level variable can then be modeled in the equations found in the *%BETWEEN%* section of the *MODEL:* statement. To model using a school-level variable in LISREL, the user will need to set up the multilevel model using LISREL programming tools and not the SIMPLIS syntax.

To address missing data in multilevel structural modeling contexts, the user has several options. When modeling data using the manual setup, the researcher supplies the covariance matrices and thus must address the missing data when calculating those matrices. If listwise or pairwise deletion is used, the researcher must ensure that the appropriate sample size is provided in the SEM syntax to obtain more accurate standard errors and test statistics. Researchers might want to use a preferred imputation technique to produce the covariance matrices under conditions of missing data (Graham & Hofer, 2000; see also Enders, Chapter 10, this volume). When using Mplus, including the statement *MISSING* = 99 (or the appropriate missing data value) without requesting any special missing data estimation techniques will result in listwise deletion of cases. However, if the user includes *MISSING* in the *ANALYSIS:* statement, the estimation uses maximum likelihood under the missing at random assumption. The user's manual specifies that observations with data on only the independent variables or missing data on one or more of the independent variables will not be used in the analysis. Note that a researcher cannot use MUML or MUMLM estimation with the *MISSING* analysis option; the estimation needs to use full information estimation such as ML or MLR; MLR is a robust estimator for missing data. When using LISREL, again, the SIMPLIS syntax options do not accommodate missing data modeling, however, the user can accommodate missing data using PRELIS or LISREL syntax.

All of the multilevel SEM approaches that have been discussed until now have been limited to random effects modeling. We assumed that the same relation among variables or constructs within schools held at all of the schools (refer back to Figure 11.1d). The Mplus software has included the ability to model random slopes in an SEM context in its version 2.14 release (Muthén & Muthén, 2003). Although LISREL does allow modeling of random slopes in a linear regression framework, it does not currently support random slopes in structural equation models. In Mplus, however, there are some restrictions on random slopes modeling: only paths from an exogenous, observed variable at level 1 are allowed to be random, that exogenous variable cannot exist at the between level, and only between-level variables can predict the slope. For example, we could not model the TV variable as being comprised of both within- (TV_W) and between-school processes (TV_B) and then model the between-school process of TV (TV_B) to predict the slope of GENDER_W on ACH_W because TV exists at both the within-school and between-school levels. Similarly, we could not model the path from TV_W to ACH_W to be random as the TV variable exists at both the within- and between-school levels. Savvy analysts should be able to use centering and newly created variables to represent school means to overcome some of these restrictions.

ADDITIONAL RESOURCES

In this chapter I attempted to provide some guidance to researchers interested in using multilevel modeling in an SEM framework. This limited treatment has only touched on the details of estimation, conceptualization, and interpretation. There are a great many resources for those who are interested in learning more about multilevel SEM and multilevel linear regression. Researchers should, of course, turn to the software user's manuals and technical addenda for their software of choice. In addition, publications by Muthén (1994) and Muthén and Satorra (1995) are particularly useful in understanding the estimation process. Also, chapters on multilevel SEM methods can be found in texts on multilevel and advanced SEM analysis by Hox (2002), Heck and Thomas (2000), and Heck (2001); the multilevel SEM chapter by Heck and Thomas provides the manual setup pseudo-balanced approach using LISREL programming syntax and the multilevel SEM chapter by Heck provides examples of Mplus syntax and output using MLM (now called MUMLM) estimation for both a confirmatory factor analysis and a path analysis. Two nice examples of multilevel path analysis can be found in articles by Kaplan and Elliott (1997a, 1997b), and examples of multilevel confirmatory factor analysis are provided in Muthén (1991) and Harnqvist, Gustafsson, Muthén, and Nelson (1994). For readers who are interested in learning about multilevel regression modeling, texts by Snijders and Bosker (1999), Kreft and deLeeuw (1998), and Raudenbush and Bryk (2002) are excellent places to start the learning process. An additional focus of this chapter has been to introduce readers to the issues of complex sampling and the analysis of such data. The text by Lee and colleagues (1989) is an excellent brief introduction to the issues in sampling and the problems that these issues present in any analysis.

SUMMARY

In this chapter I reviewed several concepts regarding data that are collected via complex sampling procedures and introduced three methods of modeling such data in an SEM framework. Readers should remember that traditional, single-level statistical procedures rely on an assumption that observations are independent. Usually, with cluster samples, that independence has been compromised. Diagnostic tools, such as the calculation of intraclass correlations, are available to assess the level of dependency and if observations are found not to be independent the researcher must then choose an appropriate course of action. However, a sample that was collected using cluster sampling does not necessarily require a multilevel analysis. The research question specifies the level of analysis required and therefore should dictate the analysis tool.

If one chooses multilevel SEM as the appropriate tool with which to answer a specific research question, a variety of software is available. Currently, Mplus version 2.14 offers the most flexibility and choice of estimation techniques; however, new versions of software are being released at a rapid pace and users should determine whether new versions of SEM software have increased capabilities in these types of analyses. Although not easy to replicate with the automatic multilevel modeling functions found in newer versions of software, the steps outlined in this chapter and developed by Hox (2002) allow the researcher to test, in a hierarchical fashion, the existence of within- and between-group variance and covariance. These steps offer a logical progression in multilevel model testing and researchers are encouraged to address each of the questions regarding the level at which (co)variation exists whenever multilevel analyses are undertaken.

As software becomes more advanced, allowing for such modeling approaches as random slopes, methodological exploration of these methods and the robustness of the resultant statistics will be required. It is hoped that this chapter has introduced a new set of researchers to various issues in analyzing complex sample data with multilevel SEM methods and that these researchers can extend the use of multilevel SEM in applied settings and further methodological study of the technique.

REFERENCES

Arbuckle, J. L. (2003). *Amos 5.0 update to the Amos user's guide* [Computer software and manual]. Chicago: Small Waters.

Bentler, P. M. (1995). *EQS Structural Equations Program manual*. Encino, CA: Multivariate Software, Inc.

Bentler, P. M., & Liang, J. (2002). Two-level mean and covariance structures: Maximum likelihood via an EM algorithm. In S. Reise & N. Duncan (Eds.) *Multilevel modeling: Methodological advances, issues, and applications* (pp. 53–70). Mahwah, NJ: Erlbaum.

du Toit, M., & du Toit, S. (2001). *Interactive LISREL: User's guide*. Lincolnwood, IL: Scientific Software International.

Goldstein, H. I., & McDonald, R. P. (1988). A general model for the analysis of multilevel data. *Psychometrika, 53*, 455–467.

Graham, J. W., & Hofer, S. M. (2000). Multiple Imputation in Multivariate Research. In T. D. Little, K. U. Schnabel, & J. Baumert (Eds.), *Modeling longitudinal and multilevel data* (pp. 201–218). Erlbaum.

Harnqvist, K., Gustafsson, J. E., Muthén, B., & Nelson, G. (1994). Hierarchical models of ability at class and individual levels. *Intelligence, 18*, 165–187.

Heck, R. H. (2001). Multilevel modeling with SEM. In G. A. Marcoulides & R. E. Schumacker (Eds.) *New developments and techniques in structural equation modeling* (pp. 89–127). Mahwah, NJ: Erlbaum.

Heck, R. H., & Thomas, S. L. (2000). *An introduction to multilevel modeling techniques*. Mahwah, NJ: Erlbaum.

Hox, J. (2002). *Multilevel analysis: Techniques and applications*. Mahwah, NJ: Erlbaum.

Hox , J. J., & Maas, C. (2001). The accuracy of multilevel structural equation modeling with pseudobalanced groups and small samples. *Structural Equation Modeling: A Multidisciplinary Journal, 8*, 157–174.

Kalton, G. (1977). Practical methods for estimating survey sampling errors. *Bulletin of the International Statistical Institute, 47*, 495–514.

Kalton, G. (1983). Models in the practice of survey sampling. *International Statistical Review, 51*, 175–188.

Kaplan, D., & Elliott, P. R. (1997a). A didactic example of multilevel structural equation modeling applicable to the study of organizations. *Structural Equation Modeling: A Multidisciplinary Journal, 4*, 1–24.

Kaplan, D., & Elliott, P. R. (1997b). A model-based approach to validating education indicators using multilevel structural equation modeling. *Journal of Educational and Behavioral Statistics, 22*, 323–347.

Kaplan, D., & Ferguson, A. J. (1999). On the utilization of sample weights in latent variable models. *Structural Equation Modeling: A Multidisciplinary Journal, 6*, 305–321.

Kenny, D. A., & Judd, C. M. (1986). Consequences of violating the independence assumption in analysis of variance. *Psychological Bulletin, 99*, 422–431.

Kish, L., & Frankel, M. R. (1974). Inference from complex samples. *Journal of the Royal Statistical Society, 36*(Series B), 1–37.

Korn, E. L., & Graubard, B. I. (1995). Examples of differing weighted and unweighted estimates from a sample survey. *The American Statistician, 49*, 291–295.

Kreft, I., & de Leeuw, J. (1998). *Introducing multilevel modeling*. Newbury Park, CA: Sage.

Lee, E. S., Forthofer, R. N., & Lorimor, R. J. (1989). *Analyzing complex survey data*. Newbury Park, CA: Sage.

Longford, N. T. (1995). *Model-based methods for analysis of data from 1990 NAEP trial state assessment*. Washington, DC: National Center for Education Statistics, 95-696.

McDonald, R. P., & Goldstein, H. (1989). Balanced versus unbalanced designs for linear structural relations in two-level data. *British Journal of Mathematical and Statistical Psychology, 42*, 215–232.

Muthén, B. O. (1994). Multilevel covariance structure analysis. *Sociological Methods & Research, 22*, 376–398.

Muthén, B. O. (1997). Latent variable modeling of longitudinal and multilevel data. In A. E. Raftery (Ed.), *Sociological methodology* (pp. 453-480). Washington, DC: Blackwell.

Muthén, B. O., & Satorra, A. (1995). Complex sample data in structural equation modeling. In P. V. Marsden (Ed.), *Sociological methodology* (pp. 267–316). Washington, DC: American Sociological Association.

Muthén, L. K., & Muthén, B. O. (1998). *Mplus User's Guide*. Los Angeles: Authors.

Muthén, L. K., & Muthén, B. O. (2003). *Mplus Version 2.13 addendum to the Mplus user's guide*. Los Angeles, CA: Authors.

Raudenbush, S. W., & Bryk, A. S. (2002). *Hierarchical linear models: Applications and data analysis methods.* Newbury Park, CA: Sage.

Satorra, A., & Bentler, P. M. (2001). A scaled difference chi-square test statistic for moment structure analysis. *Psychometrika, 66*, 507–514.

Scariano, S. M., & Davenport, J. M. (1987). The effects of violations of independence assumptions in the one-way ANOVA. *The American Statistician, 41*, 123–129.

Skinner, C. J., Holt, D., & Smith, T. M. F. (1989). *Analysis of complex surveys*. Chichester, UK: Wiley.

Snijders, T. A. B., & Bosker, R. J. (1999). *An introduction to basic and advanced multilevel modeling*. Newbury Park, CA: Sage.

Stapleton, L. M. (2002). The incorporation of sample weights into multilevel structural equation models. *Structural Equation Modeling: A Multidisciplinary Journal, 9*, 475–502.

U. S. Department of Education. (2001). *ECLS-K, Base Year Public-Use Data File, Kindergarten Class of 1998–99: Data Files and Electronic Code Book (Child, Teacher, School Files).* Washington, DC: National Center for Education Statistics.

CHAPTER 12

THE USE OF MONTE CARLO STUDIES IN STRUCTURAL EQUATION MODELING RESEARCH

Deborah L. Bandalos

Users of quantitative methods often encounter situations in which the data they have gathered violate one or more of the assumptions underlying the particular statistic(s) they wish to use. For example, data may be non-normally distributed when the statistic a researcher plans to use requires normality of the variables' distributions. In cases such as these, the focus turns to the *robustness* of the statistic; that is, will use of the statistical procedure still yield relatively unbiased values of parameter estimates, standard errors, and/or test statistics if certain assumptions, such as normality, are violated? In structural equation modeling, the inclusion of many observed variables, the concomitant estimation of numerous model parameters, the need for large sample sizes, the inclusion of coarsely categorized and/or non-normally distributed observed variables, and the possible misspecification of the model will almost guarantee assumption violations in the majority of published studies. And even in the rare situation in which data perfectly meet the requirements of the statistical estimator, the lack of known sampling distributions for most of

Structural Equation Modeling: A Second Course, 385–426

the commonly used goodness-of-fit statistics in structural equation modeling (SEM) has resulted in the need for Monte Carlo studies to determine values at which one should infer that the data fit—or do not fit—the hypothesized model. Other challenges include the analysis of a correlation matrix rather than covariance matrix and the presence of nonlinear relationships. Finally, although the maximum likelihood (ML) and generalized least squares (GLS) estimators commonly used in SEM studies do have known sampling distributions, these distributions are known only asymptotically. This means that, although the properties of these estimators can be shown to hold as sample size approaches infinity, it is not always clear how large a sample is required for them to hold in practice.

In order to investigate the questions raised by problems such as these, simulation or Monte Carlo studies are used. According to Harwell, Stone, Hsu, and Kirisci (1996), the term "Monte Carlo" was first applied to simulation studies by Metropolis and Ulam (1949). Such studies were used during World War II in research involving atomic energy. I distinguish Monte Carlo studies from simulation studies by noting that Monte Carlo studies are statistical sampling investigations in which sample data are typically generated in order to obtain empirical sampling distributions. I use the term "simulation study" to refer to studies that involve data generation of some type but do not necessarily involve the generation of sample data. For example, a simulation study might entail the generation of population data, or may simply generate a set of data to demonstrate a statistical analysis. In this chapter I focus on Monte Carlo studies in which a model of some particular type (e.g., measurement or structural) is specified and many samples of data are then generated based on that model structure. The data in these samples are generated in such a way that they violate one or more of the assumptions of the estimator being used. The samples are then analyzed using SEM software and the results are aggregated across samples to determine what effect the assumption violations had on the parameter estimates, standard errors, and/or goodness-of-fit statistics. Serlin (2000) described the utility of Monte Carlo studies as "provid[ing] the information needed to help researchers select the appropriate analytical procedures under design conditions in which the underlying assumptions of the procedures are not met" (p. 231). In SEM studies, however, the lack of viable nonparametric alternatives may often result in estimators being used even when assumptions are known to have been violated. This introduces an additional situation in which Monte Carlo studies can assist applied researchers: that of determining the degree to which the obtained parameter estimates, standard errors, and fit statistics may be compromised if assumptions are violated.

ALTERNATIVES TO MONTE CARLO STUDIES

In some cases, Monte Carlo studies may not be the best way to address the types of problems raised above. One of these situations is when an analytical solution is available for the problem. As noted by Harwell and colleagues (1996), "A Monte Carlo study should be considered only if a problem cannot be solved analytically, and should be performed only after a compelling rationale for using these techniques has been offered" (p. 103). For example, if it were possible to derive the sampling distribution of a fit index such as the Comparative Fit Index (CFI) analytically, it would be preferable to obtain regions of rejection in that manner rather than by conducting a Monte Carlo study to determine the critical value at which it produces the desired rejection rates. However, in many cases an analytical solution is not possible. One reason for this is that statistical tests are typically developed using algorithms based on the properties of known mathematical distributions such as the normal distribution. If the data do not conform to such a distribution then the sampling distribution no longer holds. In SEM studies the sampling distributions of estimators also depend on the correct specification of the model. If correct model specification (or the degree of misspecification) of the model cannot be assumed, then exact sampling distributions cannot be derived. In some situations, Monte Carlo studies are used to provide insight into the behavior of a statistic even when mathematical proofs of the problem being studied are available. One reason for this is the fact that theoretical properties of estimators do not always hold under real data conditions. Nevertheless, results of Monte Carlo studies should always be related back to exact statistical theory as much as possible.

Another situation in which Monte Carlo studies may not be appropriate is where a population study is sufficient. In a population study, a model (or models) is specified in order to study the effect of certain conditions *in the population*. This type of study is useful for situations in which the degree of sampling variability and/or the stability of results such as parameter values is not of interest. For example, Kaplan (1988) was interested in the differential impact of population model misspecification on maximum likelihood (ML) and two-stage least squares (2SLS) estimation. Because he was interested in the effects in the population it was not necessary to generate sample data.

ADVANTAGES AND DISADVANTAGES OF MONTE CARLO STUDIES

As noted previously, Monte Carlo studies can be illuminating in situations such as the following:

- Violations of assumptions such as:
  - Normality
  - Continuousness of observed variables
  - Model misspecification
  - Nonlinearity of effects
- Investigation of issues such as:
  - Small sample behavior of asymptotically based statistics
  - Effects of estimating large numbers of model parameters
  - Effects of analyzing correlation matrices rather than covariance matrices
  - Properties of goodness-of-fit statistics for which no mathematical distribution exists.

Perhaps most importantly, Monte Carlo studies make it possible to study the potential interactions of assumption violations with other problems.

On the other hand, some disadvantages of Monte Carlo studies should be noted. First, Monte Carlo studies are essentially experimental studies in which the variables of interest are manipulated. As such, they share the limitations of all experimental studies. More specifically, Monte Carlo studies are dependent on the representativeness of the conditions modeled. If these conditions are not similar to those found in real data, the usefulness of the study will be severely limited. This problem is exacerbated by the extreme complexity of most real-world data. In many cases, real data sets include such complications as hierarchically nested structures (see Stapleton, Chapter 11, this volume), data that are missing in a nonrandom fashion (see Enders, Chapter 10, this volume), and the existence of heterogeneous subgroups (see Gagné, Chapter 7, this volume). Unfortunately, the inclusion of all factors known to affect the behavior of the statistic under study could render the Monte Carlo study impossibly complex. Thus, as in any study, Monte Carlo studies must balance practicality with fidelity to the real-world phenomenon one is attempting to model.

In the next section I discuss planning Monte Carlo studies, including the study design and the choice of dependent and independent variables. In the following sections I present information relevant to data generation, analysis of data from Monte Carlo studies, and issues of data management. These sections are illustrated with an example of a Monte Carlo study currently in the planning stages that investigates the behavior of the WLSMV (weighted least squares–mean, variance) estimator available in the Mplus program. I conclude by offering suggestions for those engaged in or contemplating Monte Carlo studies. Readers interested in Monte Carlo studies should also review work by Paxton, Curran, Bollen, Kirby, and Chen (2001) and by Muthén and Muthén (2002). The former article provides an illustration of the design and implementation of a Monte Carlo study while the latter demonstrates how to conduct a Monte Carlo study to determine the sample size needed to obtain unbiased parameter estimates and standard errors under non-normality and missing data conditions.

PLANNING MONTE CARLO STUDIES

In this section I focus on the decisions to be made in planning a Monte Carlo study. First, factors involved in determining a research question are presented. Next, the selection of the independent variables in the study and appropriate levels for these are discussed, followed by suggestions for the determination of appropriate dependent variables and their formulations. The actual data generation process is discussed and illustrated in the next section. Finally, issues involved in data analysis and data management are discussed.

Determining the Research Question

Essentially, the determination of a research question in Monte Carlo research is no different from other areas. Good research questions should address gaps in the literature and be designed to advance understanding of a particular topic. In Monte Carlo research, additional stipulations are that no analytical solution is available to answer the research question of interest and that sampling stability or variability is of interest so that a population study would not be sufficient to answer the question. Also, as noted previously, the research questions should be informed as much as possible by statistical theory.

Research questions in Monte Carlo studies are often developed in answer to problems that occur in either applied or methodological research. For example, in my work I frequently encounter coarsely categorized data such as that arising from the use of Likert scales. The data are, by virtue of their categorical nature, non-normally distributed. Because popular SEM estimators such as ML require the assumption of continuousness and normality of the data, the question of proper estimation of model parameters, standard errors, and fit indices arises. Recently, the Mplus computer package has introduced a new weighted least squares estimator with corrections to the model χ^2 based on means and variances (WLSMV). This method was designed to yield more accurate estimates of parameters, standard errors, and model χ^2 values than normal theory (NT)–based estimators. As explained by Arnold-Berkovits (2002), the

WLSMV, or robust weighted least squares estimator, should result in more efficient estimation than earlier WLS estimators because estimation of the parameters is based on a diagonal weight matrix rather than the full weight matrix required by previous WLS estimators. Although based on strong statistical theory, these estimators have not been studied extensively in practice. Thus, while they are theoretically superior for applications involving categorized and/or non-normally distributed data, there is still some uncertainty regarding the conditions under which this superiority will be manifested. In a study comparing ML, GLS, and WLS estimators, Muthén and Kaplan (1992) found that performance of WLS deteriorated with larger models. Studies of the WLSMV estimator by Muthén, du Toit, and Spisic (in press), Yu and Muthén (2002), and Arnold-Berkovits (2002) have indicated that estimates of fit indices, standard errors, and parameter values based on this estimator are biased when small sample sizes are combined with non-normality of the data. The study by Muthén and colleagues also revealed that analysis of a confirmatory factor analysis (CFA) model resulted in less standard error bias than did analysis of a full latent variable path model. However, because the CFA model included fewer variables, the effect may have been due to model size rather than to the type of model. One research question of interest in my example study is therefore whether the WLSMV estimators display a sensitivity to model size and/or model type.

Related to the issue of the research question is the specification of research hypotheses. In most research studies research hypotheses flow naturally from the review of the literature. Essentially, the point of one's literature review is to demonstrate why the study is needed. In the case of Monte Carlo research, this should involve a discussion of why analytical or population studies would be unsatisfactory and how the research questions are related to statistical theory. However, it has not been typical for specific research hypotheses to be stated in Monte Carlo research. While these hypotheses may be implicit in the design of the study, my feeling is that these should be stated explicitly. Doing so makes it much easier for readers to judge whether the hypotheses are reasonable, both in terms of being addressable by the study design and of being related to statistical theory and previous research. It also allows for a more direct link with the study results and discussion. In the context of my example study, one such hypothesis would be that the new WLSMV estimator will display less sensitivity to model size than earlier categorical variable methodology (CVM) estimators. This is because the method of estimation for these estimators is not as complex as that used for the earlier estimators, and should therefore result in more stable estimation, even in the presence of larger models.

CHOICE OF INDEPENDENT VARIABLES[1] AND THEIR LEVELS

Choice of a Model and Model Characteristics

A key decision Monte Carlo researchers must make is the determination of what type of model to use. Four aspects of model choice are typically of interest: model type, model size, model complexity, and model parameters. The most basic choice with regard to model type is that between the use of a measured variable path, CFA, or full structural (i.e., latent variable path) model. Researchers may be specifically interested in only one of these, or may want to investigate different types of models, thus making the type of model one of the independent variables in the study. Model size is usually operationalized as the number of parameters to be estimated in the model, or as model degrees of freedom, both of which are based in part on the number of observed variables in the model. A closely related determination is that of model complexity. Within any of the model types, complexity can be increased through the addition of cross-loading indicators, reciprocal paths, correlated disturbances, or nonlinear effects. Finally, values must be chosen for all parameters to be estimated. These values represent the population values of interest, although when sample data are generated from these parameters their statistical values will, of course, vary to some extent because of sampling error. These four aspects of model choice are discussed in more depth in the following sections.

Model type. The research question(s) might govern the choice of a model. For example, investigation of the effects of cross-loading indicators could be of most interest in CFA models. However, researchers should keep in mind that the full information nature of commonly used estimators in SEM means that facets of the model influencing estimates of one parameter may be propagated to other, possibly unexpected, parts of the model (Kaplan, 1988). In most situations it is probably best to include more than one type of model in Monte Carlo studies because this allows for greater generalization to applied settings. However, the nature of the research question should govern this choice. As in any study, there is a tradeoff between the inclusion of many independent variables with few levels of each, or fewer independent variables with more levels of each. For areas in which little previous research has been conducted, it may be more informative to choose one representative model but to include a wider variety of other independent variables, and/or more levels of these variables. For areas in which a particular model has been studied extensively under many different conditions, a researcher may choose to broaden the research base by studying other models under some of the same conditions. For example, the bulk of the Monte Carlo studies in

SEM has been conducted with CFA models; in a review of 62 Monte Carlo studies in SEM, Hoogland and Boomsma (1998) found that 89% used CFA models. This suggests that future Monte Carlo studies should incorporate more structural models.

There are two basic ways in which researchers obtain population models. One way is to review the applied and methodological literature to determine types of models and parameter values that are commonly encountered in practice. A population model is then constructed artificially as a sort of composite of models and parameter values found in these studies. The advantage of this method is that the model and its characteristics can be manipulated experimentally in order to investigate the conditions of interest. The disadvantage is that the constructed model may not reflect real-world conditions. For examples, models are often constructed with equal factor loadings or uncorrelated errors, conditions that are unlikely to be found in practice, but which allow for unconfounded investigations of other conditions of interest.

The second, and probably less common way of obtaining a population model is to base it on an actual data set. In this method, the researcher would treat an existing dataset as the "population," and define a model that is fit to those data as the population model. The population parameters would be defined as the estimated model parameters from that dataset. The advantage of using this method is that it is more likely to reflect real-world conditions and therefore to produce results that are generalizable to these conditions. The disadvantage is that the researcher may not be able to manipulate all model characteristics that may be of interest. This approach has recently been endorsed by MacCallum (2003) because it does not make the unrealistic assumption that a model fits exactly in the population. MacCallum further argued that Monte Carlo studies based on perfectly fitting population models "are of only limited value to users of the methods" (p. 135) because they do not address the question of how methods perform when the model of interest is incorrect in the population. In his paper, MacCallum suggested other methods of incorporating model error into a population model, such as including several minor nuisance or method factors in a CFA model.

In the context of my example study, I was interested in both CFA and full structural models. Because only one previous study (Muthén et al., in press) has investigated both CFA and SEM models, I wanted to expand on that study by including both types of models.

Model size. Another important consideration is the size of the model. This is typically based on the number of observed variables. In their review of the literature, Hoogland and Boomsma (1998) found that the number of observed variables in SEM Monte Carlo studies ranged from 4 to 33, with an average of 11.6. A way to conceptualize model size is the model degrees of freedom, which is a function of the number of parameters to be estimated and the number of observed variables. This was the method used in a meta-analysis of Monte Carlo studies on the robustness of the χ^2 statistics in SEM by Powell and Schafer (2001), who found that model degrees of freedom ranged from 3 to 104. In this meta-analysis, model degrees of freedom were found to have large effects on the maximum likelihood and asymptotically distribution-free (ADF) χ^2 statistics. Model size is therefore an important model characteristic in SEM Monte Carlo studies, and should be chosen carefully. In most cases, the best approach would be to vary model size as one of the independent variables in the study design. Failure to take model size into account in this way can result in misleading conclusions. For example, in an early study of ADF estimators with four observed variables, Muthén and Kaplan (1985) concluded that ADF-based χ^2 estimates showed little bias. However, in a later study that included models with up to 15 variables (Muthén & Kaplan, 1992), these authors found that the sensitivity of ADF χ^2 tests increased with sample size.

Because of these findings, I was interested in including models with large numbers of variables in the example study. I was also interested in studying the issue of whether model size or model type (CFA or structural model) have differential impacts on model fit. As the largest number of variables included in previous studies was 15 (Arnold-Berkovits, 2002; Muthén et al., in press; Yu & Muthén, 2002), I included four levels of number of variables, ranging from 8 to 24 variables. As noted previously, within each of the four levels of number of variables I also included two models: one CFA and one full structural model. This resulted in a total of eight different models. Figure 12.1 shows two illustrative models: an eight variable, two-factor CFA model and a full SEM model with two exogenous and two endogenous factors, each measured by six variables. For estimation with categorical data, the smallest model has 17 parameters to be estimated (6 factor loadings + 1 factor correlation + 8 threshold values + 2 factor variances). However, note that this is only if the variables have two categories; with more categories, more thresholds would need to be estimated.

Model complexity. Related to model size is model complexity. Model complexity has often been operationalized as the number of free parameters in a model, with more free parameters indicating greater complexity. Model complexity can be introduced by adding such things as cross-loading indicators, reciprocal paths, correlated disturbances or measurement error terms, or nonlinear effects such as interactions or quadratic terms. Readers may wonder why a researcher would want to complicate her or his model by including such parameters, as these may confound the effects of other independent variables. One reason is that many of the datasets analyzed in practice include such effects, so models that include these may be more generalizable to conditions encountered by applied researchers. A related reason is that, because ignoring such effects may bias the estimation of other model parameters, it may be of interest to determine the amount of bias that could be introduced by such omissions. In the current example study I did not include complex models because the behavior of the WLSMV estimator has not been studied extensively even in the context of simple models. I therefore felt that the inclusion of complicating factors was premature at this stage of investigation.

Model parameter values. Choice of parameter values is an important consideration because these are related to the reliability of factor indicators as well as the strength of relationships among both observed and latent variables. These in turn affect the levels of power of tests of individual parameter estimates. In choosing the size of factor loadings, the researcher should keep in mind that indicator reliability may be defined as:

$$\frac{\lambda_i^2}{\lambda_i^2 + \theta_{ii}}, \tag{1}$$

where λ_i and θ_{ii} are the unstandardized factor loading and error variance of indicator *i*, respectively. Thus, indicator reliability depends not just on the value of the loading, but also on the ratio of loading size to the size of the error variance. This is also equivalent to the square of the standardized loading.

Hoogland and Boomsma (1998) reported standardized factor loading sizes ranging from .3 to 1.0 and factor correlations from 0 to .5 in their review of Monte Carlo research in SEM. These authors did not provide values for structural parameters, but common standardized values range from .3 to .7. The values chosen should represent what is typically seen in applied research in the field of study to which one hopes to offer guidance. A review of applied studies in the literature of that field should be done to determine if one's choices are reasonable from this point of view. One issue regarding factor loading values is whether all loadings for a particular factor should be set to the same value, although possibly different across factors, or whether loadings should have mixed values. If mixed loadings are chosen it is still possible to vary loading magnitude across factors by choosing sets of loadings whose average values differ within high, low, or medium ranges. Of course, researchers could choose to include sets of both mixed- and single-value loadings.

Because previous studies of the WLSMV estimator have included only fairly high loading conditions (i.e., loadings of .7 to .8), I decided to include both low (.4 to .5) and high (.7 to .8) loading conditions in the example study. As discussed by Enders and Finney (2003), loadings in the range of .4 to .5 are not uncommon in applied studies. These researchers found, not surprisingly, that lower loadings resulted in less power to detect model misspecifications for NT-based estimators.

Other Independent Variables

As with the choice of a model, the choice of other independent variables to include in the design of a Monte Carlo study should be based on a review of the literature. Aside from choices about aspects of the model(s) to be investigated, other common independent variables in SEM Monte Carlo studies include:

- Sample size
- Level of non-normality
- Degree of model misspecification
- Degree of categorization of observed variables
- Method of estimation

Here again, Monte Carlo research is not essentially different from other quantitative research. Determining the appropriate independent variables and their levels should be based on a review of the applied and/or methodological research.

Sample size. Powell and Schafer (2001) found that sample size affected values of some types of estimators but not of others, so the choice of whether to include sample size as an independent variable may depend on the type(s) of estimator to be studied. It would also be important to determine if there was any statistical theory to suggest how sample size would affect the estimators of interest so that appropriate research hypotheses could be posed. Sample size would also be an important factor to consider if sampling stability was of interest, because stability will obviously improve with larger samples. The interest in this case may be in determining the smallest sample size at which reasonably stable estimates can be obtained. Researchers should keep in mind that fit statistics, parameter values, and standard errors will typically have different thresholds for stability. Powell and Schafer found that sample size varied from 25 to 9,600 in the Monte Carlo studies they reviewed. A typical range of values is from 100 to 1,000, unless the researcher is interested in focusing on extremely large or small sample sizes.

Previous studies of the WLSMV estimator incorporated sample sizes ranging from 100 to 1,000. The results of these studies indicated poor performance of smaller sample sizes with non-normally distributed and/or small numbers of categories. WLSMV performance also deteriorated with larger models at the smaller sample sizes. Because the number of variables in the models for the example study varied quite a bit, I chose different sample size levels for the different models. For the smallest model with eight variables I varied sample size at 150 and 300, while for the largest model with 24 variables I used sample sizes of 450 and 950. This maintained the same ratio of number of subjects to number of estimated parameters.

Type of distribution. Distributional type was included as an independent variable in 76% of the studies in Hoogland and Boomsma's (1998) review of robustness studies in SEM. Given the assumptions of normality underlying common estimators such as ML and GLS, this is not surprising. Powell and Schafer (2001) found that researchers used several methods of generating non-normal data. One of these was to generate multivariate normal data and then categorize certain variables in such a way as to yield the desired degrees of skew and kurtosis. Over half of the studies investigated by these authors used this approach. This is also the method used in Mplus MONTECARLO procedures and in the example study. A less commonly used variation on this approach was to censor variables' data from above or below to obtain the chosen degree of non-normality. One issue to consider if non-normality is induced using these methods is the form of the underlying distribution. While most studies begin with normally distributed data and categorize the data in such as way as to yield non-normal distributions, another approach is to assume that the underlying distribution itself is non-normal and categorize the data so that it matches that distribution. This may be a more realistic approach in some situations (see Bandalos & Enders, 1996, for a discussion).

Another common approach to generating non-normal data involves creating a covariance matrix with a specified structure from which sample data can be generated, and inducing the desired levels of univariate skew and kurtosis for the observed variables using procedures such as those developed by Vale and Maurelli (1983). Data from a multivariate normal distribution can also be transformed into a non-normal multivariate χ^2 or t distribution (see, e.g., Hu, Bentler, & Kano, 1992). More recently, Mattson (1997) has proposed a method in which the distributions of the latent common and unique factors are transformed in such a way as to yield the desired levels of non-normality in the observed variables. This method was reviewed by Reinartz, Echambadi, and Chinn (2002) who found it to work well when implemented in three commonly used software packages. Finally, Muthén and Muthén (2002) have proposed a method in which non-normally distributed data are obtained as a mixture of two normally distributed subpopulations having different means and variances for the observed variables. Because the majority of these methods control the univariate, and not the multivariate, distributions, the level of multivariate normality should be computed and reported after the data generation process has been completed. A related strategy would be to vary the multivariate skew and kurtosis rather than the separate univariate values. This method would be preferable because it is the multivariate and not the univariate parameters that are assumed to be within normal limits for SEM NT estimators. However, there currently exist no published guidelines regarding the levels of multivariate skew and kurtosis to which these estimators are robust (although see Finney & DiStefano, chapter 9, this volume).

One problem in generating data from populations with the prespecified levels of skew and kurtosis is that across samples these characteristics have high levels of sampling variability. Thus, although the desired levels may be obtained for the population, for any given sample they may deviate considerably from these values. This is especially true for kurtosis, and with higher levels of both skew and kurtosis. A final point has to do with the choice of skewness and kurtosis values. Micceri (1989) found that the distributions he studied "exhibited almost every conceivable type of contamination" (p. 162), making it difficult to generalize about common levels of skew and kurtosis. Because I have no reason to think that things have changed substantially since his study, the best course of action in choosing distributional levels may be to consult both the applied and Monte Carlo literature most relevant to the applied area to which the study will be targeted. Researchers might also want to explore the limits of NT estimators by increasing the level of non-normality systematically in order to determine the degree of non-normality that can be tolerated without severe consequences to parameter estimates, standard errors, or overall fit.

Level of non-normality was of interest in my example study. Previous studies of the WLSMV estimator included levels of absolute skewness (sk) in the range of .316 to 2.03, and kurtosis (k) ranging from -1.91 to 1.41. In my study I wanted to extend the research base by including more severe levels of nonnormality and chose the following target levels for "mild," "moderate," and "severe" non-normality, respectively: sk = .4, k = 2.0; sk = 1.25, k = 3.5; and sk = 3.0, k = 7.0.

Model misspecification. MacCallum (2003) has recently argued that Monte Carlo researchers should focus more on models that contain some level of population misspecification. Population lack of fit has been referred to as "error of approximation" and is arguably a more realistic conceptualization for most phenomena in the behavioral sciences (Browne & Cudeck, 1992; Cudeck & Henly, 1991). From this point of view, consideration of models with some degree of error of approximation is likely to provide more useful information to applied researchers who are attempting to model such phenomena. Estimation methods have also been found to be differentially sensitive to model misspecification (Olsson, Foss, Troye, & Howell, 2000). Model misspecification is therefore an important independent variable in many Monte Carlo studies. One way in which lack of fit is studied is through generating sample data from a population model with a perfect fit, misspecifying the model in some way, and then fitting the generated sample data to the misspecified model. Model misspecifications are often categorized as errors of omission and errors of inclusion. As the names imply, the first type involves omitting the estimation of model parameters that were present in the population model, while the second refers to estimating parameters that were not present in the population. Although some researchers argue that errors of inclusion do not really constitute errors in a strict sense, because they do not result in parameter estimate bias or lack of fit as measured by the χ^2 statistics, they are commonly included in discussions of misspecification because they do affect model degrees of freedom and can result in inflated standard errors. Commonly omitted parameters include secondary factor loadings (i.e., cross-loadings), correlations between residuals, structural paths, and even entire factors. Incorrectly included parameters are typically modeled as extra structural parameters. Errors of omission have been found to be most serious and are therefore the object of most study.

Another method of obtaining misspecified data is to use an actual large dataset as the "population" and to generate samples from it. Because no model will fit perfectly even in a large sample, some degree of model misspecification will be present in both the "population" and sample data. Samples can be generated by either taking repeated samples of a given size with replacement or by obtaining a covariance matrix based on the "population" data and generating random samples from that parent correlational structure. This method was recommended by MacCallum (2003), who also suggested introducing lack of fit into a simulated population model by including such things as extra minor factors or method effects.

One problem with studies involving misspecification is that the degree of misspecification is often not quantified, making it difficult to model the effect of misspecification on values of fit indexes, parameters, and standard errors. In some cases model misspecifications that are assumed to be serious might not have as large an effect on these values as do presumably less serious misspecifications. This is due to the fact that statistical tests of model parameters have differential levels of power, depending on their correlations with other model parameters (Kaplan & Wenger, 1993).

While many Monte Carlo researchers simply categorize misspecifications as small, medium, or large, a more informative method of quantifying misspecification due to errors of omission is to estimate the power associated with the test of the omitted parameter, using methods such as those proposed by MacCallum, Browne, and Sugawara (1996). Alternatively, because the root mean square error of approximation (RMSEA) is a measure of model lack of fit per degree of freedom, this index could be used to quantify the degree of model misspecification. Procedures such as these not only ensure that misspecifications correspond to the desired levels but allow for more accurate modeling of the effects of model misspecification.

The degree to which the model χ^2 and other fit indexes based on a particular estimator are able to detect model misspecification is one of the most important criteria for evaluating that estimator. Given this, I gave serious consideration to including misspecified models in the example study. However, because it appeared to me that the behavior of this estimator had not been studied sufficiently even in the context of correctly specified models, I decided not to include model misspecification as a design facet in the study. This decision was also motivated by concerns about the size of the study, which at this point consists of 8 (models) × 2 (loading sizes) × 2 (sample sizes) × 3 (levels of non-normality), or 96 cells, and there was still one more condition I wanted to vary. I therefore decided to retain the issue of model misspecification for a future study.

Categorization. Because many of the measures used in the social sciences are often not at the interval level, many applied researchers are interested in the degree to which the inclusion of categorical or ordinal-level data will affect the results of SEM analyses. Categorization of generated data is usually accomplished through generating continuous data and categorizing it through the imposition of cutpoints at different places along the continuum. This method can also be used to obtain non-normal univariate distributions, and is the method that was used to create the categorical, nonnormally distributed data in the example study, as noted previously. In this method the level of non-normality is controlled through the choice of cutpoints. For example, to obtain data with two categories, I used a single cutpoint. Use of a cutpoint of 1.39, as shown in the Mplus program in a later section, resulted in skewness and kurtosis levels of approximately 3.0 and 7.0, respectively. In this context it should be noted that categorized data do not allow for the generation of perfectly normal distributions, especially when the number of categories is small. Because NT-based SEM results are affected most severely by distributions with smaller numbers of categories, Monte Carlo research in this area commonly focuses on variables with two, three, or four categories. With five or more categories, the effects of categorization on parameter estimates and standard errors are generally negligible (Dolan, 1994). For more discussion of the analysis of categorical data, see Finney and DiStefano (Chapter 9, this volume).

Because the WLSMV estimator of interest in my example study has been recommended for use with categorical data, I was particularly interested in including this as a design facet. In the study by Arnold-Berkovits (2002), data based on only two categories were found to be especially problematic in terms of bias in parameter and standard error bias. In general, studies of categorical variables have found that performance does not improve substantially when the number of categories is increased beyond five. I therefore decided to use data with two, three, and five categories.

Estimation method. As mentioned previously, the various estimators used in SEM are not equally sensitive to the effects of the other possible independent variables that have been discussed. Some estimators, such as the ADF estimator, have been formulated specifically to be less sensitive to the effects of non-normality. Other estimators commonly studied include the scaled or corrected χ^2 obtained by dividing the ML or GLS χ^2 value by a factor derived from estimates of the level of non-normality present in the data. Finally, the robust estimators used in Mplus (see Muthén & Muthén, 1998, pp. 357–358) were developed to yield unbiased χ^2 statistics and standard errors in the presence of non-normality and coarse categorization. For details on these estimators, the reader is referred to Chapter 9 by Finney and DiStefano in this volume. As noted previously, estimators have also been found to be differentially sensitive to model misspecification. Clearly the choice of which estimators to study would be based on the type of data and model(s) included in the study. Typically, however, the ML estimator is included as a basis of comparison even when it is not the estimator of interest. In the current study I followed this convention and included the ML estimator along with the WLSMV estimator.

Interaction effects. It should be clear from reading the previous sections that the various independent variables that might be included in an SEM Monte Carlo study do not operate in isolation but interact with each other. For example, estimation methods are differentially sensitive to sample size, model size, level of misspecification, and non-normality. Therefore, possible interactive effects should be anticipated when designing Monte Carlo studies and the relevant variables should be included. Of course, as in any study, the researcher must make compromises between the inclusion of multiple design characteristics and resources available. For Monte Carlo studies, time and computer capacity are two important resources that usually limit the scope of a study. This means that the researcher must choose the independent variables and their levels carefully. Researchers should also keep in mind that the interpretation of high-level interaction effects among several variables with several levels each may be very difficult both for researchers to explain and for readers to comprehend. Focusing on fewer carefully chosen variables may be more elucidating in the long run.

To sum up the research design thus far, there are eight models, two magnitudes of factor loadings, two sample sizes, three levels of non-normality, and three levels of categorization, resulting in $8 \times 2 \times 2 \times 3 \times 3$, or 288 cells, in the design. Data from each cell of the design will be analyzed using both ML and WLSMV estimation methods and results will be saved. In the next section I discuss choices for dependent variables.

CHOICE OF DEPENDENT VARIABLES

The primary focus in many Monte Carlo studies is on the effects of the independent variables on parameter estimates, standard errors, and/or fit indexes. For each of these outcomes, bias and efficiency are typically of most interest. Bias refers to a systematic difference between a sample estimate and the corresponding population value. Efficiency has to do with the sampling variability of a statistic, with statistics that vary least across samples being preferred.

Parameter Estimates

Bias. Parameter estimate bias is typically calculated in a relative manner as the average deviation of the sample estimate from its population value, relative to the population value, or:

$$Bias\left(\hat{\theta}_i\right) = \sum_{j=1}^{n_r}\left(\frac{\left(\hat{\theta}_{ij} - \theta_i\right)}{\theta_i}\right) / n_r, \qquad (2)$$

where $\hat{\theta}_{ij}$ is the jth sample estimate of the ith population parameter θ_i, and n_r is the number of replications within the cell. This quantity can be multiplied by 100 to obtain percentage bias values. If certain parameter values, such as factor loadings or error variances, are affected in the same way by the study conditions, relative bias across these parameters is sometimes averaged as well. If these parameter estimates are affected differentially, however, bias should be reported separately for the individual parameters of interest. Similarly, it may be the case that some study conditions result in large amounts of positive bias while others result in large negative bias for a particular parameter or set of parameters. Across all study conditions this bias may average to zero, even though bias is substantial under some conditions. For this reason, Equation 2 represents the calculation of separate bias values for each cell of the design. However, if similar amounts and directions of bias are found across levels of a design characteristic, bias values can be averaged across cells. Finally, Equation 2 can also be modified by taking the absolute value of the quantity in the numerator to obtain a measure of the average amount of unsigned bias.

Hoogland and Boomsma (1998) suggested as a criterion that the absolute value of $Bias(\hat{\theta}_i)$ be less than .05 for parameter estimates in a particular condition, or across conditions in a study, to be considered unbiased. Kaplan and Hollis (1987) offered the more lenient criterion that bias of less than .10 to .15 might be considered negligible.

Efficiency. The efficiency of parameter estimates is usually measured as the standard deviation of the sample estimates from their average value, or:

$$\sqrt{\frac{\sum_{j=1}^{n_r}\left(\hat{\theta}_{ij}-\bar{\hat{\theta}}_i\right)^2}{n_r-1}}. \quad (3)$$

This quantity is also known as the empirical standard error of the parameter and can be calculated either within a cell of the design or across all conditions, depending on whether the study conditions are expected to affect values differentially across cells.

Standard Errors

Relative standard error bias is assessed in a manner similar to parameter estimate bias, that is, as the deviation of each sample standard error from its population value relative to the population value, averaged across replications. However, as Hoogland and Boomsma (1998) pointed out, there are different methods of estimating the population values. These are:

- Obtain estimates from the diagonal of the asymptotic covariance matrix of the parameter estimates using the population values for the parameter estimates;

- Calculate the empirical standard errors as the standard deviation of parameter estimates over a large number of replications as in Expression 3.

The first method is only accurate if the assumptions of the estimation method are met. If these assumptions are violated, as when non-normally distributed data are used for ML or GLS estimation methods, the obtained standard errors will be incorrect. The accuracy of the second method is not dependent on any distributional assumptions, but is affected by the number of replications that are used. Because of this, the researcher should use a large number of replications to obtain the empirical standard error in Expression 3. Although it is difficult to provide guidelines for a sufficient number of replications that would be appropriate for all situations, it should be at least 10,000 and possibly as many as 100,000 for non-normally distributed data. The best way of deciding this issue would be to increase the number of replications systematically until values of the empirical standard errors change only minimally.

After the population value is decided upon, the relative standard error bias can be calculated as:

$$Bias\left(\hat{SE}(\hat{\theta}_i)\right) = \sum_{j=1}^{n_r}\left(\frac{\hat{SE}(\hat{\theta}_i)_j - SE(\hat{\theta}_i)}{SE(\hat{\theta}_i)}\right) / n_r \tag{4}$$

where $SE(\hat{\theta}_i)$ is an estimate of the population standard error of $\hat{\theta}_i$, and $\hat{SE}(\hat{\theta}_i)_j$ is the estimated standard error of $\hat{\theta}_i$ for the jth replication. As with parameter estimate bias, standard error bias can be calculated either within or across cells of the design. Hoogland and Boomsma (1998) suggested that "acceptable" levels of bias not exceed 5% of the absolute value of Equation 4.

Fit Indexes

Values of the χ^2 statistic have been studied most often as outcome variables in Monte Carlo studies. Effects on actual χ^2 values as well as rejection rates are typically studied. Rejection rates are obtained in Monte Carlo studies by calculating the number of replications for which the χ^2 value for a model would be larger than the critical value at a given nominal level for Type I error. Relative bias of actual χ^2 values can be calculated as the average difference between the sample values and the population value, expressed as a proportion of the population value.

Because the population value of the model χ^2 statistic is equal to its degrees of freedom, this can be expressed as:

$$Bias\left(\hat{\chi}^2\right) = \sum_{j=1}^{n_r} \frac{\hat{\chi}_j^2 - df}{df} / n_r . \qquad (5)$$

Because most fit indexes other than χ^2 have unknown sampling distributions, values of these indexes are often described by comparing their average values across replications to the cut-off criteria suggested in the literature (e.g., Hu & Bentler, 1999).

Equation 5 provides a way of quantifying average values of the model χ^2 statistic. However, high average χ^2 values do not necessarily mean that there will be a large number of model rejections, because the high mean values could result from only a few cases. Researchers are therefore also interested in the "tail behavior," or percentage of sample χ^2 values that fall into the region of rejection. This is often quantified as the proportion of Type I errors that are made. Bradley (1978) suggested as stringent criterion and liberal criterion that the empirical Type I error rate, π, lie in the range $\alpha \pm 0.1\alpha$ or $\alpha \pm .5\alpha$, respectively. Robey and Barcikowski (1992) have outlined more complicated procedures that can be used to determine critical values to test the hypothesis that $\pi = \alpha$. Most recently, Serlin (2000) has amended these procedures so that the range for α would depend on its value, with the distance between α and π becoming more liberal with smaller values of α.

Other Dependent Variables

Other choices for dependent variables in SEM Monte Carlo studies are as varied as are the studies themselves. Other commonly used variables include power levels and Type II error rates, proportions of convergent and/or admissible solutions, and values or patterns of modification indexes.

For my example, I was interested in bias in parameter estimates and standard errors as well as bias and Type I error rejection rates for the WLSMV estimator.

GENERATING DATA FOR MONTE CARLO STUDIES

Once the research questions have been developed and the study design has been mapped out, the researcher is ready to begin simulating the data to be used in the study. This process typically begins with the creation of a population covariance matrix. If more than one model is to be used in the study, a population covariance matrix is created for each, and data are generated from each of these. The sample data are then analyzed and the results from these are saved and evaluated. In some software packages, such as Mplus and EQS, both the generation and analysis of the sample data can be accomplished within the same program. In others such as LISREL, different (although possibly companion) programs must be used to generate the sample data and conduct the analyses. However, before samples can be generated, two important decisions must be made: the choice of a random seed and the choice of the number of replications. These issues are addressed in the following sections.

Generating a Population Matrix

A common method of creating population covariance matrices is to fix all parameter values to the desired population values in an SEM software package. The covariance matrix implied by these values can then be saved and used as a basis for generating the sample data. This method is used in LISREL and other programs. In Mplus and EQS the population and sample matrices are generated and analyzed using the same program. If population covariance matrices are created separately they should always be checked for fit to the parameter values used to create them. This is done by reading the population covariance matrix into a program that specifies the same model as was used to create it. This analysis should result in parameter values that match exactly to those used to create the data, a χ^2 value of exactly zero, optimal values of all other fit indexes, and values of zero for all residuals.

Here I demonstrate the data generation process using the smallest of the models described in the previous section, and shown on the top of Figure 12.1. The unstandardized population parameter values for this model were as follows: loadings were all specified as 1.0, measurement error variances were all set at .50, factor variances were .50, and the factor covariance was set to .25.

The program in Figure 12.2 was used to generate data for samples of size 150 on the eight-variable model. Using the MONTECARLO command in Mplus allows for the data to be both generated and analyzed in a single program. The printed output contains a summary of the results for parameter estimates, standard errors, and χ^2 values. The generated data and results can also be saved into files and analyzed

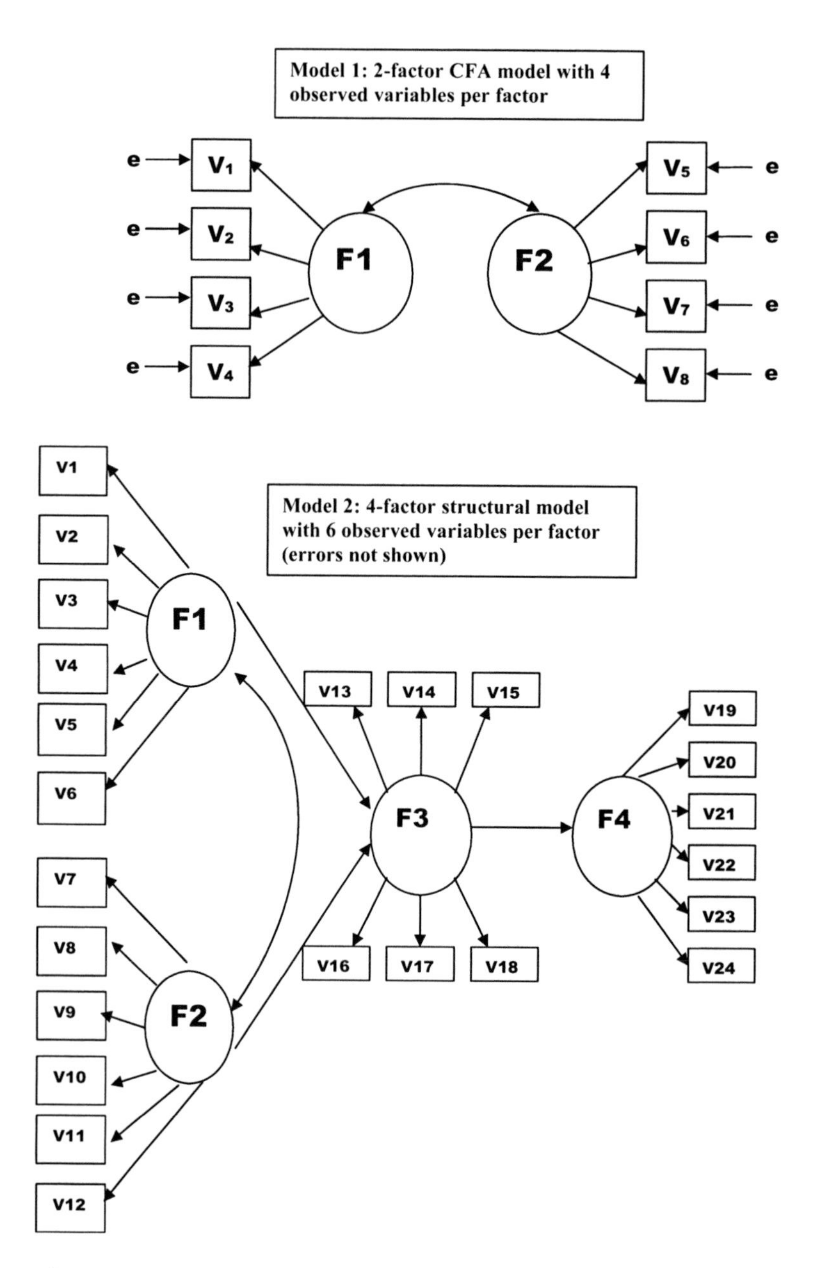

Figure 12.1. Path diagrams of two models from the example study.

```
MONTECARLO:
 NAMES = y1-y8;
 NOBSERVATIONS = 150;
 NREPS = 500;
 SEED = 741355;
 GENERATE = y1 - y8 (1);
 CATEGORICAL = y1 - y8;
 REPSAVE = ALL;
 SAVE = M1N150CAT1*.DAT;
 RESULTS = M1N150CAT1RES;
MODEL MONTECARLO:
 f1 BY y1-y4*1;
 f2 BY y5-y8*1;
 f1-f2*.50;
 f1 WITH f2*.25;
 y1-y8*.50;
 [y1$1-y8$1*1.39];
MODEL:
 f1 BY y1-y4;
 f2 BY y5-y8;
 f1-f2*.50;
 f1 WITH f2*.25;
 [y1$1-y8$1*1.39];
OUTPUT: TECH9;
```

Figure 12.2. Input for Mplus MONTECARLO program.

using external programs using the SAVE= and RESULTS= commands, respectively.

The MONTECARLO command signals that a Monte Carlo study is to be done and specifies the variable names, the sample size for each replication (NOBSERVATIONS), the number of replications (NREPS), and the random seed (SEED) to be used. The MODEL MONTECARLO command provides the population model and parameter values on which the samples are to be generated. The symbol "*" is used here to give the value of each population parameter (the "@" symbol can also be used). The MODEL command describes the model to be estimated in the samples. In this case it is the same as the population model, but if the researcher wanted to introduce some type of model misspecification, the misspecified model would be described here.

In this study I was interested in the behavior of the WLSMV estimators with categorical data. Because WLSMV is the default estimator for categorical data, use of this estimator does not need to be specified in the program. To obtain categorical data the GENERATE and CATEGORICAL options are added to the MONTECARLO command. The GENERATE option indicates the number of thresholds to be used for data generation.

In this example, it is specified as (1), indicating that variables y1 – y8 will be generated as categorical variables with two categories each. The CATEGORICAL option indicates which variables are to be analyzed as categorical. Although the variables specified in the GENERATE and CATEGORICAL options will typically be the same, having both options allows for model parameters to be generated as categorical but analyzed as continuous. To do this you would specify GENERATE = y1-y8 (1) as above but leave out the CATEGORICAL option.

The next three lines have to do with saving data and results. The REPSAVE option indicates the replications for which raw data should be saved. If all of the raw data are to be saved, as in the current example, this is indicated by specifying REPSAVE = ALL. The SAVE option specifies the name of the file(s) into which the data should be saved. If data from more than one replication are to be saved, separate files will be created for each replication. In this situation, use of an asterisk in the name of the data file will cause the files to be given the name specified plus consecutive integers indicating the replication number. In this example, I have used the name "M1N150CAT1" to indicate the data are from the first model with n = 150 using the first categorization method. The 500 data files will be named "M1N150CAT11.DAT, M1N150CAT12.DAT,...M1N150CAT1500.DAT." The RESULTS option can be used to save selected output from each replication in an external ASCII file. In this case, results from all replications will be saved in a single file, under the name specified in the option (in this example, all results will be saved in the file "M1N150CAT1.RES." Results are saved in the following order: replication number, convergence indicator (0=convergence; 1=nonconvergence), χ^2 value, parameter estimates, and standard errors. The order of parameter estimates and standard errors is given in the output from the MONTECARLO run. These results are saved in free format, separated by spaces, in a format of E15.8 (i.e., with data in scientific notation and with each variable consisting of 15 columns, 8 of which are to the right of the decimal point).

Under the MODEL MONTECARLO and MODEL commands, the line "[y1$1-y8$1*1.39] specifies that the thresholds, designated with "$", should be set to 1.39. For the population data this will result in a fixed threshold value of 1.39. For the sample data, thresholds will have start values of 1.39, but estimated values will vary due to sampling error. This value represents the cutpoint described previously that will yield non-normally distributed variables with skew and kurtosis of approximately 3.0 and 7.0, respectively. Finally, the option "TECH9" in the OUTPUT command will cause the program to print error messages in the output for replications that fail to converge.

Selected output from this program is shown in Figure 12.3. The first portion of the output gives information on fit indexes, and indicates the number of replications for which estimation was successful. In this example, only 481 out of 500 replications ran successfully. This is probably due to the combination of small sample size, non-normality, and small number of categories (2). For WLSMV estimation, fit indexes given include the χ^2 statistic, RMSEA, and WRMR. WRMR, or the weighted root mean square residual, is a measure of the difference between the sample and model-implied matrices based on the WLS estimation. Yu and Muthén (2002) have suggested a cutoff of 1.0 for this index when applied to non-normaility distributed data.

In this section, two tables, labeled "Proportions" and "Percentiles," are shown for each fit index. Under each of these headings there are columns labeled "expected" and "observed." For the χ^2 statistic, the "expected" proportions and percentiles columns contain the probability values and critical values, respectively, from a central χ^2 distribution with degrees of freedom associated with the model being tested. For example, the value of 0.010 in the "expected proportion" column corresponds to the value of 21.666 in the "expected percentile" column and indicates that, for the degrees of freedom associated with this model, a proportion of .01 of the sample χ^2 values would be expected to equal or exceed a critical value of 21.666. Values in the columns labeled "observed" contain the proportion of actual sample values at or beyond the value of χ^2 shown in the corresponding "observed percentile" column. For example, the lowest observed proportion of .091 corresponds to a χ^2 value of 32.199, indicating that .091 or 9.1% of the sample χ^2 values were greater than or equal to 32.199.

Following information on fit statistics, summaries of parameter estimates and standard errors are given under "Model Results." The first column contains the population values specified in the program. The mean and standard deviation of each parameter estimate across the successful replications are given in columns two and three. The fourth column contains the average standard error across the completed replications, while the fifth column contains the mean squared error (MSE). The latter is the average of the squared deviations of the sample parameter estimates from their population value.

Error messages from replications that failed to converge are included under "Technical 9 output." These messages indicate the sequence number of the replication and the reason for the unsuccessful run. In this example, nine replications failed to run successfully. The error message for the first two read:

Chi-Square Test of Model Fit

Number of successful computations 481

Proportions		*Percentiles*	
Expected	*Observed*	*Expected*	*Observed*
0.990	1.000	2.088	3.746
0.980	0.998	2.532	4.527
0.950	0.998	3.325	5.194
0.900	0.996	4.168	6.079
0.800	0.952	5.380	7.552
0.700	0.894	6.393	9.018
0.500	0.773	8.343	11.720
0.300	0.588	10.656	15.008
0.200	0.478	12.242	18.162
0.100	0.331	14.684	21.744
0.050	0.225	16.919	25.182
0.020	0.129	19.679	28.709
0.010	0.091	21.666	32.199

RMSEA (root mean square error of approximation)

Mean 0.062

Std Dev 0.030

Number of successful computations 481

Proportions		*Percentiles*	
Expected	*Observed*	*Expected*	*Observed*
0.990	1.000	-0.007	0.005
0.980	1.000	0.001	0.006
0.950	0.952	0.013	0.013
0.900	0.869	0.024	0.020
0.800	0.792	0.037	0.042
0.700	0.682	0.046	0.044
0.500	0.482	0.062	0.056
0.300	0.314	0.078	0.071
0.200	0.204	0.087	0.083
0.100	0.104	0.100	0.095
0.050	0.052	0.111	0.104
0.020	0.017	0.123	0.121
0.010	0.015	0.132	0.133

Figure 12.3. Selected output from Mplus MONTECARLO program.

```
REPLICATION 9:

    NO CONVERGENCE. NUMBER OF ITERATIONS EXCEEDED.

REPLICATION 19:

    WARNING: THE RESIDUAL COVARIANCE MATRIX (PSI) IS NOT
    POSITIVE DEFINITE. PROBLEM INVOLVING VARIABLE F2.
```

WRMR (weighted root mean square residual)
Mean 0.731
Std Dev 0.181
Number of successful computations 481

Proportions		*Percentiles*	
Expected	*Observed*	*Expected*	*Observed*
0.990	1.000	0.310	0.385
0.980	0.996	0.359	0.434
0.950	0.983	0.433	0.450
0.900	0.906	0.499	0.500
0.800	0.773	0.579	0.550
0.700	0.669	0.636	0.603
0.500	0.470	0.731	0.691
0.300	0.299	0.826	0.776
0.200	0.212	0.884	0.858
0.100	0.114	0.964	0.966
0.050	0.056	1.029	1.056
0.020	0.027	1.103	1.084
0.010	0.012	1.153	1.136

Figure 12.3. Selected output from Mplus MONTECARLO program (continued).

In addition to the chi-square values and parameter estimates, it is also possible to save all of the fit indexes produced by the program as well as estimates of parameter standard errors. In order to do this, however, the saved data sets must be reanalyzed using a separate program. This is referred to as an "external" Monte Carlo study in Mplus. In an external study, data created by a program such as that in the example in Figure 12.2 is read by another Monte Carlo program that reads in each data set sequentially, estimates the specified model, and saves the output into an external results file. External studies are also useful for situations in which data for a simulation are generated by a program other than Mplus, but the researcher wants to use Mplus to analyze the data. An example of an Mplus program that reads in the data created by the program in Figure 12.2 is shown in Appendix A.

Setting the Random Seed

One important consideration in generating sample data is the choice of a random seed. The seed value is used by the program as a starting point for the random draws that create the samples. One seed is needed

MODEL RESULTS

	Estimates			*S. E.*	*M. S. E.*	*95%*	*% Sig.*
	Population	*Average*	*Std. Dev.*	*Average*		*Cover*	*Coeff*
F1 BY							
Y1	1.000	1.0000	0.0000	0.0000	0.0000	1.000	0.000
Y2	1.000	1.2701	3.8987	2.3897	15.2415	0.892	0.794
Y3	1.000	1.1199	1.3495	1.0222	1.8317	0.909	0.821
Y4	1.000	1.4511	6.2867	8.0346	39.6442	0.909	0.807
F2 BY							
Y5	1.000	1.0000	0.0000	0.0000	0.0000	1.000	0.000
Y6	1.000	1.2860	2.5821	1.7941	6.7352	0.881	0.815
Y7	1.000	1.2700	1.9479	1.0736	3.8592	0.900	0.821
Y8	1.000	1.1729	1.3715	0.7642	1.9071	0.890	0.805
F1 WITH							
F2	0.250	0.2384	0.1688	0.1134	0.0286	0.721	0.526
Thresholds							
Y1$1	1.390	1.4042	0.1617	0.1505	0.0263	0.940	1.000
Y2$1	1.390	1.4078	0.1552	0.1507	0.0243	0.963	1.000
Y3$1	1.390	1.4057	0.1521	0.1504	0.0233	0.963	1.000
Y4$1	1.390	1.4003	0.1523	0.1499	0.0232	0.956	1.000
Y5$1	1.390	1.4032	0.1596	0.1504	0.0256	0.960	1.000
Y6$1	1.390	1.3932	0.1545	0.1494	0.0238	0.960	1.000
Y7$1	1.390	1.3980	0.1546	0.1498	0.0239	0.958	1.000
Y8$1	1.390	1.4021	0.1542	0.1501	0.0239	0.944	1.000
Variances							
F1	0.500	0.5635	0.3437	0.2469	0.1219	0.877	0.663
F2	0.500	0.5604	0.3435	0.2302	0.1214	0.842	0.640

Figure 12.3. Selected output from Mplus MONTECARLO program (continued).

for each set of sample data that is generated. Most programs allow two ways of setting the seed. One is to allow the computer's internal clock to derive the seed. Using this method will result in a different random seed, and thus randomly different samples, to be drawn each time the program is run. The second method is to provide a value for the starting seed using a random numbers table. This method is preferable in most situations for two reasons. The first is that, by keeping a record of the random seeds used to create the samples for different cells of the design, the data for a particular cell can be regenerated if the data are lost or damaged in some way. Another reason I prefer this method is that it allows for the seed to be tested. This is done by generating a series of random normal variates, or *z*-scores, using the seed and applying the various tests of randomness to the data. Such tests include the Runs test, the one-sample Kolmogorov-Smirnov test, or the Sign test (see Siegel & Castellan, 1988, for more details). While this is possible with seeds generated by the computer, it involves extra steps unless raw data are generated instead of covariance matrices. Studies have evaluated some of the random number generators available in commercially available software (e.g., Bang, Schumacker, & Schlieve, 1998; Fishmann & Moore, 1982) and should be consulted when choosing a software package.

One final issue with regard to seed selection is whether the same seed should be used for generating data for different cells of the design. For example, if a researcher is interested in studying the effects of 5, 10, and 15 indicators per factor on the power of the model χ^2 test in SEM, data for the three levels could be generated independently using separate seeds or the same seed could be used to generate data for all three levels. The advantage of the latter method is that it reduces the amount of sampling error across the three levels. However, it also introduces a degree of dependence among them. In terms of data analysis, this dependence could be modeled through the use of a repeated measures or other dependent-samples design. In my view, the most important consideration in choosing between the two methods is the degree to which it corresponds to the real world situation being modeled. For example, in the example involving the number of indicators, use of the same seed to generate data for all three levels would correspond to a situation in which an investigator had access to 15 total items, but may have chosen to use only 5 or 10 of them in a given study. In this case the indicators are considered to be from the same indicator pool and should be generated using the same seed. In contrast, use of separate seeds would correspond to the situation in which different sets of items were used for each of the three levels of number of indicators.

Choosing the Number of Replications

As noted by Harwell and colleagues (1996) in the context of item response theory Monte Carlo studies, the choice of the number of replications depends on the purpose of the study, the desire to minimize sampling variance, and the need for adequate power. The purpose of the study will influence the number of replications needed because some effects of interest are less stable than others. For example, more replications may be needed to study the behavior of standard errors than parameter estimates because the former generally have a greater sampling variance. However, if the object of the study is to compare the results of different parameterizations of a model, or of different software packages, it is not necessary to obtain an empirical sampling distribution so a large number of replications might not be needed. In some cases, the study design may require researcher judgment for some steps. For example, the researcher may base some decisions on several competing criteria, as in the decision to eliminate or retain items in scale development. Simulating this process would require such a judgment to be made for each replication, thus severely limiting the number of replications that is feasible. Powell and Schafer (2001) reported numbers of replications ranging from 20 to 1,000 with a median of 200 in their review of χ^2 robustness studies in SEM.

Robey and Barcikowski (1992) have argued that Monte Carlo researchers do not typically have a strong rationale for their choice of the number of replications to be used, resulting in either a lack of power or in excessive power. As in applied research, the sample size (or number of replications in the case of Monte Carlo research) will depend on the desired levels of power and Type I error, and on the sizes of the effects to be detected. One advantage of Monte Carlo studies in determining power levels of individual parameter estimates is that the population parameters are generally known, and various authors have suggested criteria for robustness of Type I error rates (e.g., Bradley, 1978; Serlin, 2000), thus making it much easier to obtain effect sizes. Robey and Barcikowski and Serlin also provided formulae for determining the number of replications necessary to detect departures from the nominal Type I error rate in statistical tests.

For my example study I chose to run 500 replications per cell. This is somewhat large for SEM Monte Carlo studies, but I felt it was necessary to obtain stable estimates of standard errors and χ^2 values based on nonnormally distributed data.

Generating the Sample Data

If the program being used does not generate both the population and sample matrices, the sample data would be generated as the next step in the study. This would be the case if the LISREL program were used to conduct the Monte Carlo study. An example of the necessary SIMPLIS and PRELIS code is included in Appendix B. Unless raw data are needed, covariance matrices are typically generated because they require less storage space. However, some estimators for non-normally distributed data as well as some missing data treatments require raw data. This is just as easy to generate but requires more storage space.

Once the sample data are generated some final checks are necessary. First, the sample data should be checked to make sure they match with theoretical expectations. For example, if sample size is varied, standard errors of parameter estimates should decrease as sample size increases. If increasing levels of model misspecification are included in the study, values of the fit indexes should be examined to confirm that these increase with the level of misspecification. If distributions with varying levels of non-normality were created, the obtained values of skew and kurtosis should be compared with the target values.

It is also important to verify that the expected amount of output was actually obtained. The primary reason that some cells may not contain the full number of replications is that some samples may result in nonconvergence and/or improper solutions. The issue of whether or not to include nonconvergent solutions is a matter of debate, but in general researchers tend to exclude these samples from further analysis. This seems consistent with practice in applied studies because nonconvergence is usually an indication of a problem with the model or data (or both) and interpreting the results of nonconvergent solutions is therefore generally considered to be poor practice. Improper solutions present more of a problem because these are not typically screened out by computer programs. However, under some conditions it is not unusual to obtain parameter values that are clearly out of range, such as negative error variances or standardized factor loadings exceeding ± 1.0. Although Mplus, LISREL, and EQS all generate warning messages when nonconvergent solutions are found, these programs may not always provide warnings for improper solutions. For this reason it is a good practice to screen parameter values to determine if they are within acceptable limits (e.g., Enders & Bandalos, 2001). If improper or out-of-range values are obtained in practice, one would hope that applied researchers would recognize them as indicative of a problem with the analysis and not interpret the results. If this is the case, eliminating such samples from the study would yield results that are more representative of the real world, which should be one's goal. Finally, researchers must decide if they will generate other samples to replace those with nonconvergent or improper solutions, or simply use the samples that do converge. Replacing the samples with new ones has the advantage of maintaining a balanced design. Whichever method is used, the number or percentage of nonconvergent or improper solutions for each cell of the design should be reported.

DATA ANALYSIS FOR MONTE CARLO STUDIES

Because Monte Carlo studies typically include several independent variables with several levels of each, the number of design cells and of replications can quickly become overwhelming. Although the use of descriptive statistics and graphical techniques can help to illustrate the results of Monte Carlo studies, especially when there are complicated interactions present, the complexity of most Monte Carlo studies requires a more powerful method of detecting the degree to which the dependent variables are affected by the independent variables. For this reason, many researchers recommend the use of appropriate inferential statistical methods in analyzing results of these studies (Harwell, 1992, 1997; Hauck & Anderson, 1984; Hoaglin & Andrews, 1975). Use of inferential statistics is also necessary to quantify and compare the effects of the various independent variables and their interactions. How many researchers would be able to detect three-way or even complex two-way interactions from perusing tables of means? Although this point may seem obvious, Harwell and colleagues (1996) found that only about 7% of Monte Carlo studies published in the psychometric literature reported the use of inferential statistics.

One common argument against the use of inferential statistics in analyzing the results of Monte Carlo studies is that the large numbers of replications typically used in these studies renders even the smallest of effects statistically significant. Several counterarguments have been advanced, however (see Harwell, 1997). First, any presentation of the results of significance testing should be accompanied by estimates of effect size such as ω^2 for ANOVA designs. While it is the case that the large numbers of replications included in most Monte Carlo studies result in very high levels of power, researchers can use effect sizes to gain some perspective on the practical significance of the results. In some cases a decision may be made to interpret only those effects reaching some prespecified level of effect size. I typically use Cohen's (1988) "large" effect size of .14 as a criterion, but others may wish to interpret smaller effects. As a second argument against this criticism, it should be pointed out that guidelines for determining the number of replications to obtain given levels of power have recently been offered (Robey & Barcikowski, 1992; Serlin, 2000). By following guidelines such as these, researchers can ensure that their studies will have adequate but not excessive levels of power. Another argument for the use of inferential statistics in Monte Carlo studies is that these studies do incorporate sampling error through the inclusion of multiple replications. The use of inferential statistics allows for this sampling error to be taken into account when determining the size of effects. Finally, the use of inferential statistics in Monte Carlo studies allows for study results to be integrated quantitatively through the use of meta-analysis, as illustrated most recently by Harwell (2003).

To summarize, I recommend the use of both inferential and descriptive or graphical methods in analyzing and conveying the results of Monte Carlo studies. While the use of inferential statistics, along with reporting of effect sizes and/or confidence intervals, is invaluable in detecting complicated effects, including interactions, descriptive and graphical techniques are useful in elucidating the nature of these effects.

With regard to the type of inferential statistics to be used, this should be determined by the study design and research questions. If the design factors are primarily categorical, an ANOVA design is more appropriate. This design is preferred by many researchers because it lends itself to the use of tables of means in presenting results. If design factors are primarily continuous, regression analyses may be used. These analyses offer the advantage of allowing for the estimation of values of the dependent variables for various values of the independent variables within the range of those studied. For example, if sample size were varied at 100, 300, and 500, values of the dependent variables for sample sizes of 200 and 400 could also be estimated using the obtained regression equation. Of course, these values must be interpreted with caution as no data for those conditions were actually observed.

Issues of Data Management

It is perhaps an understatement to say that Monte Carlo studies result in a great deal of output that must somehow be sorted into files, saved, and analyzed. Paxton and colleagues (2001) provided an excellent discussion of issues involved in the actual execution of Monte Carlo studies and the attendant data management. These matters will be discussed briefly here; readers desiring a more thorough treatment should consult the Paxton and colleagues article.

What Should Be Saved? The answer to this question will of course depend on the research questions. Most SEM software packages now incorporate Monte Carlo capabilities that allow for parameter estimates, standard errors, and various fit indexes to be output to files. In most software packages results from the different replications are concatenated and saved into one file, or the average values across a set of replications may be saved. Information about each run, such as whether estimates converged and were admissible and the replication number, are also commonly included, either in the same file or as a separate file.

There are two basic options for analyzing this output. All three software packages discussed in this volume have limited data analysis capabilities and allow for the calculation of basic descriptive statistics for the output data. However, most Monte Carlo researchers prefer to read the output data into statistical software packages such as SPSS or SAS and conduct their data analyses within one of those programs. The advantage of this is that the various measures of bias and variability described previously can be calculated and saved in these programs, and values designating the categorical design factors and their levels can be inserted to allow for inferential statistics to be used.

Because SEM software packages typically save the results of interest, one may question the necessity of saving the original raw data or sample covariance matrices. If the researcher has kept a log of the seeds used to generate the data, they can always be regenerated if any questions arise. However, this may be extremely time-consuming if a complicated design was used or if time-intensive analyses such as bootstrapping or ADF estimation was used. This decision will therefore depend to some extent on the type of study that was done. However, if the raw data are not stored, provision should be made for regenerating the data should the need arise.

Finally, the issue of documentation should be addressed. As Paxton and colleagues (2001) pointed out, it is not possible to have too much documentation in a Monte Carlo study. In addition to saving the raw and summary data, copies of all programs used should be annotated liberally and archived. Annotation of programs is particularly important because it is often difficult to remember the rationale for program code even after a short period of time. Also, if the research is being carried out collaboratively, program annotation provides an easy way to communicate the purpose of each program to other members of the research group. Paxton and colleagues recommended keeping multiple copies of each program, and based on my experience in Monte Carlo research I would second this recommendation. Programs should also be dated and corrections or changes noted in the program so that there is no confusion over which version of a program is correct. I also like to keep written records of every analysis from the generation of the population matrices to final analysis of the data, with dates to indicate when each was completed. If analyses are rerun or data regenerated, as sometimes proves necessary, it is vital to be able to identify the most current or corrected data and analyses. Finally, I would recommend keeping all of the raw data as well as several versions of the parameter estimates, standard errors, fit indices, or other output data of interest.

SUMMARY

Monte Carlo studies in SEM involve the generation of sample data in order to study issues such as the robustness of statistical estimators to assumption violations, the small sample behavior of asymptotic estimators, or the behavior of ad hoc fit indexes. These studies are essentially the same as other research studies in that the research questions should be based on theory and previous research and should be clearly stated. Similarly, the choice of independent variables and their levels and of dependent variables should be based on a thorough review of the available literature and theory. In this chapter I have introduced the most commonly used independent and dependent variables in SEM Monte Carlo studies and have provided information that I hope will be useful in making choices among these. I have also illustrated the use of the Mplus and LISREL computer packages for conducting Monte Carlo studies in SEM. However, although SEM packages are most commonly used for these studies, it is also possible to use the matrix features of such statistical packages as SAS, SPSS, or S+, or computer languages such as C++ or FORTRAN, to generate and even analyze data for Monte Carlo studies. Finally, I have emphasized that the analysis of data from Monte Carlo studies should include inferential statistics chosen to provide answers to the initial research questions as well as tables and/or graphical displays to clarify or elucidate aspects of the results. I have ended by discussing the types of output that should be saved and by stressing the importance of clear and thorough documentation of all syntax and data files. Although the example study used in this chapter is still in the planning stages, I plan to follow my own recommendations when data generation and analyses begin in earnest.

NOTE

1. I use the term *independent variables* here to designate facets of the research design. I use the term *dependent variable* to designate outcome variables. To avoid confusion the terms *exogenous* and *endogenous variables* are used to designate independent and dependent variables within a measured variable or latent variable path model.

APPENDIX A

Mplus Program to Read in External Data

TITLE: Example of Reading in External Data for CFA Model

DATA: FILE = MM1N150CAT1list.dat;

! Comment: *The file specification in this line will cause Mplus to sequentially read in the 500 previously created data files. When these files are created Mplus also creates a "list" file that is literally a list of the names of all data files created. In the current program, the names of the data files are taken from this list and then read in one at a time.*

TYPE = MONTECARLO;

VARIABLE: NAMES = y1-y8;

CATEGORICAL ARE y1-y8;
! Comment: *This line specifies that the variables are all categorical. The number of categories need not be specified here as the program will determine this from the raw data.*

MODEL: f1 BY y1 y2-y4*1;
f2 BY y5 y6-y8*1;
f1-f2*.50;
f1 WITH f2*.25;
[y1$1-y8$1*1.39];
! Comment: *The lines in the MODEL command specify the model to be run. Note that the numerical values given in these lines are starting values, not fixed values as in the previous MONTECARLO program. In this example the model is the same as the model specified in the program that created the data. However, a mis-specified model could also be indicated here. Note that by default Mplus defines the factor metric by setting the values of the first loading for each factor equal to 1.0 These loadings should therefore not be given starting values as this would indicate they should be estimated.*

OUTPUT: TECH9;

SAVEDATA: results are M1N150CAT1.res;
! Comment: *The results from all 500 replications will be saved in the file specified here. The order in which results are saved is given in the printed output file.*

APPENDIX B

SIMPLIS Program to Create a Population Covariance Matrix

creating a population covariance matrix
observed variables: y1-y8
****the observed variables command identifies the number of observed variables and their names****
covariance matrix from file id.cov
****the matrix id.cov contains an eight variable identity matrix, but any 8-variable matrix could be used****
sample size 100000
latent variables: f1 f2
**** the latent variables command identifies the number of latent variables and their names ****
relationships:
  y1 –y4 = 1.0*f1
  y5-y8 = 1.0*f2
  Set the error variances of y1-y8 to .5
  Set variance of f1 - f2 To .5
  Set covariance of f1 - f2 To .25
**** the relationships command identifies the structure of the model. Here all parameter values are set to their prespecified population values through the use of the "let" and "set" commands. For factor loadings, the population value is simply specified by statements such as y1-y4 = 1.0*f1, indicating that the loadings of y1-y4 should be fixed to 1.0. ****
options: nd = 3 si = m1.cov
**** the option "nd" allows for specification of the number of decimals in the output. The option "si=m1.cov" causes the model-implied matrix to be saved under the name "m1.cov."****

SIMPLIS Program to Obtain Coefficients for Data with a Specified Correlation Matrix

Title: obtaining the t-matrix for Cholesky decomposition
observed variables: y1-y8
covariance matrix from file m1.cov
sample size 100000
latent variables: f1-f8
relationships:
  y1 = (1)*f1
  y2 = (1)*f1 + (1)*f2

```
y3 = (1)*f1 + (1)*f2 + (1)*f3
y4 = (1)*f1 + (1)*f2 + (1)*f3 + (1)*f4
y5 = (1)*f1 + (1)*f2 + (1)*f3 + (1)*f4 + (1)*f5
y6 = (1)*f1 + (1)*f2 + (1)*f3 + (1)*f4 + (1)*f5 + (1)*f6
y7 = (1)*f1 + (1)*f2 + (1)*f3 + (1)*f4 + (1)*f5 + (1)*f6 + (1)*f7
y8 = (1)*f1 + (1)*f2 + (1)*f3 + (1)*f4 + (1)*f5 + (1)*f6 + (1)*f7 +
   (1)*f8
```

*** *Here the numbers in parentheses are starting values****

```
set the covariances of f1-f8 to 0.0
set the variances of f1-f8 to 1.0
set the error variances of y1-y8 to 0.0
options: nd = 6 ad = off
```

*** *The number of decimals (ND) is set to 6 for greater precision. The admissibility index (AD) is turned OFF because the unusual nature of this analysis often triggers warnings.* ***

PRELIS Program to Generate Multivariate Normal Data With a Specified Covariance Matrix

```
Prelis program to generate multivariate normal data with a specified
covariance matrix
da no = 150 rp = 500
```

*** *"da" stands for data. "No" is the number of observations (sample size) and "rp" is the number of replications.****

```
ne v1 = NRAND
ne v2 = NRAND
ne v3 = NRAND
ne v4 = NRAND
ne v5 = NRAND
ne v6 = NRAND
ne v7 = NRAND
ne v8 = NRAND
```

*** *The eight lines above will create 8 new ("ne") variables (v1-v8) that will be distributed as normal random deviates (NRAND)* ***

```
ne x1 = 1.0000*v1
ne x2 = 0.500001*v1 + 0.866025*v2
ne x3 = 0.500000*v1 + 0.288675*v2 + 0.816497*v3
ne x4 = 0.500000*v1 + 0.288675*v2 + 0.204124*v3 + 0.790569*v4
ne x5 = 0.250000*v1 + 0.144337*v2 + 0.102062*v3 + 0.0790569*v4 +
   0.948683*v5
ne x6 = 0.250000*v1 + 0.144337*v2 + 0.102062*v3 + 0.0790569*v4 +
   0.421637*v5 + 0.849837*v6
```

ne x7 = 0.250000*v1 + 0.144337*v2 + 0.102062*v3 + 0.0790569*v4 +
0.421637*v5 + 0.261488*v6
ne x7 = x7 + 0.808597*v7
ne x8 = 0.250000*v1 + 0.144337*v2 + 0.102062*v3 + 0.0790568*v4 +
0.421637*v5 + 0.261488*v6
ne x8 = x8 + 0.190261*v7 + 0.785905*v8
*** *The lines above will transform the 8 normal random deviates into correlated variables with the correlational structure specified in the previous program. The coefficients of each variable are the values obtained for factor loadings from that program.****
co all
*** *The command "co all" specifies that all variables are continuous.* ***
sd v1-v8
*** *The "sd" statement will delete the original variables v1-v8 (which have been replaced by the new, correlated variables x1-x8.* ***
ou cm=m1n150.cm ix = 741355 XM XT
*** *The specification "cm = m1n150.cm" will cause the 500 new covariance matrices to be saved in the file "m1n150.cm" – I have chosen this name to indicate the matrices are from Model 1, n = 150. The random seed to be used is specified in the "ix" specification. XM and XT suppress some of the output.****

SIMPLIS Program to Analyze Data and Save Output

Analyzing data for model 1, *n* = 150
observed variables: y1-y8
covariance matrix from file m1n150.cm
*** *The matrix "m1n150.cm" contains the 500 covariance matrices created by the previous program.****
sample size 150
latent variables: f1 f2
relationships:
y1-y4 = f1
y5-y8 = f2
options: rp = 500 lx = m1n150.lx td = m1n150.td ph = m1n150.ph gf=m1n150.gf
*** *The option "rp" specified the number of replications; this tells the program it will be reading 500 consecutive covariance matrices. The next 4 specifications indicate that the factor loadings (lx) should be saved to the matrix "m1n150.lx", error variances (td) to the matrix "m1n150.td", factor covariances (ph) to the matrix "m1n150.ph" and goodness of fit indexes (gf) to the matrix "m1n150.gf." These matrices will contain 500 consecutive sets of parameter estimates or goodness of fit indexes.* ***

REFERENCES

Arnold-Berkovits, I. (2002). *Structural equation modeling with ordered polytomous and continuous variables: A simulation study comparing full-information Bayesian estimation to correlation/covariance methods.* Unpublished doctoral dissertation, University of Maryland, College Park.

Bandalos, D. L., & Enders, C. K. (1996). The effects of nonnormality and number of response categories on reliability. *Applied Measurement in Education, 9,* 151–160.

Bang, J. W., Schumacker, R. E., & Schlieve, P. (1998). Random-number generator validity in simulation studies: An investigation of normality. *Educational and Psychological Measurement, 58,* 430–450.

Bradley. J. V. (1978). Robustness? *British Journal of Mathematical and Statistical Psychology, 31,* 144–152.

Browne, M. W., & Cudeck, R. (1992). Alternative ways of assessing model fit. *Sociological Methods & Research, 21,* 230–258.

Cohen, J. (1988). *Statistical power analysis for the behavioral sciences* (2nd ed.). Hillsdale, NJ: Erlbaum.

Cudeck, R., & Henly, S. J. (1991). Model selection in covariance structure analysis and the "problem" of sample size: A clarification. *Psychological Bulletin, 109,* 512–519.

Dolan, C. V. (1994). Factor analysis of variables with 2, 3, 5, and 7 response categories: A comparison of categorical variable estimators using simulated data. *British Journal of Mathematical and Statistical Psychology, 47,* 309–326.

Enders, C. K., & Bandalos, D. L. (2001). The relative performance of full information maximum likelihood estimation for missing data in structural equation models. *Structural Equation Modeling: A Multidisciplinary Journal, 8,* 430–457.

Enders, C. K., & Finney, S. J. (2003, April). *SEM fit index criteria re-examined: An investigation of ML and robust fit indices in complex models.* Paper presented at the annual meeting of the American Educational Research Association, Chicago, IL.

Fishmann, G. S., & Moore, L. R. (1982). A statistical evaluation of multiplicative congruential random number generators with Modulus 2^{31-1}. *Journal of the American Statistical Association, 77,* 129–136.

Harwell, M. R. (1992) Summarizing Monte Carlo results in methodological research. *Journal of Educational Statistics, 17,* 297–313.

Harwell, M. R. (1997). Analyzing the results of Monte Carlo studies in item response theory. *Educational and Psychological Measurement, 57,* 266–279.

Harwell, M. R. (2003). Summarizing Monte Carlo results in methodological research: The single-factor, fixed-effects ANCOVA case. *Journal of Educational and Behavioral Statistics, 28,* 45–70.

Harwell, M. R., Stone, C. A., Hsu, T.-C., & Kirisci, L. (1996). Monte Carlo studies in Item Response Theory. *Applied Psychological Measurement, 20,* 101–125.

Hauck, W. W., & Anderson, S. (1984). A survey regarding the reporting of simulation studies. *The American Statistician, 38,* 214–216.

Hoaglin, D. C., & Andrews, S. (1984). The reporting of computation-based results in statistics. *The American Statistician, 29,* 122-126.

Hoogland, J. J., & Boomsma, A. (1998). Robustness studies in covariance structure modeling: An overview and meta-analysis. *Sociological Methods & Research, 26*, 329–367.

Hu, L. T., & Bentler, P. M. (1999). Cutoff criteria for fit indexes in covariance structure analysis: Conventional criteria versus new alternatives. *Structural Equation Modeling: A Multidisciplinary Journal, 6*, 1–55.

Hu, L. T., Bentler, P. M., & Kano, Y. (1992). Can test statistics in covariance modeling be trusted? *Psychological Bulletin, 112*, 351–362.

Kaplan, D. (1988). The impact of specification error on the estimation, testing, and improvement of structural equation models. *Multivariate Behavioral Research, 23*, 69–86.

Kaplan, D., & Wenger, R. N. (1993). Asymptotic independence and separability in covariance structure models: Implications for specification error, power, and model identification. *Multivariate Behavioral Research, 28*, 467–482.

MacCallum, R. (2003).Working with imperfect models. *Multivariate Behavioral Research, 38*, 113–139.

MacCallum, R. C., Browne, M. W., & Sugawara, H. M. (1996). Power analysis and determination of sample size for covariance structure modeling. *Psychological Methods, 1*, 130–149.

Mattson, S. (1997). How to generate non-normal data for simulation of structural equation models. *Multivariate Behavioral Research, 32*, 355–373.

Metropolis, N., & Ulam, S. (1949). The Monte Carlo method. *Journal of the American Statistical Association, 44*, 335–341.

Micceri, T. (1989). The unicorn, the normal curve, and other improbably creatures. *Psychological Bulletin, 105*, 156–166.

Muthén, B. O., du Toit, S. H. C., & Spisic, D. (in press). Robust inference using weighted least squares and quadratic estimating equations in latent variable modeling with categorical outcomes. *Psychometrika*.

Muthén, B. O., & Kaplan, D. (1985). A comparison of some methodologies for the factor analysis of non-normal Likert variables. *British Journal of Mathematical and Statistical Psychology, 38*, 171–189.

Muthén, B. O., & Kaplan, D. (1992). A comparison of some methodologies for the factor analysis of non-normal Likert variables: A note on the size of the model. *British Journal of Mathematical and Statistical Psychology, 45*, 19–30.

Muthén, L. K., & Muthén, B. O. (1998). *Mplus: The comprehensive modeling program for applied researchers: User's guide*. Los Angeles: Authors.

Muthén, L. K., & Muthén, B. O. (2002). How to use a Monte Carlo study to decide on sample size and determine power. *Structural Equation Modeling: A Multidisciplinary Journal, 9*, 599–620.

Olsson, U. H., Foss, T., Troye, S. V., & Howell, R. D. (2000). The performance of ML, GLS, and WLS estimation in structural equation modeling under conditions of misspecification and nonnormality. *Structural Equation Modeling: A Multidisciplinary Journal, 7*, 557–595.

Paxton, P., Curran, P. J., Bollen, K. A., Kirby, J., & Chen, F. (2001). Monte Carlo experiments: Design and implementation. *Structural Equation Modeling: A Multidisciplinary Journal, 8*, 287–312.

Powell, D. A., & Schafer, W. D. (2001). The robustness of the likelihood ratio chi-square test for structural equation models: A meta-analysis. *Journal of Educational and Behavioral Statistics, 26*, 105–132.

Reinartz, W. J., Echambadi, R., & Chinn, W. (2002). Generating non-normal data for simulation of structural equation models using Mattson's method. *Multivariate Behavioral Research, 37*, 227–244.

Robey, R. R., & Barcikowski, R. S. (1992). Type I error and the number of iterations in Monte Carlo studies of robustness. *British Journal of Mathematical and Statistical Psychology, 45*, 283–288.

Serlin, R. C. (2000). Testing for robustness in Monte Carlo studies. *Psychological Methods, 5*, 230–240.

Siegel, S., & Castellan, N. J., Jr. (1988). *Nonparametric statistics for the behavioral sciences*. New York: McGraw-Hill.

Vale, C. D., & Maurelli, V. A. (1983). Simulating multivariate non-normal distributions. *Psychometrika, 48*, 465–471.

Yu, C.-Y., & Muthén, B. O. (2002, March). *Evaluation of model fit indices for latent variable models with categorical and continuous outcomes.* Paper presented at the annual meeting of the American Educational Research Association, New Orleans, LA.

ABOUT THE AUTHORS

Deborah Bandalos is a professor of educational psychology and director of the Research, Evaluation, Measurement, and Statistics Program at the University of Georgia, where she teaches courses in structural equation modeling, measurement theory, and scale development. Her research interests include the use and misuse of item parceling in structural equation modeling, applications of structural equation modeling in scale development, validity studies, and assessment and accountability.

Christine DiStefano is an assistant professor at the University of South Carolina where she teaches courses in assessment and statistics. Her research interests include survey methodology, cluster analysis, and the application of structural equation modeling and measurement techniques to issues of psychological testing and child behavior.

Craig K. Enders is an assistant professor in the Department of Psychology at Arizona State University. His primary research focus is in the area of missing data algorithms and structural equation modeling, and he has authored numerous journal articles on these topics.

Sara J. Finney is an assistant professor in the Department of Graduate Psychology at James Madison University where she teaches courses in structural equation modeling and multivariate statistics. The majority of her research involves the use of structural equation modeling to gather construct validity evidence for various self-report measures. Her other research interests include student achievement motivation and statistics education.

Phill Gagné is an assistant professor at Georgia State University. He teaches applied statistics courses in the Department of Educational Policy Studies and conducts primarily simulation research in structural equation modeling and other statistical techniques.

Samuel B. Green is a professor of measurement, statistics, and methodological studies in the Division of Psychology in Education at Arizona State University. He is leader of the Educational Psychology Program in the division. His research interests include structural equation modeling, reliability theory, and mixed modeling as they apply to educational and psychological research.

Gregory R. Hancock is a professor in the Department of Measurement, Statistics, and Evaluation at the University of Maryland, College Park. His research is primarily in structural equation modeling (SEM), with specific focus on latent variable experimental design and latent growth modeling. He is the former chair of the SEM special interest group of the American Educational Research Association, serves on the editorial board of several methodological and applied journals, and teaches workshops on SEM to national and international audiences.

Kit-Tai Hau is a professor and chair of the Educational Psychology Department at the Chinese University of Hong Kong. His research centers on motivation, structural equation modeling, psychometrics, computerized testing, and adolescent suicide. He has run a great number of national-level workshops on structural equation modeling in China and has played a very active role in Hong Kong government policy formation on ability segregation of students, computerized assessment, medium of instruction, and public examination systems.

Scott L. Hershberger is a professor of psychology in the Department of Psychology at California State University, Long Beach. He has published extensively in the areas of structural equation modeling, psychometric theory, behavior genetics, and sexual orientation and behavior. He is a fellow of the Royal Statistical Society and International Statistical Institute.

Rex B. Kline is an associate professor of psychology at Concordia University in Montreal. Since earning a PhD in psychology, his areas of research and writing have included the psychometric evaluation of cognitive abilities, clinical child assessment, structural equation modeling, reform of data analysis methods in the behavioral sciences, and usability engineering in computer science. He is the author of over 40 empirical studies, three books, and is the coauthor of a teacher-informant questionnaire of student behavioral and scholastic adjustment.

Frank R. Lawrence is an assistant professor in the College of Health and Human Development, Pennsylvania State University. Previously he was a research professor in the College of Education, University of Alabama at Birmingham. He teaches graduate courses in regression and longitudinal data analyses. His recent and current research involves examining the behavior of general growth mixture models.

Herbert W. Marsh is a professor of educational psychology and founding director of the SELF Research Centre, University of Western Sydney, Australia, is widely published (250 journal articles, 22 chapters, 8 monographs, 275 conference papers, and widely used tests measuring self-concept, motivation, and evaluations of teaching effectiveness) and a "highly cited researcher" on ISI's list of the "world's most cited and influential scientific authors." His research interests include self-concept and motivational constructs, evaluations of teaching effectiveness, developmental psychology, quantitative analysis, sports psychology, peer review, peer support, and antibullying interventions.

Ralph O. Mueller is a professor of educational research and of public policy and public administration at The George Washington University, Washington, D.C., and currently serves as chair of the Department of Educational Leadership. He is the former chair of AERA's SIG/Structural Equation Modeling and SIG/Professors of Educational Research and has taught (and co-taught with Greg Hancock) many national and international SEM workshops for universities, professional associations, and software companies. He is the author of an introductory textbook and other didactic writings on SEM.

Laura M. Stapleton is an assistant professor in the Department of Psychology at the University of Maryland Baltimore County. Her research interests include variance estimation and statistical modeling of data that arise from complex sampling designs, as well as various issues in multilevel modeling.

Marilyn S. Thompson is an assistant professor of measurement, statistics, and methodological studies in the Division of Psychology in Education at Arizona State University. Her research interests include methodological techniques for large data set analysis, including structural equation modeling and multilevel modeling. She also studies the use and misuse of achievement data to inform education policy. She is Director of the

EDCARE laboratory, which provides educational statistics, measurement, and evaluation services.

Zhonglin Wen is a professor in the Department of Psychology, South China Normal University. Mainly educated at universities in China (mainland) and the Chinese University of Hong Kong, he studied as a visiting scholar at the University of Manchester (UK) and the University of Western Sydney (Australia) for one year, respectively. His major research interests include mathematical statistics and research methods in psychology and education, especially structural equation modeling.